Cognitive Science and Technology

Series Editor

David M. W. Powers, Adelaide, SA, Australia

This series aims to publish work at the intersection of Computational Intelligence and Cognitive Science that is truly interdisciplinary and meets the standards and conventions of each of the component disciplines, whilst having the flexibility to explore new methodologies and paradigms. Artificial Intelligence was originally founded by Computer Scientists and Psychologists, and tends to have stagnated with a symbolic focus. Computational Intelligence broke away from AI to explore controversial metaphors ranging from neural models and fuzzy models, to evolutionary models and physical models, but tends to stay at the level of metaphor. Cognitive Science formed as the ability to model theories with Computers provided a unifying mechanism for the formalisation and testing of theories from linguistics, psychology and philosophy, but the disciplinary backgrounds of single discipline Cognitive Scientists tends to keep this mechanism at the level of a loose metaphor. User Centric Systems and Human Factors similarly should inform the development of physical or information systems, but too often remain in the focal domains of sociology and psychology, with the engineers and technologists lacking the human factors skills, and the social scientists lacking the technological skills. The key feature is that volumes must conform to the standards of both hard (Computing & Engineering) and social/health sciences (Linguistics, Psychology, Neurology, Philosophy, etc.). All volumes will be reviewed by experts with formal qualifications on both sides of this divide (and an understanding of and history of collaboration across the interdisciplinary nexus).

Indexed by SCOPUS

More information about this series at http://www.springer.com/series/11554

Dennis Sale

Creative Teachers

Self-directed Learners

 Springer

Dennis Sale
Singapore Polytechnic
Singapore, Singapore

ISSN 2195-3988   ISSN 2195-3996   (electronic)
Cognitive Science and Technology
ISBN 978-981-15-3471-3   ISBN 978-981-15-3469-0   (eBook)
https://doi.org/10.1007/978-981-15-3469-0

This Springer imprint is published by the registered company Springer Nature Singapore Pte Ltd.
The registered company address is: 152 Beach Road, #21-01/04 Gateway East, Singapore 189721,
Singapore

To my wife, Jane and daughters,
Adele & Lydia

Preface

In my previous book, *Creative Teaching: An Evidence-Based Approach*, I applied current research on human learning from the cognitive sciences to demystify the underpinning syntax of creative teaching, specifically identifying what creative teachers do and how they do it. This enables any motivated teaching professional to develop expertise through acquiring the necessary knowledge, understanding and skills through deliberate practice. That work established the basis for an *Evidence-Based Creative Teaching* (EBCT) *framework*, which provides the means for designing and facilitating effective, efficient and engaging learning experience for students, irrespective of delivery mode (e.g., face-to-face, blended or fully online).

This book applies the EBCT framework to major educational challenges that teachers face now, especially that of developing students' capability to be self-directed lifelong learners, equipped with twenty-first-century competencies. Students must be able to survive these turbulent times—euphemistically referred to as the *VUCA* world—as well as have opportunities to prosper, contribute to work and the community, and find purpose and meaning—experience well-being—in life. That is a tough challenge.

The book contains an extensive synthesis of the research literature across all fields of applied psychology, as well as related works in biology, philosophy and futurism. It provides a current evidence-based resource for helping teaching and training professionals to tackle today's curriculum and professional development challenges—as best as we can frame them and as best as we might thoughtfully address them.

It has also been written for persons who are interested in understanding how learning really works in terms of psychological processes and brain functioning in an easier to read format than is typically the case in this genre. Hence, I use a more informal narrative style, with many stories and some humour to illustrate key facts about human learning, specifically identifying factors that enhance—as well as inhibit—our competence for this essential capability. Such understandings and practices will help one's personal learning, self-regulation and well-being.

Chapter 1 frames the context for what follows in subsequent chapters, which is the systematic analysis and evaluation of the pedagogic issues and necessary core competencies for facilitating both teaching expertise and self-directed learning for students.

Chapter 2 provides a comprehensive synthesis of the extensive research relating to human learning, captured in terms of universal cognitive scientific principles, which I frame as the *Core Principles of Learning*. These constitute an evidence-based pedagogic framework—the essential knowledge bases underpinning teaching expertise—what I refer to as *Pedagogic Literacy*.

Chapter 3 argues that *Metacognitive Capability* is the superordinate twenty-first-century competence, essential for both creative teachers and self-directed learners. I extensively analyse the components of metacognition—how they work as a dynamic system in human psychological functioning—and, most importantly, how this unique human capability can be used to maximize learning, self-regulation and personal well-being.

Chapter 4 addresses the challenge of enhancing students' intrinsic motivation in school-based learning and, in essence, human motivation generically. Motivation underpins learning, in that without motivation, people do not bother to learn—or think—too much.

Chapter 5 tackles the psychological capability for creative thinking and how this is contextualized to everyday practical teaching and learning contexts—what I refer to as *Creative Teaching Competence*.

Chapter 6, using an evidence-based approach, demonstrates how the affordances of EdTech can be used to positively impact specific aspects of the learning process, hence providing better student attainment and engagement opportunities.

Chapter 7 critically analyses and evaluates what constitutes twenty-first-century competencies, how these are best derived from an evidence-based approach and the implications for framing educational aims and outcomes. I take as a core valuation that while *curriculum* must support industry requirements and employability, there is also a need to accommodate competencies for wider issues of well-being and citizenship.

Chapter 8 focuses on assessing twenty-first-century competencies. Different competencies, as well as different aspects of a competence, require different assessment methods. Assessing complex competencies such as metacognition in valid and reliable ways, in real application contexts, will be a big challenge for curriculum planners and teachers.

Chapter 9 provides an evidence-based framework for implementing professional development that is both practically viable in the real world of educational institutions and most likely to be perceived as such by teaching professionals themselves. While we increasingly know what to do and how to do it, high-quality professional learning comes at a cost in terms of time and resources.

Singapore   Dennis Sale

Acknowledgements

I have been privileged to have worked and lived in Singapore for the past 24 years. Now acknowledged as the best educational system in the world, and much is to do with the approaches taken, not least placing great value on teachers and their professional development. I have been part of Singapore's educational development, both in terms of unlimited opportunities for personal learning and professional growth, as well as a contributor to many of its educational innovations—so thank you Singapore.

Writing books on the *realities* of teaching is challenging, as it requires a wide range of participating professionals to open up their classrooms for me to experience their practice and engage in much mediation relating to its impact on learning. Acknowledgement and thanks go to the thousands of teaching professionals, from many educational/training sectors, countries and cultural contexts, who have shared their experiences, thoughts and feelings with me. It has been enriching and enjoyable.

Special appreciation goes to my excellent Research Team (Ngoh Shwu Lan, Cheah Sin Moh, Mark Wan, M. Fikret Ercan, M. Thiyagarajan, Roland Soh, Zhou Shang Ping, Ng-Soo Geok Ling and Wong Yunyi) who allowed me full access to observe their lessons, talk extensively to their students, as well as conducting their own Evidence-Based Reflective Practice.

I would also like to specifically acknowledge the following individuals who have made significant contributions to my thinking, research work and writing:

Geoff Petty, one of Britain's leading experts on teaching methods and author of Teaching Today and Evidence-Based Teaching: A Practical Approach. Apart from being inspired by Geoff's pioneering work in this area, I am especially grateful to him for his feedback on my work.

Ochan Kusuma-Powell and Bill Powell, veteran global international educators, and authors of numerous books on teaching and educational leadership. Their feedback and friendship over the years were invaluable in framing the style and direction of my writing.

Helene Leong (director of Educational Development at Singapore Polytechnic) and Mark Niven Singh (deputy director of Educational Development at Singapore Polytechnic) for their exceptional leadership and full support over the many years of research and professional development that has underpinned this work.

Johnmarshall Reeve, professor in the Institute of Positive Psychology and Education at the Australian Catholic University, for his input and support on research design in the field of intrinsic motivation.

My two research associates (Melissa Ng and Yiren Chua) and assistant in putting the book together (Vanessa Lee) for doing both good work and being such fun to work with.

Contents

About the Author

Dennis Sale has taught across all sectors of the British educational system, and for the past 24 years trained and coached over 10,000 teaching/training professionals in Singapore and most countries in the Asian region. He has invented curriculum and pedagogic models in the areas of metacognitive capability, creative teaching, and blended learning. His most recent books include Creative Teaching: An Evidence-Based Approach (Springer, 2015) and The Challenge of Reframing Engineering Education (Springer, 2013). While in Singapore, Dennis worked as Senior Education Advisor at Singapore Polytechnic for over 15 years, conducted research for the Ministry of Education, and provided consultancy and training for both private and public institutions across Asia. His work focuses on how humans learn best, translating evidence-based findings from cognitive science and related fields into professional development programs that offer high impact learning outcomes for both teachers and students. Dennis has set himself the personal goal of enhancing teacher expertise and learner capability globally.

Chapter 1
Making Sense of Teaching: From Mystery to Heuristics

Abstract This introductory chapter frames the context for establishing an evidence-based pedagogic framework for developing *Creative Teachers* and *Self-Directed Learners*. Self-directed learning is now considered an educational priority for students who will need new skills and attributes to successfully navigate the challenges and uncertainties of today's rapidly changing technological society. Equally, teachers will need heightened expertise, most notably in pedagogy and technology related skills, as many teaching roles will be challenged in an educational landscape where expectations of value are rising, but financial resources are decreasing. The chapter summarizes both the reasons why teaching has lacked the evidence-based approaches to practice that are the norm in other professions such as medicine and engineering and how it is now, from extensive research in the cognitive sciences and on what teaching methods work best, in a *learning revolution*.

1.1 Introduction

Teaching is the only major occupation of man for which we have not yet developed tools that make an average person capable of competence and performance. In teaching, we rely on the 'naturals', the ones who somehow know how to teach.

(Peter Drucker)

The title of the book 'Fifteen Thousand Hours' by Rutter (1982) is based on the approximate number of hours pupils spend in school. So, what did I learn in my 15,000 h? Well, I certainly acquired two useful skill-sets, football and boxing. What makes football such a good learning experience and a useful skill set to learn? Travel practically anywhere in the world and you will easily find soccer-players, who play in teams either in organized leagues or social set-ups. They love the game and the 'crack' (I think this term has Irish origins, 'craic'—loosely means fun; apologies if I'm wrong). The essential point is that this offers the opportunity for immediate membership and friendship in the local community.

D. Sale, *Creative Teachers*, Cognitive Science and Technology,
https://doi.org/10.1007/978-981-15-3469-0_1

To play football, at whatever level, requires the ability to function as a team-player (i.e. collaboration), which is a much muted so-called twenty-first-century competency in the educational and business literature. It's a necessary skill of human survival per se—didn't the cave-dwellers need it? It may have been more useful in their world to keep dangerous creatures and hostile human tribes from killing them, as compared to us soccer-players kicking a leather ball (it was made of leather in the 1960s) between wooden posts into a net. I would argue the same logic for good communication skills, which are important in soccer and essential in all aspects of human life. It may have been less complex and sophisticated in the caves than in the modern context, but it has been and will remain a key factor in our survival and progress as a species.

You may ask, can the same argument be made for the educational value of boxing? At the secondary school I attended in Hoxton, East London, boxing was a compulsory subject for boys and, as I had learned to box earlier, I was quite competent in the boxing ring by then. This meant that I was not a target for bullies and was able to keep my school lunchbox and eat its contents—which figures highly in terms of any notion or hierarchy of human needs, for me anyway. There were many of my peers who were not so privileged at school. Apart from my daily nutritional needs being met, I feel—even now—that I did learn to be resilient, self-disciplined and respecting of others from my boxing experience. The boxing instructor, a Second World War veteran was a tough man but had core values of honesty, integrity and fairness. Apart from the rigorous training for physical skill development, strong ethical components relating to human conduct were also emphasized. To be frank, and with no conceit, life at university was easy. In contrast to the experience of getting up at 5 am, doing what seemed an eternity of a run, often swimming across a freezing cold lake, followed by 100+ sits-ups—and how I hated that relentless skipping—a 9 am start at university, 4–5 h of lectures a day, and studying something I was actually interested in, was nothing compared to facing the 'tough boys' in the boxing ring and on the streets of Hoxton. Could I deal with exam stress and put in a hard shift at university and now at work? No sweat.

Sadly, the formal curriculum at school was a pretty tedious experience. Weeks were spent learning about a plethora of irrelevances; I can still recall in the biology class learning about spirogyra, a hermaphroditic pondweed. As an East London youth in the 1960s, was I interested in or see any practical use for such knowledge? Similarly, in music, I was occasionally caned for messing about while being taught (but failing to learn) scales in music and how to play the flute. On Top of the Pops (a weekly music show in Britain in the 1960s), the Beatles, the Rolling Stones and other favourite bands played the guitar, and sang pop music. There was no connection between the music I enjoyed out of school and what we did in music class. This was the typical learning experience in school. We never really thought about what we were learning or why; it was just school and we went there from Monday until Friday.

1.2 What I Learned About Teachers: The Good, the Bad and the Ugly

Certainly, there were things I learned about teaching and teachers. Most significantly, the teachers were not alike, far from it. The 1966 film "The Good, the Bad and the Ugly", which starred one of my favourite actors, Clint Eastwood, comes readily to mind when thinking about my teachers. Let's lump the "bad and the ugly" together and this was my maths teacher for 'O' level. Unintelligible on all counts, the lesson may just as well have been taught in a Malagasy dialect. I sat the 'O' level mathematics exam in June 1968 and achieved the undistinguished grade of 9 (6 being the lowest passing grade at that time). You may be wondering what a grade 9 means in the context of mathematical competence? Well, if my Jack Russell would have sat the same exam, he could not have fared worse. Jack (what else would you call a Jack Russell) is a clever dog but still hasn't worked out how to open the fridge door and get to his chicken meal autonomously.

Fortunately, in the following academic year, I had a change in maths teacher, Mr. Edrich, and he represented an example of 'the good'. We could understand what he was saying, and he recognized that we were far from confident or competent in maths, which was not rocket science to ascertain. However, most importantly, instead of communicating negative and pessimistic views, he communicated to us that with effort and some hard work we could learn this stuff. In terms of his teaching, as I remember it, he slowed down the pace and kept providing examples and non-examples (I did not recognize this strategy then) and we gradually began to understand and eventually do those basic factorization procedures. I was able to make some sense of how simultaneous and quadratic equations worked, and I could increasingly solve the questions set. However, it was not a deep understanding as I re-sat the maths 'O' level in the November series and still failed it—but only just, a grade 7. The happy ending was that in the following June exam series in 1969, I passed with a grade 3, which was very credible in context. If I am honest, I don't think I ever achieved a deep understanding of maths at that time, but passing it was crucial as it was a high stakes exam; without it, I could not have got a place at university on a B.Sc. Programme. Mr. Edrich will never know his specific and positive impact on my life, and this is the norm for many teachers. Unfortunately, I did not consider thanking him at that time, as it's only retrospectively, and much later, as I came to fully realize how important such teachers are.

In my 40 plus years as a teaching professional of various genres, similar experiences emerge and play out in terms of different perceptions of teacher's abilities. For example, as a classroom teacher, there were many instances of parents asking me something akin to, "Is there any way I can avoid my daughter Linda being taught by Mr. Lee next term? Everybody knows he's dead boring and students can't understand what he's on about." Parents certainly know that teachers vary greatly in competency (however defined) and that their children's performance is not just a reflection of a fixed innate capability in the subject, but varies considerably depending on who teaches them. Similarly, as a parent, it was very apparent that both my daughters' enjoyment and grades reflected, in no small part, the experiences that certain teachers created for them throughout the course programme. In one situation, for my youngest daughter, I ended up paying for private tuition, on top of the already expensive expatriate private school fees. In the previous year, she was meeting the attainment targets comfortably and enjoyed the subject. However, in this particular year, she found the teacher less friendly, not easy to follow and the experience generally dull. Her interest waned and her grades fell significantly. Whatever one's views on educational equity, and I favour providing as much equality of opportunity as possible for all, by not paying for this extra tuition from a different (maybe better) teacher could have resulted in an outcome similar to my first two attempts at the 'O' level maths exam, back in the 1960s. I was lucky I had Mr. Edrich to teach me that darn math, and he was not on extra salary.

It is not surprising, therefore, that an increasing body of research shows the massive impact that teachers have on student attainment. Izumi and Evers (2002), from an overview of research on the impact of teachers on student achievement, summarized:

> …nothing is as important to learning as the quality of a student's teacher. The difference between a good teacher and a bad teacher is so great that fifth-grade students who have poor teachers in grades three to five score roughly 50 percentile points below similar groups of students who are fortunate enough to have effective teachers. (ix)

Hargreaves and Fullan (2012), which is even more damning, documented that:

> … the Los Angeles Times reporters gained access to 7 years of value-added test performance data for 6,000 third through fifth-grade teachers in English and Mathematics in the Los Angeles Unified Public-school District – one of the poorest districts in the United States. They passed the data to expert economists, who came up with an even more remarkable finding. There were differences of up to 41% in value-added performance between teachers of the same kind of children in the very same school. (p. 15)

At the school level, Rowe and Rowe (1993) argued:

> Based on our findings to date it could be argued that effective schools are only effective to the extent that they have effective teachers. (p. 15)

Petty (2009) fully contextualized the importance of good teachers in real-life terms when he wrote:

> Good teachers touch people's lives forever. If you teach well, some of your students will only succeed because of your excellent teaching. They then might go on to get more advanced qualifications and skills, again just because of your expert teaching. Then they might get a career, indeed a whole life, built on your excellent teaching. No other profession is that consequential and enabling. (v)

As a pupil at school, a practising teacher, a parent, and a teacher educator in many educational sectors and contexts, I have experienced the good, the bad and the ugly, and I don't particularly feel good about the bad and the ugly. Most importantly, as there is now a solid evidence base to validate the high impact of teachers on attainment and that's without the impacts on socio-psychological aspects such as self-efficacy, self-esteem and well-being, we should not shirk the responsibility of enhancing teaching expertise as a priority goal in educational policy. It is shocking, though perhaps not that surprising, that there has been a lack of clarity on what constitutes highly effective creative teaching and how this can be systematically incorporated into professional development programmes. The basis of highly effective teaching, let alone creative teaching, has long been debated in the educational literature. For example, Ornstein (1995) from reviewing the literature, suggested that "…few facts concerning teacher effectiveness have been established" (p. 77). In the following sections, I will make the case that while much has changed in terms of our understanding of teaching effectiveness since such reviews, much of actual practice still seems rooted in confusion over what constitutes effective teaching. To understand this better, let's take a short tour into the nether regions of Educational Jurassic Park.

1.3   A Short Tour into Educational Jurassic Park

Much of the confusion about what is or should be good teaching can be explained in large part by Sallis and Hingley's (1991) assertion that "…education is a creature of fashion" (p. 9). I like this analogy as it is so grounded in my experience of fashion. While I have relatively little interest in fashion now, my teenage years were spent in the 1960s, the era of great musical bands, full employment, a real sense of optimism about the future and, of course, the famous fashion icon, the mini-skirt. This was the world as I knew it, and this was ladies fashion as I experienced it, and it seemed an objective reality of what was natural. I was, of course, unaware of such notions as 'socially constructed realities' (Berger and Luckman 1967). The reality was exactly as I perceived it, what else could it be? Equally, I never considered what it might have been like for the ladies wearing such attire in the winter months. However, one evening, my mother was showing me pictures of herself when she was young, and the thing I noticed was the long skirts she wore. I vividly remember commenting that this seemed strange and I was glad that evolution had moved on from then. It never dawned on me that fashion was the product of a deliberate industry ploy that systematically creates, manages and periodically changes images of desirable attire. After all, it must do this once the marketed item is saturated—how many pairs of flair bottom trousers can you fit in a typical male wardrobe? In defence of my lack of understanding on such matters, I don't think many 14-year-olds at that time were well versed in such sociological imagination either. It's no big deal that fashion in clothes is manipulated to ensure new revenue is generated and novelty is added to an aspect of human experience. However, I am far less comfortable when this is applied to professions, especially where human well-being is concerned, and teaching meets that criteria full on. It is inevitable that any profession, indeed any aspect of human activity, can only be as good as the most current knowledge bases—conventional wisdom—of the time. Let's not blame Ptolemy for thinking that the Earth was the centre of the universe, it would have made perfect sense at his time. Similarly, before we could have known otherwise, the simple notion of gravity, it makes perfect sense to believe that the Earth is flat. In the case of teaching, I take the stand that much professional activity has not sufficiently and consistently reflected knowledge bases that could have improved practice at the level of student learning opportunities and outcomes (e.g., attainment, engagement, well-being). For example, for those of us who have been in, or around, the profession over the past 30–40 years, one could not have failed to notice such major shifts in teaching focus from 'traditional' to 'progressive' education and then 'back to basics', as well as, more recently, the teacher's role allegedly changing from 'sage on the stage' to 'guide on the side.'

A negative consequence of this contested nature and periodic radical reframing of what constitutes good teaching is that it does little to convince anybody that teaching is truly a profession with well-constituted bases of professional knowledge, as is the case for medicine or engineering. This is not to say that the medical profession, or other well-established professions, have not gone through similar epochs of fads masquerading as practice, as Thomas' (1979) depiction of the medical profession before the drive towards evidence-based practice portrayed:

> It is hard to conceive of a less scientific enterprise among human endeavours. Virtually anything that could be thought up for treatment was tried out at one time or another, and, once tried, lasted decades or even centuries before being given up. It was, in retrospect, the most frivolous and irresponsible kind of experimentation, based on nothing but trial and error, and usually resulting in precisely that sequence. (p. 159)

One would probably be both shocked and frightened if, on a visit to a modern medical centre, the doctor produced a saw, some leeches and asked you to drink a large dose of alcohol. We now see increasing sophistication of practice through a whole range of complex technology infrastructure. This is not to argue that all is well in the medical profession and there are probably still some 'dodgy' practices. However, it feels like the profession, in most modern societies anyway, is now largely driven by established and rigorous standards of research and validation, which seems to be relatively lacking in the context of education. Indeed, one may argue that this is visibly apparent as many classrooms look pretty similar to what they were decades or even centuries past. However, the major reason for the slower acceleration towards accepted high professionalism in teaching is that much of practice is still largely driven by dominant paradigms or perspectives in psychology and pedagogy, rather than a solid empirical base. Paradigms are ways of looking at things in the world (e.g. the meaning of life, human conduct, educational aims and practice) and contain certain premises and methodologies relating to those particular domains of reality. These, in turn, shape how we perceive and orientate ourselves to such realities. Kuhn (1996) famously noted that when socialized into a paradigm, it becomes a prerequisite to perception itself:

> What a man sees depends both upon what he looks at and also upon what his previous visual-conceptual experience has taught him to see. (p. 113)

World religions are other notable examples of a paradigm in that they typically contain explicit assumptions about the nature of reality (e.g., a belief in a metaphysical being, absolute codes of conduct, building a relationship with that being) which shape the thinking and behavioural aspects of adherents to specific faiths. In education, much has been similar, though lacking adherence to a metaphysical being, only the occasional psychological guru, which may have had similar impacts in terms of practice. Prominent paradigms in education have included 'behaviourism', 'cognitivism' and, probably the most dominant one in terms of 'current vogue', 'constructivism'. These paradigms do offer insights relating to aspects of the learning process and provide some useful overall framing for approaching teaching. However, they are far from constituting a comprehensive evidence-based framework that has strong predictive value in terms of enhancing student attainment. The danger of limiting practice to one paradigm is well captured by Pratt (2002):

> Perspectives are neither good nor bad. They are simply philosophical orientations to knowledge, learning and the role and responsibility of being a teacher. Therefore, it is important to remember that each of these perspectives represents a legitimate view of teaching when enacted appropriately. Conversely, each holds the potential for poor teaching. (p. 14)

Anderson et al. (1998) are more explicit in identifying the problem when they argued that:

> What is needed more than a philosophy of education is a science of education. Modern attempts at educational improvement point back to theorists (Piaget, Vygotsky, and Dewey) whose theories are vague by current psychological standards and lack the strong connection to empirical evidence that has become standard in the field. (p. 237)

Mayer (2004) is even most blunt in advocating the necessity of making the kind of changes in approach to practice that have occurred in other more established professions. He argued that we must:

> ...move educational reform efforts from the fuzzy and unproductive world of ideology - which sometimes hides under the various banners of constructivism – to the sharp and productive world of theory-based research on how people learn. (p. 18)

Finally, the problem appears systemic, both in terms of policy and practices, and shapes the socialization of recruits into the profession, which is the hallmark of a paradigm. Stone's (2000) criticism of some teacher education programmes further illustrates the continuation of paradigms rather than evidence-based practice in the training of teachers:

> What teachers are told, however, is that student differences are important and if their teaching is truly creative, energetic and engaging, they will succeed in individualising and bringing forth the best from all students. In effect, teachers are being taught to make diagnoses that heighten their awareness of differences without advancing their ability to teach. (p. 43)

In consequence, this has created much confusion for many teaching professionals as to what is good pedagogy (indeed, what is pedagogy) and what are truly useful knowledge bases from which we can design and facilitate instructional strategies with high predictive value in terms of meeting desired learning outcomes. It is unlikely that many in the teaching profession believe that this is the result of limited available literature on teaching and how to teach , just as there is no shortage of writings on other topics of educational relevance, such as parenting. However, the confusion does not seem to abate. Hattie (1999), for example, stated:

> A glance at the journals on most shelves of most libraries, my colleagues' shelves, and on web pages would indicate that the state of knowledge in the discipline is healthy. The worldwide picture is certainly one of plenty. (p. 1)

However, in the same address, he argued that:

1. Teachers/researchers have models of learning that are rarely externally elaborated or asked for
2. Teachers/researchers seek evidence to buttress their models of learning and thus rarely seek to refute them or introduce major changes.

> We all seek positive evidence in that which we love. Teachers/researchers, like lovers, are often blind. (p. 2)

As a result, to quote Hattie again in this context, this results in:

> …a school community peopled with teachers with self-fulfilling prophecies, all believing they are doing a good job, and with models of learning rarely based on any other evidence than that "it works for me". As well, we have an educational research community peopled with academics chasing their pet theory, promoting their methodology while passing each other in corridors, and rarely asking for negative evidence, and pushing with a passion that "if only the teachers would do this, or know that". Both educational communities work behind closed doors, coming out to discuss kids, curricula, accountability, and each other, but rarely discussing the fundamental tenets about their teaching that leads to positive impacts on student learning. (p. 2)

A particularly notable example, that fully illustrates the above analysis, is that of *Learning Styles* which has shaped the thinking and practices of many teachers worldwide. Over the years, I have had many heated debates on this topic and always refused to conduct workshops or seminars on it, as I felt it was, at best, an ephemeral entity in the learning and attainment stakes. From an evidence-based point of view, it now seems little more than 'folk psychology', and I can take some solace in that. As Hattie (2009) summarized:

> One of the more fruitless pursuits is labelling students with 'learning styles'. This modern fad for learning styles, not to be confused with the more worthwhile notion of multiple learning strategies, assumes that different students have differing preferences for particular ways of learning.
>
> Often, the claim is that when teaching is aligned with the preferred or dominant learning style (for example, auditory, visual, tactile, or kinesthetic) then achievement is enhanced. While there can be many advantages by teaching content using many different methods (visual, spoken, movement), this must not be confused with thinking that students have differential strengths in thinking in these styles. (p. 89)

1.4 Moving Out of Educational Jurassic Park

Much is changing as far as teaching is concerned and it may, as Petty (2009) argued, be ready to:

> …embark on a revolution, and like medicine, abandon both custom and practice and fashions and fads, to become evidence-based (cover page).

In terms of paradigms, there is a significant shift towards a more evidence-based approach to learning and teaching—what Petty framed as 'Evidence-Based Teaching'. One may argue that this is simply another paradigm shift and may not constitute a more valid or verifiable base of knowledge from which to design and enact the practices we call teaching. Indeed, this is very much the standpoint of more radical forms of constructivism. For example, Lincoln (1990) pointed out:

> The constructivist paradigm…has as its central focus, not the abstraction (reduction) or the approximation (modelling) of a single reality but the presentation of multiple, holistic, competing, and often conflicting realities of multiple stakeholders and research participants (including the inquirers). (p. 73)

Invariably, one cannot escape the essential subjectivity of experience, and suggestions of a value-free science are untenable. However, I feel it is necessary to retain at least a critical operational notion of objectivity as a 'regulatory ideal'; otherwise, there is little point in conducting an inquiry, whether it be about good teaching or good soccer, or whatever. As Phillips (1990) argued:

> If we abandon such notions, it is not sensible to make inquiries at all. For if a sloppy inquiry is as acceptable as a careful one, and if an inquiry that is careless about evidence is as acceptable as an inquiry that has taken pains to be precise and unbiased, then there is no need to inquire… (p. 43)

In this context, to argue that there are no better nor worse ways in which to design student learning experiences is both absurd and dangerous. As Ramsden (1992) wrote:

> It is a folly to suggest that there are no better or worse ways of teaching, no general attributes that distinguish good teaching from the bad. (p. 87)

It is now firmly established that there is a strong evidence base relating to how best to design and facilitate the various practices we call teaching that can significantly enhance student learning opportunities, attainment levels and the experience of learning (e.g. intrinsic motivation). This change is an inevitable result of our increasing knowledge relating to how humans learn, what teaching methods and practices work best and why, and the unpacking of what the best teaching practitioners do and how. Much of this significant research on learning has already been documented in the literature (e.g., Bransford 1999; Marzano 2007; Mayer and Alexander 2010; Hattie and Yates 2014). Collectively, the research evidence is now providing us with a heightened pedagogic understanding of the various facets of highly effective teaching and, when this is used creatively in context, it will optimize attainment and engagement for a wider range of student groups. In most basic terms we can now engage in useful dialogue about professional knowledge and practices in teaching from a validated empirical base, much as is the norm for the more established professions (e.g., medicine and engineering). Indeed, even two decades ago, Marzano (1992) argued:

> ...over the past 3 decades, we have amassed enough research and theory about learning to derive a truly research based-model of Instruction. (p. 2)

More recently, Darling-Hammond and Bransford (2005), from surveying the research findings, concluded that:

> There are systematic and principled aspects of effective teaching, and there is a base of verifiable evidence of knowledge that supports that work in the sense that it is like engineering or medicine. (p. 12)

There is no doubt that our understanding of how humans learn is rapidly increasing, especially as the fields of cognitive and social neuroscience provide further insights into brain functioning at the neurological level, and how this plays out in terms of human cognition and behaviour relating to learning. Equally, and fully consonant with this heightened understanding of human learning, is the accumulation of extensive and rigorous research activity, which is uncovering from a strong empirical base what teaching methods tend to work best and on what basis. Perhaps most publicized in this area is the work of Hattie (e.g., 2009, 2012), though many others have been providing significant contributions over recent years (e.g., Bransford 1999; Marzano 2007; Mayer and Alexander 2010; Petty 2009). Mansell (2008) referred to Hattie's seminal work on the effectiveness of different teaching methods and strategies as:

> ... perhaps education's equivalent to the search for the Holy Grail - or the answer to life, the universe and everything.

Also, Hattie's work was a definitive landmark in educational research, providing a key push towards evidence-based practice in teaching. Hattie synthesized over 800 meta-analyses of the influences on learning and, most significantly, he was interested not just in what factors impacted learning, but the extent of their impact—referred to as Effect-Size. Effect size is a way to measure the effectiveness of a particular intervention to ascertain a measure of both the improvement (gain) in learner achievement for a group of learners and the variation of learner performances expressed on a standardised scale. By taking into account both improvement and variation it provides information on which interventions are most worth implementing.

Hattie firstly identified the typical effect sizes of schooling without specific interventions, for example, what gains in attainment are we likely to expect over a one-year academic cycle? Typically, for students moving from one year to the next, the average effect size across all students is 0.40. Hence, for Hattie, effect sizes above 0.40 are of particular interest. As a baseline an effect size of 1.0 is massive and is typically associated with:

Table 1.1 Examples of effect sizes in learner attainment from Hattie's Meta-analysis

Influence	Mean effect size
Whole-class interactive teaching (direct instruction) A specific approach to active learning in class, which is highly teacher led, but very active for students. This involves summaries reviews and a range of active learning methods, including questioning	0.81
Feedback Students getting feedback on their work from the teacher or from themselves (self-assessment or from peers or some other sources	0.73
Metacognitive strategies Explicit teaching and use of metacognitive strategies (e.g., conscious planning, monitoring, and evaluating of thinking and learning	0.69
Challenging goals for students Goals that students can meet through effort on their part—they should be a specific as possible, and meaningful to the students involved	0.59
Advance organizers Giving students a summary in advance and a purposes for the learning	0.41

- Advancing the learner's achievement by one year
- Improving the rate of learning by 50%
- A two-grade leap in GCSE grades.

Selected examples of high effect size methods are presented in Table 1.1.

However, as Hattie notes, it is important to balance effect size with the level of difficulty of interventions. For example, providing advance organizers (summaries in advance of the teaching) have an effect size of 0.41 which is pretty average, but they only take up a few minutes at the beginning of the lesson, and potentially offer the equivalent of moving up a year in terms of a student's achievement.

He goes on to make relative comparisons of intervention use, which enables us to go beyond identifying the effect sizes for particular innovations (e.g., deliberative intervention involving a strategy/method, used for a group of students), and ascertain whether the effects of a particular innovation were better for students than what they would achieve if they had received alternative innovations.

Of particular significance is the fact that it is not just the effect size of one intervention that is important, but how several effective methods can be strategically and creatively combined to produce powerful instructional strategies that significantly impact student attainment. As Hattie (2009) pointed out:

> …some effect sizes are 'Russian dolls' containing more than one strategy. For example, 'feedback' requires that the student has been given a goal, and completed an activity for which the feedback is to be given; 'whole-class interactive teaching' is a strategy that includes 'advance organisers' and feedback and reviews. (p. 62)

For readers not familiar with 'Russian Dolls', they are a set of different sized dolls, usually around 5, and they fit one inside another from the smallest to the biggest. Figure 1.1 provides a visual example of high effect method combination.

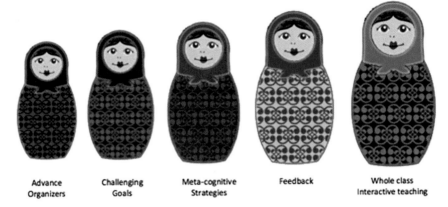

| Advance | Challenging | Meta-cognitive | Feedback | Whole class |
| Organizers | Goals | Strategies | | Interactive teaching |

Fig. 1.1 Illustration of Hattie's Russian Doll analogy

The Russian Doll analogy provides an easy to remember generic advance organizer for planning lessons as it should easily evoke the key question of what strategy or method combination is likely to be most effective for the particular student group. However, as will be explored in subsequent chapters, some methods may have a better overall impact on student attainment but in learning, as in all aspects of life, too much of a good thing often leads to habituation and boredom and subsequently loses its impact. Also, in designing the overall instructional strategy, we must take into account the learner profile (especially prior competence and motivational status), the learning outcomes to be attained, and the available resource facilities that can be accessed.

The interested reader can refer to Hattie's original works (e.g., Hattie 2009, 2012) for the extensive detailed coverage of the research methodology employed and the full range of effect sizes for different instructional methods and learning strategies.

1.5   Moving Teaching from Mystery to Heuristics

In summary to this introductory chapter, I offer an analogy between recent developments in knowledge bases relating to how humans learn and the effectiveness of different teaching methods with Martin's (2009) depiction of the *Knowledge funnel* (Fig. 1.2). In developing our understanding of the nature and working of things in the world, he depicts a process in which phenomena can move from being a *Mystery* (experienced in some way but not understood) to a *Heuristic* (understandable in good part) and finally to *Algorithmic* (fully understood, predictable and controllable). A highly significant global example of moving down the knowledge funnel can be seen in terms of our understanding and subsequent response to the discovery of the Acquired Immune Deficiency Syndrome (AIDS). AIDS was only formally identified as a new disease in 1981 and remained a mystery until 1983 when the

Fig. 1.2 Martin's
knowledge funnel

Human Immunodeficiency Virus (HIV), the virus that causes AIDS, was discovered. As is well documented, the origins of the virus, as well as the specific means and rates of transmission, evoked much controversy and debate at that time. Also, treatments, despite the initial optimistic hype of azidothymidine (AZT), were controversial and driven more by desperation rather than hard science which, in context, was understandable. Today, HIV is clearly in the realm of heuristics, in that we have a comprehensive understanding of the behaviour of the virus and its impact on the immune system, as well as how to mitigate its most deleterious effects—in an increasing number of cases—to the progression into full-blown AIDS. At present, our knowledge is not Algorithmic, as this level of knowledge would require both a sterilizing cure (where the virus is completely eradicated from the body), as well as an effective vaccine. However, as Avert (2019), which provides global information on HIV and Aids, summarizes:

> There is no cure for HIV, although antiretroviral treatment can control the virus, meaning that people with HIV can live long and healthy lives. Most research is looking for a functional cure where HIV is reduced to undetectable and harmless levels in the body permanently, but some residual virus may remain.

In terms of Martin's knowledge funnel, I am suggesting that, as far as teaching is concerned, we have moved a long away from it constituting a Mystery to one of clearly identifiable and understandable Heuristics. Will teaching ever be heuristic? A decade or so ago, I would probably have said 'never'. However, with the growth of knowledge bases in the fields of biology and artificial intelligence, and notions of 'Radical Evolution' (e.g., Garreau 2005), which will be explored in context later—who knows? Presently, we may see some powerful learning environments being produced by what is referred to as 'mixed reality' technologies (e.g. (merging of real and virtual worlds to produce new environments and visualizations where physical and digital objects co-exist and interact in real-time), but I would suggest that this is far from algorithmic, working with the human brain as it is.

For now, we are firmly working in the realm of heuristics, which provides an optimistic frame in terms of better understanding the present educational landscape, the challenges we need to address, and ways to enhance the practices of teaching. As Martin (2009) notes, heuristics:

> …represent an incomplete yet distinctly advanced understanding of what was previously a mystery. But that understanding is unequally distributed. Some people remain stuck in the

world of mystery, while others master its heuristics. The beauty of heuristics is that they guide us toward a solution by way of organized exploration of possibilities. (p. 12)

It is to be noted that heuristics in this context retains the more generic notion of 'rules of thumb' that enables people to solve problems and make judgments quickly and efficiently, but extends the concept to include existing (but as-to-yet, incomplete) knowledge about phenomena in the world. In this way, good heuristics will enable teaching professionals to design and facilitate learning experiences effectively and efficiently from a sound pedagogic base but, as the term denotes, not with the certainty of outcomes in all situations. How this works in practice will be explained and illustrated in the forthcoming chapters. It certainly constitutes a significant shift towards a more substantive evidence-based profession, and reflects very strikingly the description by Perkins (1992) of the 'unequal distribution of knowledge' concerning what we know about learning and teaching and what happens in many classrooms:

...we do not have a knowledge gap – we have a monumental use-of-knowledge gap. (p. 2)

1.6 Summary

In many ways, this book represents a convergence on differing conceptions of teaching as 'art', 'craft' or 'science' in the research literature (e.g., Eisner 1995). We may be finally moving towards a situation in which there is both increasing understanding and capability to develop the practices and tools of effective—even creative teaching, which Drucker suggested were only previously known by "the naturals, the ones who somehow know how to teach".

References

Anderson JR et al (1998) Radical constructivism and cognitive psychology. In: Ravitch D (ed) Brookings papers on education policy: 1998. Brookings Institution, Washington, D.C., pp 227–255

Avert (2019) Global information on HIV and Aids. https://www.avert.org/. Last accessed 30 Nov 2019

Berger PL, Luckmann T (1967) The social construction of reality: a treatise in the sociology of knowledge. Anchor Books, New York

Bransford J et al (1999) Brain, mind, experience & school. National Academy Press, Washington, D.C.

Darling-Hammond L, Bransford J (2005) Preparing teachers for a changing world: what teachers should learn and be able to do. Jossey-Bass, San-Francisco

Eisner EW (1995) The art and craft of teaching. In: Ornstein AC, Behar LS (eds) Contemporary issues in curriculum. Allyn & Bacon, Massachusetts

Garreau J (2005) Radical evolution: the promise and peril of enhancing our minds, our bodies—and what is means to be human. Doubleday, New York

Hargreaves A, Fullan M (2012) Professional capital: transforming teaching in every school. Teachers College Press, New York

Hattie J (1999) Influences on student learning. Inaugural lecture. University of Auckland

Hattie J (2009) Visible learning. Routledge, New York

Hattie J (2012) Visible learning for teachers: maximizing impact on learning. Routledge, London

Hattie J, Yates GCR (2014) Visible learning and the science of how we learn. Routledge, New York

Izumi TL, Evers WM (2002) Teacher quality. Hoover Institutional Press, San Francisco

Kuhn TS (1996) The structure of scientific revolutions. The University of Chicago Press, Chicago

Lincoln YS (1990) The making of a constructivist: a remembrance of transformations past. In: Guba EG (ed) The paradigm dialogue. Sage, London

Mansell W (2008) Research reveals teaching's Holy Grail. TES Newspaper on 21 November

Martin R (2009) The design of business. Harvard Business Press, Massachusetts

Marzano RJ (1992) A different kind of classroom. ASCD, Alexandria, VA

Marzano RJ (2007) The art and science of teaching: a comprehensive framework for effective instruction. ASCD, Alexandria, VA

Mayer RE (2004) Should there be a three-strikes rule against pure discovery learning?. Am Psychol 59(1):14–19

Mayer RE, Alexander PA (2010) Handbook of research on learning and instruction. Routledge, London

Ornstein AC (1995) Teaching: theory into practice. Allyn & Bacon, Needham Heights, Massachusetts

Perkins DN (1992) Smart schools. The Free Press, London

Petty G (2009) Evidence-based teaching: a practical approach. Nelson Thornes, Cheltenham

Phillips DC (1990) Postpositivistic science: myths and realities. In: Guba EG (ed) The paradigm dialogue. Sage, London

Pratt DD (2002) Good teaching: one size fits all? In: Ross-Gordon JM (ed) Contemporary viewpoints on teaching adults effectively, no 93. Spring 2002. Jossey-Bass, San Francisco, pp 5–15

Ramsden P (1992) Learning to teach in higher education. Routledge, London

Rowe KJ, Rowe KS (1993) Assessing student behavior: the utility and measurement properties of a simple parent and teacher-administered behavioural rating instrument for use in educational and epidemiological research. Paper presented at the annual conference of the Australian association for research in education. Fremantle, WA

Rutter M (1982) Fifteen thousand hours: secondary schools and their effects on children. Harvard University Press, Cambridge, MA

Sallis E, Hingley P (1991) College quality assurance systems in Mendip Paper D20. Coombe Lodge, Bristol

Stone JE (2000) Teacher training and pedagogical methods. In: Teacher quality conference. Hoover Institution/Pacific Research Institute, May 12

Thomas L (1979) The Medusa and the Snail: more notes of a biology watcher. Viking Press, New York

Chapter 2
Towards an Evidence-Based Pedagogic Literacy: The Core Principles of Learning

Abstract This chapter provides a comprehensive synthesis of the extensive research relating to human learning and its practical implications for designing and facilitating learning events that have high predictive capability for enhancing students' learning and well-being. It documents the key research areas that have led to a greater understanding of what teaching methods and strategies work best and why, as well as the underlying cognitive scientific principles that facilitate effective and efficient learning. Such understandings constitute an evidence-based pedagogic framework—the essential knowledge bases underpinning teaching expertise—what I refer to as *Pedagogic Literacy*. This forms the basis for a professional development approach that is grounded in what we know about human learning and, based on this, can be translated into high impact instructional strategies.

2.1 Introduction

Chapter 1 took you on a short tour into Educational Jurassic Park, and how EBT (Petty 2009) can close its gates forever, keeping teachers safely on the outside. I used Martin's depiction of the 'knowledge funnel' as a conceptual frame to position effective teaching as moving away from being a Mystery to one of potentially useful Heuristics. Underpinning such heuristics are increasing knowledge bases on how humans learn and extensive research on what teaching methods work best and on what basis—enabling teaching to adopt evidence-based approaches to practice as in the case of other professions. The pioneering work of Hattie (2009) outlined in that chapter opened up the possibility of a more comprehensive evidence-based approach to all aspects of pedagogy.

Pedagogy is a much-used term by educationalists and other personnel in the learning industry when talking about matters of curriculum, teaching and learning. In many curriculum-related meetings, I am still amused by the plethora of terminology that surface in this area (e.g., pedagogical approach, pedagogic practices, pedagogical content knowledge and, more recently, signature pedagogies). However, what is equally apparent is that for many there is still a high level of conceptual confusion, as I still get asked questions such as, "Is there one pedagogy or many"? This can

© Springer Nature Singapore Pte Ltd. 2020
D. Sale, *Creative Teachers*, Cognitive Science and Technology,
https://doi.org/10.1007/978-981-15-3469-0_2

be explained as just another consequence of the periodic radical reframing of what constitutes good teaching, so it is not surprising that people are confused, because there is confusion. This chapter will reduce such confusion and, more importantly, offer a pedagogical framework that is firmly grounded in the increasing evidence-bases relating to how humans learn and what teaching methods work best and why.

Historically the term pedagogy seems to have been derived from the Greek words 'paid', meaning "child" and 'agogos' meaning "leader of". This essentially frames pedagogy as referring primarily to the teaching of children. Mortimore (1999), in a comprehensive review of the literature on pedagogy, noted that approaches to pedagogy have gone through various phases, focusing on such aspects as 'teaching styles', 'paradigms of learning', 'models and methods of teaching' and 'the context of teaching'. He, not surprisingly, concluded that:

> Pedagogy has been seen by many within and outside the teaching profession as a somewhat vague concept. (p. 228)

More recent definitions (e.g., The Free Dictionary 2019) have dropped the reference to the child and applied it more generically to "the principles, practice, or profession of teaching" or "the activities of educating or instructing". Pedagogy has also been contrasted with the term *Andragogy* (Knowles 1973), which focuses on the teaching of adult learners. This has invariably led to further confusion, and questions being asked as to whether or not adults learn differently from children and should they be taught differently, and in what ways and how. The issue of differential learning between adults and children is addressed in some detail in Chap. 9, but key points are summarized here for purposes of context in this chapter.

Certainly, there are significant differences in the level of prior experience of adults, as compared to children. Adults also choose what they want to learn, and this is typically consciously directed to meet work or personal learning goals. Kids at school are largely told what to learn, at least in the earlier years. However, whilst there are important motivational and life experience differences for adults, it is questionable whether the underlying learning process is structurally different from that of children who have attained the stage of formal operational thought (Piaget 2001), typically around 12–15 years of age. At this stage of brain maturation, children can reason logically and use a range of thinking skills (e.g., analyze, compare & contrast, make inferences and interpretations, and evaluate). In some ways, this has similarities with the notion of different learning styles, which was popular in the educational literature for a couple of decades. As outlined prior, research has far from validated such theories, especially their usefulness in terms of pedagogically beneficial applications. I agree with Schank (1997) who argued that:

> Contrary to common belief, people don't have different learning styles. They do, however, have different personalities. The distinction is important because we need to be clear that everybody learns in the same way. (p. 48)

A similar frame is made by Goulston (2009), who argued that:

> While our lives and our problems are very different, our brains work in similar ways. (p. 3)

While philosophical discussions on how best to frame pedagogy will inevitably continue and this is pertinent to critical educational discourse, it has limited usefulness for busy teaching professionals seeking practical guidance on how best to design effective learning experiences and conduct their teaching practices skilfully. The present scenario is analogous to completing a large complicated jig-saw puzzle, and we don't have all the pieces (some are missing). However, we have enough pieces and the intelligence to construct a sufficiently useful picture of what evidence-based creative teaching entails, and what is required for successful enactment in practice. It is useful to have strong empirical evidence that teachers do make the most significant difference (positive or otherwise) in terms of student attainment levels and, possibly, well-being—at least in the school context. However, we must go further to frame an evidence-based pedagogic framework, based on what the most effective teachers do, and how this enhances student learning (e.g., the psychological processes involved) in specific ways. It is only from such an evidence-based approach can we produce professional development programmes which have high predictability in terms of improving teaching practices and student attainment and engagement levels. Similarly, Hattie's (2009) summary of differential teacher proficiency is salient in this context:

> Not all teachers are effective, not all teachers are experts, and not all teachers have powerful effects on students. The important consideration is the extent to which they do influence student achievements, and what it is that makes the most difference. (p. 34)

In summary, much has been learned about the effectiveness of different methods and strategies of teaching and their impact on student attainment. The big questions now centre on what makes such methods and strategies work better and how they operate in terms of productively structuring the subjective experience of learners. To put it in simpler terms, what specifically goes on inside students' heads and how does this enhance their learning processes, resulting in better attainment, engagement and well-being? The more we frame better evidence-based answers to these questions, the more we move towards a pedagogy that is practically useful in terms of how we teach, and all that this entails.

In the following sections, through an extensive synthesis of a wide range of knowledge bases relating to human learning, I outline and illustrate certain key heuristics, what I have referred to as *Core Principles of Learning* (Sale 2015) that underpin effective teaching. Together they constitute a pedagogic framework from which teaching professionals can thoughtfully plan and facilitate learning experiences from a more evidence-based perspective. The framework does not claim to be exhaustive or summative, as new knowledge and insights will continually enhance our understanding of human learning and the implications for how we teach. However, from much validation in practice across a wide range of educational sectors and cultural contexts, I see them as contributing to a much-needed *Pedagogic Literacy*.

Furthermore, while each Core Principle of Learning focuses attention on a key area or process relating to how humans learn and the specific implications for planning instruction, they are not discrete or separate in that they should be considered independently of each other. They are mutually supporting, interdependent and potentially highly synergistic. As Stigler and Hiebert (1999) highlighted:

> Teaching is a system. It is not a loose mixture of individual features thrown together by the teacher. It works more like a machine, with the parts operating together and reinforcing one another, driving the vehicle forward. (p. 75)

These core principles of learning are essential parts of this system, the underpinning knowledge of how people learn and how to use this in the planning and delivery of what we call teaching. Willingham (2009) used the term Cognitive Scientific Principles to describe such principles of learning and illustrates their implications for practice through an analogy with engineering:

> Principles of physics do not prescribe for a civil engineer exactly how to build a bridge, but they do let him predict how it is likely to perform if he builds it. Similarly, cognitive scientific principles do not prescribe how to teach, but they can help you predict how much your students are likely to learn. If you follow these principles, you maximize the chances that your students will flourish. (p. 165)

In explaining and illustrating each of these core principles of learning, calibrations will be made to Hattie's Effect Sizes of different strategies/methods, subjecting them to pedagogic analysis in the process. If certain methods have high effect sizes, there must be correlations with underpinning brain activity and the psychological functioning of the mind. While the field of cognitive neuroscience is uncovering how certain parts of the brain responds to external stimuli and how this can impact people's feelings and behaviour, the more significant insights are those from cognitive psychology. In broad terms, Cognitive Psychology is the scientific study of mental processes such as attention, language development and use, perception and belief formation, memory process and how they work, thinking and problem-solving. It can also legitimately cover other aspects of psychological functions such as emotions and volition as cognition, motivation and other aspects of the mind work as a dynamic system. However, as we are increasingly discovering, this is both a complex dynamic system and one that is far from integrated in terms of the various components working in unison. We could argue the contrary as Pinker (2003), perhaps the leading writer in the field, points out:

> Behaviour…comes from an internal struggle among mental modules with differing agendas and goals. (p. 40)

Similarly (Apter 2001) describes the mind in these terms:

> …everyday life, as it is experienced, is a tangled web of changing desires, perceptions, feelings, and emotions that filter in and out of awareness in a perceptual swirl. (p. 33)

Bandler and Grinder (1990) make the summative point:

> It's really important to understand that most people are very chaotically organised on the inside. (p. 71)

The purpose of the above emphasis on how the brain works at the level of mind is to emphasize that as teachers we are not able to simply transmit information into students brains, like we can download information on our computers. Instead, we must navigate many aspects—foibles—of human psychological functioning, much of which is systemically not functioning as we would like it in terms of facilitating desired learning outcomes in educational contexts.

However, it is through understanding how the mind works (and often doesn't work well) that can best arm us with useful knowledge for developing skills and tools to teach in ways that are both brain compatible, as well as mitigating many of its less helpful features. This is the territory that we must navigate if we are to be expert teachers and, in navigation, it helps if we know the territory well and have a half-decent map for getting from one location to another. Yes, it's *Heuristics*, but this is much better than *Mystery*.

The key message of the chapter is quite clear and simply captured by Hart (1983):

> ...designing educational experiences without knowledge about how human brains learn naturally and most efficiently can be compared to designing a glove without any knowledge of the human hand. (p. 4)

2.2 Core Principle 1: Learning Goals, Objectives and Proficiency Expectations Are Visible to Learners

I fail to recall much by the way of consciously ever considering any learning goals over my 15,000 h at school, beyond getting a regular place in the school soccer team. Even for this desirable goal, I had little idea of what I specifically needed to do to achieve it—except to be good at soccer. The physical education teacher never helped me to understand my limitations as a soccer player and what I might do to enhance specific skill areas. Indeed, my school life lacked an explicit structure for learning beyond the fact that I was supposed to be there. In terms of the subjects I studied (the word does not fit well), I had little notion of what I should be learning in terms of specific outcomes and to what level of proficiency. For the exams I took, I tried to memorize what I had written down in class. I had no benchmarks for my performance, so it was a surprise, and delight, when I passed those 'O' and 'A' levels.

On leaving school, I had little direction on what I wanted to do. After all, my school-friends had long since left school and were going out to local pubs and clubs with money in their pockets. And there was a real motivational base to this—girls. For my first 6 months or so, I worked as a labourer on a building site for the scaffolding crew. Scaffolders were a tough bunch of guys, and they had to be able to carry those 22-foot poles—which were cast iron in those days. Anyway, the money was good, and my boxing background meant I could match the scaffolders in the practices of pole carrying.

However, somewhere around this time, my father, obviously concerned about where his only son (in fact only child) was going in life, called me in for a 'father and son conversation'. These were not frequent, so I still can recall the main content of this conversation. Most significant was him pointing out that while scaffolding paid well now, I would be earning similar amounts, in real terms, some 20 years in the future, and will not find it such a physically relishing challenge as the years pass. Also, he pointed out that with my 'O' and 'A' levels, I should have plenty of choices. The problem was, I did not know what I wanted—well not in occupational terms anyway.

In response, I went to the local career office, where I must say, personnel were helpful. I had many interviews, including at accountancy and legal firms, so my options were good. Not sure which way to go, I was eventually introduced to the possibility of going to university, something I had no meaningful frame on whatever. I had never met anyone who had gone to university and my only prior knowledge in this area was a weekly TV quiz show 'University Challenge' in which different universities competed for some prize or other. It soon became apparent, however, that there were some attractive aspects in going to university, not least government grants, long holiday periods and opportunities to develop my soccer playing skills. The only missing piece of this jigsaw was that one had to study a subject. Nothing came to mind for me. Motivated to some extent to pursue this option, I browsed through the university prospectuses and—hey presto, 'psychology'. In all honesty, I did not know much about psychology, but I guessed it was a bit like sociology, which was one of my 'A' level subjects. Sociology was also my favourite school subject, again made interesting by the teacher.

In summary, serendipity rather than any thoughtful sense of direction shaped my learning and career to this point. However, studying psychology was a life-changing experience as I realized that one's learning is within a persons' locus of control. Invariably, the constraints of finance, time and commitments may significantly impact the timing of career choices, but successful learning is very much in one's own hands. However, successful learning involves in no small part knowing what it is that you want to learn and for what life goals. It also requires a strategy, and not least a fair bit of effort, which in turn is aided by a belief system that sees attainment as a product of these processes, not a predestined neurological state. There is a saying in soccer circles that, "You are only as good as your last game." That makes perfect sense. I have noticed, over many years of watching professional soccer, how fickle soccer fans are. When a player has had a few poor games, there are often sounds of derision when his name is read out on the team sheet. Three weeks prior, the same player was greeted with great applause. A similar frame plays out in life. I was once a grade 9 'O' level student in maths. I could still have been that, but I am not, and I know what changed that reality and how it works. Hopefully, that has made me a better teacher. Poor thinking, limiting beliefs and lack of competence are not existentially fixed for the mainstream population of learners, but if no change in perception occurs, they can become stable realities—even identities—for the people concerned. Learning is about change, and productive change can be greatly helped by others, but these others need to be good models. For example, in the case of thinking, as Dilts et al. (1980) illustrates:

Effective thinking strategies can be modelled and utilized by any individual who wishes to do so. (p. 193)

The key point to this heuristic for effective teaching and enhancing learner attainment is that students require structure for their learning, and this starts with having a meaningful goal. While students are ultimately responsible for their learning, helping them to frame clear and meaningful goals, as well as what is involved in meeting them, is fundamental to providing structure, direction and motivation to their learning. As Ramsden (1992) pointed out:

It is indisputable that, from the students' perspective, clear standards and goals are a vitally important element of the effective educational experience. Lack of clarity on these points is almost always associated with negative evaluations, learning difficulties and poor performance. (p. 127)

There is a strong evidence base supporting the importance of establishing clear, meaningful and challenging goals for learners, For example, Marzano (2007) found an effect size of 0.97 for Specifying Goals, and Hattie (2009) found an effect size of 0.56 for Challenging Goals. The more we can articulate learning goals, be specific about what is to be learned—make it visible (what it looks like, sounds like and feels like)—the more likely learners are to achieve these outcomes. Of course, it helps even more if the learners themselves are motivated and committed to achieving such outcomes. As Hattie (2009) highlighted:

...effective teachers set appropriately challenging goals and then structure situations so that students can reach these goals. If teachers can encourage students to share a commitment to these challenging goals, and if they provide feedback on how to be successful in learning as one is working to achieve the goals, then goals are more likely to be attained. (p. 165)

Similarly, Schank (2011) reinforces the important outcome of student buy-in:

Teaching works best when you teach students who agree that they want to learn whatever it is you have to teach. (p. 43)

There is often a need for creative teaching to facilitate such high-level student buy-in across divergent student groups, as this involves a major perceptual shift for many students in terms of motivation and learning approaches. However, if this can be attained, the focus can then be largely on the *how* of learning effectively, rather than frequently revisiting the why. What constitutes challenging is, of course, subjective in part, but most importantly, we are seeking the best contextualization to the learner profile. Providing goals that are easy to attain results in little value in the learning stakes. The idea of giving students such goals to ensure they get plenty of positive feedback regarding their successful attainment, to promote self-esteem, is naïve at best. Students know that they are being 'dumbed down' and will not be duped by such token positive self-regard. Similarly, if the goals are not realistically achievable in terms of student's prior knowledge (e.g., level of conceptual understanding; skill sets), and in the time frames defined, this creates frustration and stress which is detrimental to learning and attainment. While it is sometimes challenging for the teacher to establish meaningful and challenging goals for students, it is time well spent, as Hattie (2009) concluded:

Educating students to have high, challenging, appropriate expectations is among the most powerful influence in enhancing student achievement. (p. 60)

It is important therefore to be able to, as far as is possible, ascertain their prior learning before setting goals. This is covered in some detail in Core Principle 2: Learners' prior knowledge is activated and connected to new learning. Once the student profile is ascertained in terms of prior knowledge, always recognizing that there will be variation in almost any student group (and this should be accommodated for whenever possible), there are many ways to represent appropriate goals to learners. What is most important is that students are provided with as clear as possible a definition of what the goal entails, the level of proficiency of the performance activities and the products that are required to be produced in meeting the goal, and any other key information that provides essential structure to make it as tangible as possible. This can involve showing examples of what good performance and product outcomes look like, sound like and feel like. For example, in working with teaching professionals who are seeking to attain a goal that involves being able to use specific instructional techniques (e.g., questioning techniques to promote critical thinking in a facilitation session), I often initially provide them with a range of video exemplars of what this looks like in real teaching contexts related to their field of practice. It is much easier to work towards goals when you have a clear sense of what goal attainment looks like, and what's involved in achieving it.

A noted effective way of supporting this in the context of a lesson is through the provision of an *advance organizer*, which is a summary of what is to be learned in the coming lesson—as identified in Chap. 1. Advance organizers are used at the beginning of a lesson and serve to provide an organizing frame for the content that is to follow, as well as a means for students to monitor personal learning in meeting the stated learning objectives related to the overall learning goal. The more these organizers connect to the desired goal, the better is the guide for learning. It's a bit like using a road map. A very accurate one can make the journey easy; the converse is also true. Apart from providing clarity and structure to the learning process, advance organizers help students to see a purpose in the learning and further reinforce the meaningfulness and motivation for successful goal attainment.

2.3 Core Principle 2: Learners Prior Knowledge Is Activated and Connected to New Learning

On the first day of my appointment at Singapore Polytechnic in 1985, I was asked by a colleague if I had been able to access my email. Immediately, a sense of anxiety became apparent as I posed the question to myself, "How do I do this?" I had never used email before or even accessed the internet. The internet was at best a very fuzzy concept in my head. It became no less fuzzy after a few days when I was a participant in a one-day training programme on using the internet. At the end of the workshop, I was even more confused and could not even recognize or open the internet browser,

Netscape. Yes, I started to feel a bit silly, but this was not a concern, as I knew exactly where I was in the learning stakes—a complete novice. In this learning situation, I was very aware of my limited prior knowledge of email and the internet. Furthermore, as a novice, it's natural to experience feelings of uncertainty, even dependency, and performance will be erratic at best. That's the profile of a novice in any unfamiliar learning situation, irrespective of whether one has great expertise in other fields. Aside, I am also very much a novice as a guitar player and on the one occasion I did a public performance, fortunately in a minor venue, even my basic chord playing went out of synchronization. I have never played publicly since.

Looking back on that one-day internet training programme highlights the difficulties faced by any learner who is confronted with a learning situation in which there is little prior knowledge to connect to, and where the instruction is far too fast to build any useful understanding of what is being taught. I went back to my office tired, confused, and with no useful understanding or competence to use the internet. However, what I did know was that this was a typical and almost inevitable result given the learning context and, most importantly, I knew how to deal with it effectively.

Learners come to any new learning situation (whether it be the classroom or elsewhere) with preconceptions about how the world works based on their life experiences. Within this framing, they may have developed some generalized beliefs about themselves as learners (which may or may not have been favourable) in terms of their capability for learning. As described prior, post my grade 9 math 'O' level result and the preceding learning experience, I did not feel competent or confident in learning mathematics. The problem is that prior learning can create a whole host of misconceptions and motivational dispositions that lead people to avoid any further attempts at learning in a specific area. This takes on an added significance in that all learning, whether accurate or otherwise, exists as a relatively permanent structure in our neural architecture. I was fortunate in that my final 'O' level math teacher, Mr. Edrich, was able to challenge and disrupt my existing knowledge and beliefs relating to learning mathematics. For many, they become stuck in an abyss of misconceptions and perceived limited capability. The important point is that new learning cannot avoid being connected to prior learning. As Shulman (1991) pointed out:

> All new knowledge gains its form and meaning through its connection with pre-existing knowledge and its influence on the organization and reorganization of prior knowledge. (p. 10)

Our prior knowledge is stored in our long-term memory in organized mental models (often referred to as schemata). These schemata exist as connected neural networks in our brain architecture and this is what makes possible our understanding (or lack of it) of phenomena in the real world that they represent. The more complete and better integrated these schemata are—in terms of the phenomena they represent— the more useful they are in terms of the application of that learning. For example, an expert motor mechanic should easily be able to diagnose what is wrong with my car when it breaks down and knows how to fix it. I am only able to know that it isn't working—as I have no useful prior knowledge on what causes cars to break down, and even less on how to fix them.

Prior knowledge then is the lens through which students will perceive and react to new information provided in a learning event. If prior learning is inaccurate, incongruent or limited, it is likely to interfere with the meaningful integration of the new knowledge presented. This provides real challenges for teachers. Ausubel (1978) went as far as arguing that:

> If I had to reduce all of educational psychology to just one principle, I would say this: the most important single factor influencing learning is what the learner already knows. Ascertain this and teach him(sic) accordingly. (p. 163)

Making student's prior knowledge explicit helps not only to deal with misconceptions and facilitate better linking of new knowledge to existing knowledge structures but also saves an enormous amount of time in terms of duplicated learning (e.g., Nuthall 2005), boredom for students, as well as frustration for teachers. Finding out what students already know, understand and can do is fundamental to teaching in any context. Hattie (2012) argued that:

> …we must know what students already know, know how they think, and then aim to progress all students towards the success criteria of the lesson. (p. 44)

> There is then the challenge of designing ways to connect new knowledge to the level of knowledge and understanding that learners already have. This requires both a good understanding of the subject matter content and the students being taught, as well as some creativity in order to design the most appropriate instructional strategy to best facilitate such connectivity. Wlodkowski (2008), using the language of cognitive neuroscience, suggests that this involves the following:

> …begin with what they already know and biologically assemble them with the new knowledge or skill by connecting the established networks and the new networks. (p. 13)

This heuristic has an effect size of 0.41 (Hattie 2009) and in combination with clear goals and advance organizers, provides a strong foundation for subsequent learning, and can be seen as significant components of an impactful "Russian Doll" instructional strategy—to reiterate this metaphor introduced in Chap. 1. Once students have clarity of purpose in the learning goals, a sense of direction for meeting them, appraised their existing knowledge and dealt with any restrictive misconceptions, they are in a much better position to tackle new concepts effectively. Of course, this is an ideal scenario and it is unlikely to happen so easily for all students in all situations. However, it is a much better strategy than going straight into the new content delivery, for all the reasons outlined above.

Activation of students' prior knowledge can be done in many ways, but all involve eliciting specific feedback concerning what they already know, understand and can do (and to what level of proficiency) in relation to new learning goals and specific outcomes. This can be done through written and oral pre-tests, and by way of open discussion with students to explore more fully their mental models and ways they are thinking about the topic area to be covered. It is important to recognize that many students are unlikely to be clear on specific things they don't know and may not even be able to effectively make this explicit. For this reason, it is important to create a psychological climate in which students feel comfortable sharing their learning concerns and are not afraid of admitting to not knowing. This is explored in some detail in Core Principle 9: *A psychological climate is created which is both success-orientated and fun.*

2.4   Core Principle 3: Content Is Organized Around Key Concepts and Principles that Are Fundamental to Understanding the Structure of a Subject

Understanding is about making personal meaning of knowledge and seeing how it is used in real-world applications and problem-solving. When learners have developed a good understanding of a topic, they will have acquired an organized and accurate mental representation (schemata) of the key concepts in their minds. Once attained, understanding will facilitate effective and efficient retrieval of the relevant knowledge of the topic from long-term memory, easy explanation of what the topic is about, its key components, areas of contention, as well as its thoughtful application in real-world problem-solving. Furthermore, with a good understanding of something, whether it's the working of mechanical systems or, in the context of this book, pedagogy, it's then possible to use this knowledge effectively across the domain field, what is referred to as transfer of learning. Transfer facilitates accurate diagnosis of problem situations and the capability to create solutions with a high degree of outcome prediction because it means that the person fully understands the knowledge bases involved. For myself, I have little understanding of mechanical systems; hence I am unable to fix anything mechanical. My Jack Russell dog occasionally sits on the remote-control devices that operate the television and related systems, often resulting in picture loss on the television. It typically ends up with me ringing the technical support helpline. I don't know what most of the various buttons on the different remote-control devices mean, what aspect of the system behaviour they control, or their relationships to each other (buttons and the different remote-control devices). In a situation of picture loss, unless it is patently obvious what has happened (e.g., the on button is now off), my understanding is so limited I am effectively taking part in a lottery where there is a low probability of success; my chances of hitting the appropriate buttons on the relevant remote control devices in the correct sequences are not good.

In the literature, much is written about the nature of knowledge, types of knowledge and how knowledge and cognitive processes interact to build understanding by philosophers, educationalists, and cognitive scientists. The study of the nature, form and structure of knowledge is a recognized discipline, typically referred to as epistemology. We will avoid an extensive coverage of this area as much of the different terminology conflates and may, in this context, add more confusion rather than insight into how core concepts and principles help students to understand the key structure of a topic, and what makes this particularly important to learning and attainment.

One area of general agreement among writers on the types of knowledge, which provides a useful understanding of what knowledge entails, is the categorization of knowledge into Declarative Knowledge and Procedural Knowledge, as summarized below:

Declarative Knowledge: As the term implies, it refers to knowledge that can be clearly stated as facts, concepts, generalizations or principles within a content knowledge field. For example, once acquired, we might be able to clearly assess that a learner knows or understands:

- the concept of democracy
- the defining attributes of a dog
- the conventions of punctuation
- Cristiano Ronaldo plays football for Juventus (at the time of writing).

Procedural Knowledge: This refers to knowing how to do something, typically involving performing a process or demonstrating a skill. For example, once acquired, we might be able to clearly assess that a learner is able to:

- Add and subtract
- Write a paragraph
- Juggle
- Set up an experiment
- Read music
- Search for a database.

In many practical tasks both types of knowledge are involved, as to do something typically involves knowing something about it. For example, while the amount of declarative knowledge involved in being able to play soccer is not extensive, no amount of skill in procedural terms would be useful if one did not know what goal to score into. Invariably, there is much variation in terms of both the number of knowledge components and level of complexity involved in knowledge acquisition and deployment when procedural. For example, to acquire a single piece of factual knowledge such as England won the soccer world cup in 1966 is very straightforward. Around 5 repetitions should put it firmly into long-term memory. How memory works and its crucial role in effective learning is outlined in detail in Core Principle 7: Learning design takes into account the working of memory systems. In exceptional circumstances, a little idiosyncratic knowledge may be amazingly useful to an individual, as was powerfully illustrated in the 2008 film 'Slumdog Millionaire'. The film featured a young man (Jamal) from the slums of Mumbai who appears on the Indian version of 'Who wants to be a millionaire?' and answers all the questions correctly, though arouses suspicions that he must have cheated. However, in the film, Jamal recounts in a flashback how he knows the answer to each question, each one linked to a key event in his life. His learning of these specific bits of factual knowledge happened idiosyncratically, but through great serendipity resulted in the illusion of him being highly knowledgeable, which ran counter to his slum living existence. In most real-world contexts, we are very unlikely to get such highly favourable results from limited knowledge bases. The building of accurate organized mental models (deep understanding) of complex phenomena in the world requires much internal cognitive work on the part of the learner to negotiate and assimilate the vast knowledge bases involved. One does not need much knowledge (declarative or procedural) to ascertain why one's pencil is not working and how to fix it. However, this is unlikely to be the case in a situation of aeroplane failure, unless of course, you happen to be an expert aeroplane engineer. What this cognitive work is and how it works is the focus of Core Principle 4: Good thinking promotes the building of understanding.

Understanding is something students can achieve themselves only through the acquisition of relevant knowledge, actively making appropriate connections between the knowledge components (e.g., declarative and procedural) to build an accurate mental picture—schemata of the intended learning goal. The rote memorization of knowledge, while fundamentally important in effective learning, will not in itself result in understanding as this requires the learner to actively make the mental connections and create accurate internal representations. This involves what we refer to as 'thinking'. However, thinking without knowledge is of no value—try thinking about nothing. As Lang (2016) emphasizes:

> One of the first and most important tasks as a teacher is to help students develop a rich body of knowledge in our content areas—without doing so we handicap considerably their ability to engage in cognitive activities like thinking and evaluating and creating.

> ...such cognitive skills require extensive factual knowledge. We have to know things, in other words, to think critically about them. Facts are related to other facts, and the more of those relationships we can see, the more we will prove capable of critical analysis and creative thinking. Students who don't bother to memorize anything will never get much beyond skating over the surface of a topic. (p. 15)

Notions that today, all we need to do to get content is to search the internet and find it, and 'hey presto' we have knowledge and understanding—even expertise—is highly dubious, as Keen (2007) exposed in his book, 'The cult of the Amateur: How Today's Internet is Killing our Culture'. In the days when we used physical encyclopaedias and went to the library and read books, information was there, but it did not mean that we could understand it. Good teachers get good results for a reason and, amongst other important aspects, they can identify the key concepts essential to understanding the structure of a topic area in the context of learners' prior knowledge and experience, and then work with this for extending their learning. Willingham (2009) makes an interesting point in relation to the importance of key knowledge in the learning process:

> The very processes that teachers care about most – critical thinking processes such as reasoning and problem-solving – are intimately intertwined with factual knowledge that is stored in long-term memory (not just found in the environment). (p. 22)

> First, you should know that much of the time when we see someone apparently thinking, he or she is actually engaged in memory retrieval…memory is the cognitive process of first resort. (pp. 28–29)

In terms of our everyday learning, such cognitive science explains why we can read, understand and easily apply new knowledge in areas that we have the expertise, and do this quickly. In contrast, when our knowledge is limited and idiosyncratic, we struggle to make sense of what we are reading and feel that we are not so bright (in those situated moments). For example, I can breeze through research papers in the field of psychology and education but always struggle with any instructional booklet on how to assemble any do-it-yourself (DIY) equipment such as furniture or a mechanical device—much to my wife's annoyance. Quite simply, I have such limited knowledge or interest in this area. The same applies to gardening, which I once thought I should do, as we had a garden, and other folk often said how relaxing and enjoyable such activity was. Well, I tried to grow lettuce, and guess what, the slugs ate them; I tried to grow strawberries, which the birds ate, and my attempt at growing runner beans, as I did not secure their structures against a strong wind, was destroyed. Basically, I lacked too much basic necessary gardening knowledge. Was it relaxing and enjoyable? Well, you can infer the answer—I had the whole garden ripped up and concretized.

So, what are the pedagogic implications of this? In Willingham's words:

> Factual knowledge must precede skill. Critical thinking is not a set of procedures that can be practised and perfected while divorced from background knowledge. Cognitive science leads to the rather obvious conclusion that students must learn the concepts that come up again and again – the unifying ideas of each discipline. (p. 33)

In summary, as Resnick (1989) summarized:

> Study after study shows that people who know more about a topic reason more profoundly about that topic than people who know little about it. (p. 4)

There should be a little surprise here, after all, "knowledge is power"—right? Essentially what this means in practice is that the acquisition, organization and integration of relevant knowledge bases in one's long-term memory system are foundational to better understanding—hence learning and attainment. Berliner's (1987) description of the benefits of comprehensive and well-organized schemata (the basis of good understanding) in a specific or domain is fundamentally informative in this context:

> Individuals possessing rich, relatively complete schemas about certain phenomena need very little personal experience to learn easily, quickly and retain well information pertaining to those phenomena. A well-developed schema allows very efficient learning from a verbal and written discourse on a topic about which much is known. (p. 61)

Similarly, as Pugh and Bergin (2006) point out:

> …for students to access and apply their learning they need to possess a deep level, connected knowledge structures. That is, their knowledge needs to be conceptually deep, cohesive, and connected to other key ideas, relevant prior knowledge, multiple representations, and everyday experiences. (p. 148)

There is much we can do as teaching professionals to facilitate understanding. You will note that the two preceding core principles of learning (clear goals and activating prior knowledge) all contribute in some significant way to facilitating the process of building understanding. Through a careful analysis of the learning goals, the specific outcomes and proficiency standards that we seek to achieve with our students, it is possible to identify the key declarative and procedural knowledge (especially core concepts and principles) that underpin an understanding of the key structure of the topic areas we are teaching. Bruner (1966) identified what are essentially key evidence-based principles underpinning the importance of good structure in enhancing learning:

> The first is that understanding fundamentals makes a subject more comprehensible. (p. 23)

> The second point relates to human memory. Perhaps the most basic thing that can be said about human memory, after a century of extensive research, is that unless detail is placed in a structured pattern, it is rapidly forgotten. (p. 24)

> Third, understanding of fundamental principles and ideas…appears to be the main road to adequate "transfer of training." To understand something like a specific instance of a more general case – which is what understanding of a more fundamental principle of structure means – is to have learned not only a specific thing but also a model for understanding other things like it one may encounter. (p. 25)

Basically, understanding is a difficult learning goal, as it involves much cognitive work—which we typically refer to as thinking. As Willingham (2009) succinctly noted:

> Understanding is hard for students. After all, if understanding were easy for students, teaching would be easy for you. (p. 78)

> Bruner advocated a Spiral Curriculum in which the key concepts and principles are revisited over time to further clarify and extend the knowledge base in terms of adding new related knowledge, enhancing integration and further refining until the students mental schemata has the most accurate and appropriate mental representation, what he refers to as "the full formal apparatus that goes with them", (p. 13)

He is famously noted for asserting that:

> We begin with the hypothesis that any subject can be taught effectively in some intellectually honest form to any child at any age of development. (p. 33)

For example, in this chapter, the Core Principles of Learning constitute key concepts and principles fundamental to understanding the structure of what constitutes Good Teaching (e.g., effective, efficient, creative teaching). Once these are understood, the more specific factual content relating to how they enhance aspects of the learning process will become increasingly easier to accommodate into a solid and meaningful mental model in long-term memory. Over time, with a thoughtful application, the knowledge base becomes more refined, elaborated and practically useful. In the wider context of this book, as the key structure becomes increasingly understandable in terms of how to enhance the practice of teaching from an evidence-based approach, the more abstract notion of 'Pedagogic Literacy' starts to become a meaningful and useful proposition (he says, hopefully). Just as clear and meaningful learning goals and advance organizers provide structure to what is to be learned, this heuristic focuses our attention to the most appropriate selection of knowledge components and their best organizational structuring and sequencing for facilitating the learning experience to maximize attainment opportunities for learners. While the mind has a natural tendency to organize information into meaning wholes, as Gestalt psychology established in the early twentieth century (e.g., Koffka 1915; Köhler 1929), this is greatly aided and enhanced when there is a clear and logical structure in the presentation of knowledge in the first place. Hattie and Yates (2014) pointed out:

> The mind does not relate well to unstructured data. We find it extremely taxing to learn random lists or when coping with unrelated materials. We need to learn the organization, structure, and meaning in whatever we learn. Meaningfulness, or relatedness, stems directly from prior knowledge. We benefit enormously from being shown how to group information, how to locate patterns, how to use order, and how to schematise and summarise. (p. 115)

Furthermore, it has long been recognized that different subject areas, by their very nature, lend themselves to different teaching and learning approaches in terms of effective student learning. For example, Shulman (1991) argues that teachers require 'pedagogic content knowledge', which is the ability to fully understand how their specific disciplines are most effectively taught. This involves not only the identification of core concepts and principles essential for building understanding but also key areas where misconceptions and areas of difficulty are likely to be encountered by students. In this way, the instructional strategy can be systematically tailored to incorporate effective methods that are specifically contextualized to the nature of the discipline and how practitioners in the field conduct their practices in real-world contexts. The importance of applying not just pedagogical knowledge to the ways we teach but also supplementing this with pedagogical content knowledge is captured by Shulman when he argued:

> When was the last time you saw a problem set in the study of Hamlet? Or in Asian History? Can you have guided practice in a poem? Or for evolutionary theory? I would argue that we have reflected in the differences among the disciples, different ways of knowing that are tied to different ways of teaching. (p. 5)

It is essential, therefore, that teachers know their subjects especially well in order to be able to identify the most appropriate method combinations to effectively teach the key concepts and principles that are fundamental to understanding in the specific context of their subject topic areas. In a similar vein, McTighe and Wiggins (1998) refer to the importance of focusing content on the 'big ideas' and the 'essential-questions' in making sense of the content knowledge and its importance within the wider subject context. The big ideas relate to the more fundamental and enduring understandings relating to a topic area, as they:

- Provide a conceptual "lens" for any student
- Provide a breadth of meaning by connecting and organizing many facts, concepts and skills; serving as a lynchpin for understanding
- Point to key knowledge at the heart of the expert understanding of the subject
- Require "uncoverage" because its meaning or value is rarely obvious to the learner, is counterintuitive or prone to misunderstanding

- Have great transfer value; applying to many other inquiries and issues over time— "horizontally (across subjects) and "vertically" (through the years in later courses) in the curriculum and out of school. (p. 69)

Big ideas provide an excellent vehicle for helping students to understand both the key structure of a topic area as well as its relevance to real-life contexts. As the authors argue, they provide:

> …a conceptual tool for sharpening thinking, connecting discrepant pieces of knowledge, and equipping learners for transferable applications. (p. 70)

Essential questions are core to the subject and will stimulate thought, provoke enquiry, and spark more questions relating to the essential core structure of the topic area, further enhancing understanding. As the authors summarized:

> The best questions point to and highlight the big ideas. They serve as doorways through which learners explore the key concepts, themes, theories, issues, and problems that reside within the content, perhaps as yet unseen: it is through the process of actively "interrogating" the content through provocative questions that students deepen their understanding. (p. 106)

McTighe and Wiggins argue that a question is 'essential' if it can:

1. Cause genuine and relevant enquiry into the big ideas and core content
2. Provoke deep thought, lively discussion, new understandings and more questions
3. Require learners to consider alternatives, weigh evidence, support ideas, etc.
4. Help make connections with prior learning and personal experiences
5. Naturally reoccur, creating opportunities for transfer to other situations and subjects.

Furthermore, as knowledge is increasing almost exponentially, and it is not possible to keep adding more and more subject content in the curriculum, the selection of the most relevant content knowledge for developing key understandings that are fundamental to the structure of topic areas becomes a pedagogic prerequisite to teaching. Willingham (2009) cleverly framed this essential question in terms of, "What knowledge yields the greatest cognitive benefit" (p. 36). In more lay terms, as the maxim goes, "More is not better, better is better" and this applies particularly well to the selection of subject content in preparing to teach. Equally, research (e.g., Hattie and Yates 2014, p. 7) argues that we will invest effort more strongly when we have already built some useful foundation of knowledge (e.g., understanding), in contrast to when there is nothing to build on. Being able to quickly help students achieve a basic understanding of what a topic entails and its relevance to their learning goals not only helps the cognitive aspects of the learning process, but also the affective domain of emotions and feelings in that this is more likely to generate and maintain a better motivational base for a more sustained learning experience.

This explains why we are often reluctant to take on tasks in which we feel we have very little understanding of the competence and perceive a big gap between where we are and where we need to be in the learning stakes. Sadly, this often results in a person giving up in an area of learning that they had an initial interest in pursuing. I nearly did this with math but was fortunate to have a good teacher 'to pull me out of the pit'—so to speak, which made the difference.

There are some old sayings in this context, which provide easy to remember analogies. We need to be able to:

> "See the wood from the trees" and "Separate the wheat from the chaff".

In summary, being able to identify the key concepts and principles of a subject from the mass of tertiary information flying around is surely a core principle of learning and must be a key heuristic in planning and facilitating instruction. Equally, students need to be well informed and taught how to do this effectively. Brown's et al. (2014) reflection is worth 'reflecting upon':

> Each of us has a large basket of resources in the form of aptitudes, prior knowledge, intelligence, interests, and sense of personal empowerment that shape how we learn and how we overcome our shortcomings, some of these differences matter a lot – for example, our ability to extract underlying principles from new experiences and to convert new knowledge into mental structures. Other differences we may think count for a lot, for example, having a verbal or visual learning style, actually don't. (p. 141)

2.5   Core Principle 4: Good Thinking Promotes the Building of Understanding

In Chap. 1, I mentioned that thinking was not something I learned from my 15,000 h at school. Well, my teachers can be easily forgiven, if Wagner's (2010) conclusion is correct:

> In schools, critical thinking has long been a buzz phrase. Educators pay lip service to its importance, but few can tell me what they mean by the phrase or how they teach and test it… (p. 16)
>
> For the most part, teachers haven't been trained to teach students how to think. (xxiv)

There is often an assumption that thinking is simply common sense. Well, even if it is, and I don't think it is, it's not that common. In most basic terms thinking is goal-directed cognitive activity, which seems to occur not just at a conscious level (e.g., "I just think this through"), but also subconsciously and unconsciously. The outcome of good thinking is typically a heightened, or at least improved, understanding of something. Certainly, thinking is essential to building understanding as it involves the making of connections in the brain, and this is learning at the neural level.

However, thinking does not occur in isolation, but rather through connecting and making sense of information, which ultimately (if successful) is retained as neurologically useful mental models in long-term memory. As Willingham (2009) summarized:

> Thinking occurs when you combine information (from the environment and long-term memory) in new ways…That combining happens in working memory. (p. 11)

Hence, …we must ensure that students acquire background knowledge parallel with practising critical thinking skills. (p. 22)

There is, despite differences in perspective and terminology in the literature, strong agreement that thinking is crucial to the quality of human learning. As Paul (1993) summarized:

The thought is the key to knowledge. Knowledge is discovered by thinking, analyzed by thinking, organized by thinking, transformed by thinking, assessed by thinking, and, most importantly, acquired by thinking. (vii)

Petty (2009) puts this into a very practical context when he argued that:

It is no exaggeration to say that almost every aspect of private and public life is driven by our ability (or inability) to use these thinking skills effectively, and to 'think straight'. (p. 325)

However, while good thinking may be beneficial in the learning stakes, there are those who do not see the human mind as particularly well developed for such activity, as Willingham (2009) concluded:

Humans don't think very often because our brains are designed not for thought but for the avoidance of thought. (p. 4)

Hattie and Yates (2014) offer the following analysis:

The ability to think well, to learn efficiently, and solve problems successfully are attributes that do not figure in most descriptions of natural human adroitness. While a few of us seem to want to develop good thinking skills (however defined) – it does not seem to be typical – …humans naturally assimilate the vast bulk of their knowledge through direct social influence processes that do not make great demands on thinking capabilities. (p. 7)

There is indeed an interesting paradox as far as thinking is concerned. On the one hand, as Jensen (1996) argued:

The best thing we can do, from the point of view of the brain, is to teach our learners how to think. (p. 163)

On the other hand, the human brain for a significant proportion of the population does not seem to want to do this too willingly. Kahneman (2012) provides a powerful insight here, which has extensive implications in educational contexts and how we teach. He argues that thinking can be conceptualized in terms of two systems; System 1 and System 2. These are, of course, metaphors, but they convey something that instantly has strong face validity:

System 1 is a fast reflexive system that identifies the familiar, especially threatening elements in a situation and quickly activates automatic response patterns. This system is the most essential for survival and is the default system. It typically works well in familiar everyday life where most situations and problems are familiar, and we have long-established patterned responses to them. However, this system also results in rapid stereotypical/prejudicial judgements and action. It is the price we pay for this powerful survival system.

System 2 is a slow, analytic, reflective system that explores the more objective factual elements of a situation, compares them with previously learned elements, and then responds. However, this requires self-control, effort and time, which is essentially tiring. As Kahneman summarizes:

> System 1 is impulsive and intuitive; System 2 is capable of reasoning, and it is cautious, but at least for some people, it is also lazy. (p. 48)

The development of good thinking, then, has much in terms of similarity with other desirable outcomes sought by people. For example, few people enjoy going on a diet or working long hours of overtime. However, there is a benefit to weight loss when obese and extra money is useful and often essential for some. The same can be said for developing good thinking, as far as effective learning is concerned. We clearly recognize the longer-term benefits, but the shorter-term cognitive strain is often likely to short cut our perseverance to do this well in many situations.

If good thinking is hampered by it being a tiring activity and some of us have 'lazy' brains, this is further compounded by the impact of beliefs and emotions on our capability for rational cognitive activity. Marcus (2009), from a cognitive neuroscience perspective, highlights how our belief systems further provide challenges to the brain functioning as a good 'thinking machine':

> Our beliefs are contaminated by the tricks of memory, by emotion, and by the vagaries of a perceptual system that really ought to be fully separate – not to mention a logic and inference system that is as yet, in the early twenty-first century, far from fully hatched. (p. 67)

Indeed, as Brown et al. (2014) summarized, and this is worrying:

> The truth is that we're all hardwired to make errors in judgement. Good judgement is a skill one must acquire, becoming an astute observer of one's own thinking and performance. We start at a disadvantage for several reasons. One is that when we're incompetent, we tend to overestimate our competence and see little reason for the change. Another is that as humans, we are readily misled by illusions, cognitive biases, and the stories we construct to explain the world around us and our place within it. (pp. 104–105)

It is therefore not that surprising that good thinking is more than just common sense, or we may need to accept that common sense is a much rarer capability than is typically assumed. However, despite the many barriers to good thinking, it can be effectively modelled, understood, and improved. As Perkins (1995) pointed out, "People can learn to think and act intelligently" (p. 18). In consequence, there is little point in asking students to engage in good thinking if they have no accurate and useful prior knowledge of what this means. In the absence of useful knowledge in this area, as for any area of new learning, a whole host of misconceptions are likely to come into play, and we know what this eventually leads to—a confused and frustrated learner.

There are many models of thinking in the literature (e.g., Marzano 1988; Swartz and Parks 1994; Perkins 1985) and the keen reader can find much of interest here.

However, it is also full of different terms (e.g., 'critical thinking', 'creative think-ing', 'lateral thinking', 'analytical thinking', 'dialogical thinking', 'parallel think-ing', 'design thinking'—even 'thinking out of the box') relating to thinking, that often confuses rather than aids the development of good thinking in curriculum design and practical teaching. Having spent many years researching this elusive human quality, I have evolved a model of thinking (Sale 2014) based on extensive modelling of how professionals, across a wide range of fields, actually solve problems in their work contexts. It must be recognized at the outset that accurate conceptualization of internal cognitive processes is inherently problematic and invariably unreliable, especially across subject domains. However, without some valid practical frame on what these elusive but desirable skills are, and how they work in terms of the wider context of internal mental activity, there is little chance of the effective teaching and assessment of them.

What this means, for example, is that while psychologists may solve problems in some qualitatively different ways from engineers, both at the individual and collective level, there is similarity in the types of cognitive activity involved. For example, they will need to analyse situations (e.g., cases), make comparison and contrast with similar past cases, build up inferences and interpretations from ongoing perceptions and data accumulation, generate possible solutions and decide action based on chosen criteria. Around this swirl of cognitive activity, there will be overall monitoring of what is going on, typically referred to as metacognition. The summary model is depicted in Fig. 2.1, and the typical cognitive heuristics involved are outlined in Table 2.1. Note that the cognitive heuristics for each type of thinking are the essential framing questions that are to be negotiated in making sense of information and building understanding.

In this model, analysis, compare & contrast, inference & interpretation and evalua-tion are typically employed during critical thinking; whereas generating possibilities, as the term implies, is predominantly employed in creative thinking. However, it is

Fig. 2.1 Sale model of types of thinking

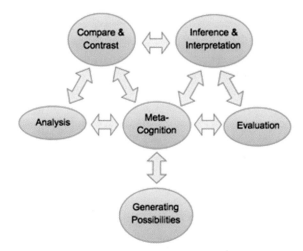

Table 2.1 Cognitive heuristics of types of thinking

Cognitive heuristics of types of thinking
Generating: possibilities
• Generate many possibilities
• Generate different types of possibilities
• Generate novel possibilities
Compare & contrast
• Identify what is similar between things (e.g.. objects/options/ideas)
• Identify what is different between things
• Identify and consider what is important about both the similarities and differences
• Identify a range of situations when the different features are applicable
Analysis
• Identify relationship or the parts to a whole in system/structure/model
• Identify functions of each part
• Identify consequences to the whole, if a part was missing or malfunctioning
• Identify what collections of parts from important sub-systems of the whole
• Identify if and how certain parts have a synergistic effect (for open systems)
Inference & interpretation
• Identify intentions and assumptions in data
• Separate fact from opinion in data
• Identify key points, connections, and contradictions in data
• Make meaning of the data/information available
• Establish a best picture to make predictions
Evaluation
• Decide on what is to be evaluated
• Identify appropriate criteria from which evaluation can be made
• Prioritize the importance of the criteria
• Apply the criteria and make decision
Meta-cognition
• Recognize the ability and usefulness of thinking in an organized manner
• Actively think about the ways in which we are thinking
• Monitor and evaluate how effective we are thinking
• Identify and manage beliefs and emotions which may hinder the quality of thinking
• Identify and utilize strategies to improve the quality and thinking

metacognition that is the higher executive capability and, as Flavell (1979) and Martinez (2006) maintain, critical thinking is subsumed under metacognition. Similarly, Halpern (1998) argued that when people are engaged in critical thinking, they need to use metacognitive skills to monitor the thinking process (e.g., checking the progress being made towards a personal goal, ensuring accuracy, and making decisions relating to time and mental effort). Brown et al. (2014) is even more direct in this analysis and argues that without metacognition, critical thinking is impossible to achieve. Also, creative thinking is not simply a question of clicking on a creativity switch in one's brain. Such notions are both naïve and dangerous. While creative thinking works along with different heuristics, it is closely linked to both critical thinking and metacognition's executive control. Without significant content knowledge in a field (and preferably more than one field) and if metacognition and critical thinking are poor, little by way of creativity is likely to manifest itself in terms of real-world usefulness.

Furthermore, and in the context of the above framing of how the mind works (or doesn't work), some writers see 'good thinking' not just in terms of cognitive processes and heuristics' but also in terms of the development of intellectual traits and standards. However, while definitions of critical thinking vary in the literature, most share a consensus that it involves certain dispositional factors as well as the cognitive heuristics outlined above. For example, Paul et al. (2006) identify the following traits as central to acquiring a high level of expertise in critical thinking:

- Intellectual humility—sensitivity to one's own biases and the limitations of knowing
- Intellectual courage—prepared to question own beliefs and those of others, even if unpopular with dominant perspectives and people
- Intellectual empathy—awareness of the need to actively entertain different views from one's own
- Intellectual integrity—holding oneself to the same intellectual standards of others (no double standards)
- Intellectual perseverance—working through intellectual complexities despite frustration
- Confidence in reason—recognizing that humankind's interests are best served by giving free play to reason
- Intellectual autonomy—thinking for oneself in relation to standards of rationality and not uncritically accepting the judgements of others
- Fair-mindedness—conscious of the need to treat all viewpoints alike and not be influenced by vested interests.

Such dispositions are certainly desirable, but the extent to which they are integral to deep-seated personality traits and trainable by pedagogic interventions is questionable and will be explored further in the following chapters.

In summary, these types of thinking run as overlapping and intertwined neural programmes, moving from foreground to background as the focus of framing a problem changes and new questions emerge. Certainly, when creativity is sought, generating possibilities is at the mind's forefront, but other types of thinking will weave in and out of consciousness and, typically run continuously in the subconscious mind. However, the good thinker will periodically take a conscious metacognitive view and attempt to make sense of (understand) what is actually going on in his/her mind, check various aspects of cognitive and affective processes (e.g., the types of thinking; impact of beliefs and emotions) and adjust when necessary. Good thinking can, therefore, be framed as the ability to navigate this 'perpetual cognitive and affective swirl', and to be able to employ the various heuristics of these types of thinking in a fluid, effective, efficient and highly synergistic manner. This is perhaps the reason that good thinking is quite rare in many situations, and why we really need to teach it to our students.

The development of Good Thinking is so central to learning and well-being, that Chap. 3: Metacognitive Capability *The Superordinate Competence for the Twenty-First Century*) extensively focuses on these essential though elusive skill sets. For now, just remember it is a *Core Principle of Learning*.

2.6 Core Principle 5: Instructional Methods and Presentation Mediums Engage the Range of Human Senses

Against boredom even gods struggle in vain.

(Friedrich Nietzsche)

When I first arrived in Singapore, I took an instant liking for a local delicacy, 'chilli crab'—you must try it if you come to Singapore. In fact, I had this, and other local dishes, almost every night. Indeed, on one occasion, I remember an elderly Chinese lady at the local hawker centre (that's a Singaporean term for food court) saying to me, "Why you always have chilli crab, lah, why not spring roll."—or something like that. Well, the answer at that point in time was easy, "I like chilli crab." However, one night, and it was inevitable in retrospect, the chilli crab was served up in its typical form, but my response was not the usual positive one. Suddenly, its appeal seemed to have vanished completely. The chilli crab was no different, but my perception had somehow changed and with this, my whole orientation to it was different. Invariably, based on my East London values, I ate it; after all, it's not right to leave good food—a punishable offence by parents in my younger years, if caught. Quite simply, in psychological terms, I had become habituated to chilli crab and its appeal had greatly diminished. In most basic terms, I had become bored with it. Sadly, as humans, we have an inherent tendency for this to happen, even for things we really like.

When asked what is the best teaching method to use and why, I tend to recite a variant of the chilli crab story as an advance organizer. Yes, some methods are more effective than others, but the overuse of any one method will create habituation and students will get bored. I can recall academic faculty at a previous educational institution attending a workshop on Project-based Learning. Many came back excited and wanting to use it in their teaching. Well, imagine the students on a Monday morning, when for the first time they get to choose aspects of their learning and be more actively involved in the learning process, it was a novelty. However, after 2–3 weeks of such activity, when they have amassed several projects, the enthusiasm for such pedagogy had long receded. Too much of the same thing gets boring, and as Willingham (2009) concluded, "Change grabs attention, as you no doubt know" (p. 17). This is often why we go on holiday—even though it often ends up stressful, especially when taking young children who keep saying, "I wish we could go home". If the Gods struggle in vain, what chance for us mere mortals with this existential nemesis?

Hence, in terms of learning and teaching, the creation of appropriate variation in the modes and mediums of delivery, which stimulate the range of senses, is highly significant for enhancing the learning experience for students. Attentional processes play a major role in what enters our consciousness and influences, at the subjective level of experience, what we see and hear. The environment at any given time can present multiple sources of stimuli impacting our senses, but most of it will be selectively (mainly unconsciously) ignored. We typically pay attention to what we are motivated/interested in, as we will then be looking out for it. My wife at social events, for example, will notice what other women are wearing and make comparisons. I never notice what other men are wearing, as I have no interest in fashion. I simply wear what my wife suggests, though often insists on. As it doesn't matter much to me, I always (ok, mostly) concur. Of course, if someone was wearing something so different, or wearing nothing at all, I would almost certainly notice, as our dominant visual sense and brain's natural intuition to respond to novelty, would largely ensure this happens without my volition. As Sylwester (1998) pointed out:

It's biologically impossible to learn anything that you're not paying attention to; the attentional mechanism drives the whole learning and memory process. (p. 6)

In a similar vein, Csikszentmihalyi (1990) argued:

The shape and content of life depend on how attention has been used …Attention is the most important tool in the task of improving the quality of experience. (p. 33)

Getting attention is of key importance for learning and, as teachers, we need to tackle this difficult existential task as best we can. We know that human attention, unless the stimulus is very striking, either in terms of strong interest (e.g., pleasurable) or necessity to avoid pain (e.g., in a perceived crisis situation), will typically drift off and/or be distracted by some other feature or activity in the environment. How long is a typical attention span? There probably is not one, as it is variable from person to person and situation to situation. In a classroom, if students expect the teacher to be boring, there may not be any attention given. I have observed classes where some students come to class with a pillow and are nicely 'tucked in' for a nap before the teacher has commenced the lesson. Over the years, I have asked students why they have done this; it's, in fact, a silly question. They have made the decision, from previous lessons with this teacher that there is no reason to give any attention whatsoever. Getting some rest, for a long day at school, is more important than being bored. Hence, the importance of a good first impression—often referred to, in the psychological literature, as a Primacy effect (the tendency of the brain to take in more information when presented with a new stimulus)—when teaching a new class. As in the world of dating, if a first date goes badly, you are unlikely to get a second one—with that person anyway. Similarly, if your first class finds your lesson boring, you are likely to struggle for attention in the second session, and that's if students turn up—which in voluntary contexts, many may not.

Hence, as Miller (2016), in terms of a pedagogic frame, but with big practical teaching implications, states:

> Optimizing your learning experiences for attention is the first step towards optimizing it for long-term memory and higher thought processes. (p. 87)

Mental activity is stimulated through our five senses, with the visual sense probably the most dominant. The relative dominance of our vision system may well be the result of our evolution, as Mlodinow (2012) captures so interestingly:

> …an animal that sees better eats better and avoids danger better, and hence lives longer. As a result, evolution has arranged it so that about a third of your brain is devoted to processing vision. (p. 35)

In many situations the greater the combination of our senses that are appropriately stimulated in a planned learning event, the more potentially effective the experience is likely to be in terms of gaining better attention and facilitating the desired learning outcomes. For example, it is estimated that when we see and hear something, this doubles the sensory impact as compared to just hearing it. Direct experience will increase the impact further and, teaching it, will enhance it further still. This should not be surprising as the act of teaching, if conducted properly, will involve much by way of preparation. Most specifically, it will involve developing a strong understanding of the key content areas, especially those concepts and principles that are fundamental to understanding the key structure of the topic area. It will also involve identifying areas of potential difficulty and where the main misconceptions are likely to be experienced by learners. Finally, there will a systematic structuring and sequencing of how best to present this content in the most effective and efficient method combination. In my experience, by assessing how well someone has learned takes this process even further. When assessing students, one must firstly be able to validly ascertain what they have learned and to what extent the key learning components (e.g., the desired learning outcomes) have been met in the performance evidence to ensure accurate judgement of performance. In making assessment decisions, especially of a summative nature (e.g., when one is making a final assessment decision or ascribing a grade), the assessor is claiming to know learners in some fundamental way that often has a significant impact on their access to future educational channels and employment opportunities. Secondly, as assessment (formative) is a key aspect of the learning processes, this requires assessors to accurately diagnose students specific areas of weakness and then provide tailored feedback to help them strategize effective future learning strategies.

Pedagogically there is logic in providing or enabling students to engage in real-world tasks that they find meaningful and interesting, that gets them highly engaged (i.e., behaviourally, emotionally, cognitively and agentically), proactively thinking and asking questions—and assessing their learning collaboratively. This is a good instructional approach. However, its rarely the reality in many classrooms, and certainly not for all students. The foibles of the human condition make this so. Many teachers often bemoan the lack of student engagement and interest and feel it's often their fault. In the case of some teachers, this may be true, to varying degrees, but teaching is not easy, and expecting full attention and engagement is wishful thinking. One can only do one's professional best. Remember teaching is heuristic not algorithmic. Water always boils at 100 °C—correct? Well, not exactly, your elevation relative to sea level can affect the temperature at which water boils, due to differences in air pressure. Hence, even algorithms don't always behave 100% and don't worry, we won't go on to discuss string theory. What we can say is that on a bad day, students may not give much attention, no matter how well you teach. So many factors can affect student behaviour, such as personal circumstances, moods, time of day, etc., and if these negatives conflate, the best-planned and delivered lesson may go nowhere useful. Just as many patients do not take 'good' medical advice, so many students may not see the point in learning what is being taught—despite your best attention and good teaching.

In terms of ascertaining the various ways in which teaching and learning arrangements impact the senses, Edgar Dales' famous 'Cone of Learning' (Fig. 2.2) is often shown to illustrate how different senses and activities affect the learning process. The percentages have a limited empirical base and are quite arbitrary; however, it provides a generalized illustration of how different combinations of sensory input may affect the type and quality of learning.

The use of audio-visual aids is common practice in seeking to enhance student learning through different sensory modalities, and it is certainly the case that the human mind responds positively to multimedia (Hattie and Yates 2014). The cinema, of course, exploits this to its fullest impact. Our brain is set up well to integrate information from different source inputs, especially from different modalities. Strong learning occurs when words and images are combined, and these effects become especially strong when the words and images are made meaningful through accessing prior knowledge. Good visual representations work because:

- Recall is almost always visually triggered; hence visual representation acts as a cue triggering the full memory system

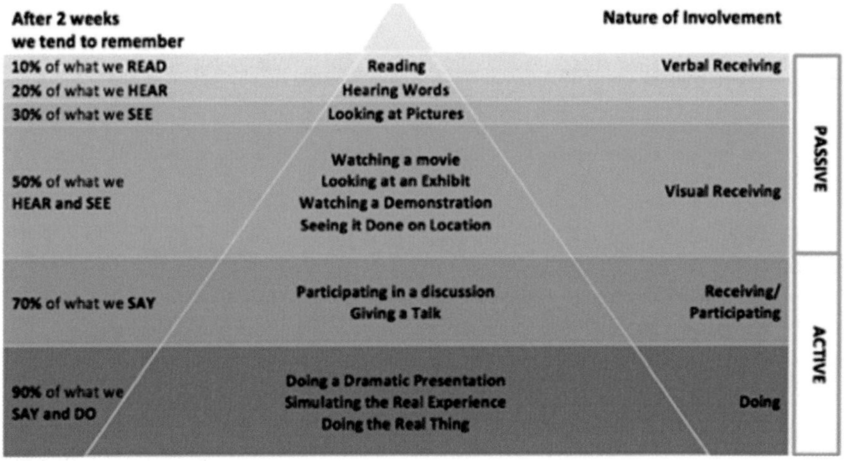

Fig. 2.2   Edgar Dales' *cone of learning*

Table 2.2 Key principles of good instructional design for audio-visual presentations

Key principles of good instructional design for audio-visual presentations
Five principles for reducing extraneous process • Coherence (reduce extraneous words and pictures) • Signalling (highlight essential words and pictures) • Redundancy (do not add onscreen text to narrated graphics) • Spatial contiguity (present printed words near corresponding graphics) • Temporal contiguity (present corresponding words and graphics simultaneously)
Three principles for managing essential processing • Segmenting (break down instruction into learner-paced segments) • Pre-training (Provide pre-training in names and characteristics of each main concept) • Modality (use spoken words for visualization rather than text)
Two principles for fostering generative processing • Personalization (put words in conversational style) • Voice (use friendly human voice for speaking words)

- Only structured information can go into Long-Term Memory, so this helps the transmission from Working Memory into Long Term Memory and subsequent recall
- They facilitate the ability of learners to see the relationship of a whole to its various parts, which fosters understanding.

However, it is important not to over-use audio-visual aids or to create too much variation in modes and mediums of presentation. I have seen many teachers using audio-visual aids and varied presentation format, all with good intentions to enhance the learning experience, but only to create confusion for students. There is now much evidence-based research on how best to present visual material to facilitate effective learning. For example, Mayer and Alexander (2011) summarized essential key principles that specifically impact the effectiveness of multimedia on learning (see Table 2.2).

As Mayer makes clear:

> These practical implications are examples of evidence-based practice – basing instructional methods on research evidence rather than on conventional wisdom, opinion, speculation, fads, or doctrine. (p. 441)

This heuristic is not difficult to understand in terms of how it can enhance student attention and attainment as it has strong face validity. For example, we have all both experienced boredom and how it affects our attention and disrupts learning, as well as being stimulated by high impact multi-media movies. I remember being amazed by the film 'Avatar' because of the multi-media effects, even though the story had some ridiculous concepts such as helicopter gunships, resembling what are used today, on a planet in another solar system many light years away—really? However, today's multi-media and internet-rich resource pool is a double-edged sword. On the one hand, it offers the creative teacher much in the way of capability for building networks of integrated resources, differentiating the learning experience and creating instructional strategies that provide better attentional, engaging and attainment opportunities for an increasingly wider cohort of learners. On the other hand, we must bear in mind that today's learners, so familiar with the internet and its diverse entertainment and communication options, will not simply give attention to 'bells and whistles' multi-media.

The ability to design creative content and effective instructional strategies may be even more necessary today than in yesteryear.

2.7 Core Principle 6: Learning Design Takes into Account the Working of Memory Systems

Human memory is a little bit like having a Maserati sports car, but only being allowed to use the first gear, except on special occasions. A Maserati will hit a top speed of 185 miles per hour, but certainly not in first gear. Our memory has two main systems, long-term memory (LTM) and working memory (WM). These are depicted in Fig. 2.3. Our LTM seems to have unlimited storage capability. It's not that our brain gets bigger as we learn more; rather it becomes denser in terms of neural connectedness, though we can never live long enough to test its full capability. However, before information can be stored in LTM, it must first pass through WM, which has limited immediate capability when processing new information. The 'magic' 7 (able to process around 7 plus or minus 2 bits of information at one go) was originally documented by Miller (1956), for what was then referred to as short term memory. However, more recent research (e.g., Van Merrienboer and Sweller 2005) suggests that in everyday situational use, this tends to be only 2–4 elements at a time. WM also needs quick rehearsal for information to be effectively captured and processed, otherwise it is

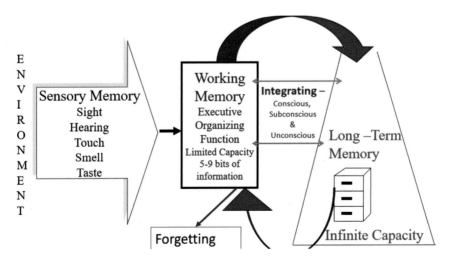

Fig. 2.3 Summary of memory systems

typically lost (forgotten) after only a few seconds. The limited capacity of working memory poses problems for learning, as Clark and Lyons (2005) point out:

> …it is in working memory that active mental work, including learning, takes place. Working memory is the site of conscious thought and processing. (p. 48)

Similarly, Ormrod (2011) summarizes the importance of this key memory system:

> Working memory is the component of our memory system in which we hold attended-to information for a short time while we try to make sense of it. More generally, it's where our thinking occurs. For example, working memory is where we think about the content of a lecture, try to decipher a confusing textbook passage, or solve a problem. Whatever our consciousness is, this is probably where it is housed. (p. 55)

Willingham (2009) further emphasizes that what we do with the information when in WM, has implications for what we end up with in LTM—for better or for worse:

> For material to be learned (that is, to end up in long-term memory) it must reside for some period in working memory – that is, a student must pay attention to it. Further, how the student thinks of the experience completely determines what will end up in long-term memory. (p. 49)

His descriptions of the internal mental processing involved demonstrate the crucial dynamic relationship of thinking and knowledge in building understanding, If knowledge is limited and thinking is poor, what we end up with in LTM will be of little or no value. Hence, to develop well-structured, comprehensive, and useful mental models of what we are learning, neurologically well-wired and cemented in LTM, there is a need for the right knowledge (e.g., key concepts and principles) and good thinking to be occurring in WM, and ongoing checking of what's 'building up' in LTM. To reinforce this process, Brown's et al. (2014) summary is spot-on:

> To be useful, learning requires memory, so what we have learned is still there when we need it. (p. 2)
>
> Practice at retrieving new knowledge or skill from memory is a potent tool for learning and durable retention. (p. 43)

That is why retrieval practice, as will be explored in more detail later, is so important in the learning process. Willingham (2009) is insightful in his framing that:

> Memory is the residue of thought. (p. 4)

You may also remember from Core Principle 4 that the human mind is, for many of us, inherently lazy in that System 2 thinking typically is draining on our cognitive resources and results in what is often referred to as 'ego-depletion' (Kahneman 2012). Quite simply, excessive cognitive activity, like excessive physical activity, is not the norm for most people—one must choose to develop these capabilities. It is also the case that, at a conscious processing level, the brain is relatively slow as a processing system, especially when compared to computer technology. If you have any doubt, do this simple exercise:

How many capital letters in the English alphabet are curved?

If you already know the answer, it would be immediate; otherwise, it would probably have taken you some 25–30 s to get the correct answer (11). However, type in Jack Russell terrier on your PC and you will get around 99,400,000 hits (Well, at 4.40 p.m., Singapore-time, November 2019). Given the limitations of WM, a largely lazy thinking system and slow processing speed, we start to get a somewhat limiting picture of human learning capability. Hattie and Yates (2014) make the point that many of us don't want to face up to when they highlighted:

> Notions such as instant experts, superfast learning, speed reading, and other magic-like programs, amount to faddish quackery in violation of known and validated principles of human learning. If only it was that simple. (p. 113)

However, the picture is not as bleak as it seems, as there are ways in which we can use our memory akin to driving the Maserati in 4th gear. This becomes possible, even easy, once we have acquired vast knowledge, understanding and expertise in a specific field. Such capability is fully encoded as highly integrated neural networks (e.g., cognitive schemata) in LTM. WM has no limitations when dealing with such information retrieved from LTM, as it dramatically alters the functionality of what is taking place within the memory systems. The two systems effectively merge into one fluent dynamic entity working towards meeting the conscious goal of desired information retrieval and solving the problem in hand. Furthermore, over time, this process becomes increasingly automated, and as Hattie and Yates (2014) summarized:

> When your knowledge becomes so automatic that you can access it quickly, with virtually no effort, then the WM system is said to be bypassed through the automaticity stage – a most desirable place to be. (p. 147)

This enhanced memory capability explains why a person very fluent in a language can always find the words they want to use and assemble them in complex sentences instantaneously. Contrast this with the novice trying to learn the days of the week in a new language. It took me more than an hour to learn (as in encoding sufficiently in LTM for later effective retrieval) the days of the week in Mandarin, and that was quite good.

It has been popular in educational circles to downplay the importance of rote learning and memorization. After all, we want flexible adaptive and creative thinkers today—right? Yes, but such high-level human capability is largely based on what we have already acquired in our LTM system. Quite simply, if there is not much information there, and it's not particularly well organized and connected, there is little chance of creative or even useful outcomes. This could not have been levelled at the neural arrangements of Einstein or Da Vinci, and it may have been a definitive factor in their genius capabilities. It is not surprising that Kircher et al (2006) concluded that:

> …long term memory is now viewed as the central dominant structure of human cognition. Everything we see, hear and think about is critically dependent on and influenced by our long-term memory. (pp. 3–4)

Research clearly shows that a major factor that differentiates experts from novices is that expert problem-solvers can draw on the vast knowledge bases in their LTM and quickly select the best approach and procedures for solving a given problem. As Kircher et al. further point out:

> We are skillful in an area because our long-term memory contains huge amounts of information concerning that area. That information permits us to quickly recognize the characteristics of a situation and indicates to us, often unconsciously, what to do and how to do it. (p. 4)

This essentially means that the more you have effectively learned and appropriately organized in LTM makes subsequent learning in that area or field more effective. As Willingham (2009) noted:

> …having factual knowledge in long term memory makes it easier to acquire still more factual knowledge. (p. 34)

> One of the main factors that contribute to successful thought is the amount and quality of the information in long term memory. (p. 17)

This goes very much against the prevalent view among many educationalists that we should not be encouraging rote learning but instead focusing on building understanding through the development of thinking. As documented earlier, understanding is important, and the development of good thinking is essential to achieving this. However, this is a bit like having a Maserati, knowing how to drive it, but not having any petrol to put in the tank. Csikszentmihalyi (1990) was correct in arguing that, "it is a mistake to assume that creativity and rote learning are incompatible" (p. 123). Memory and thinking are equally important in the development of understanding, share interdependent functionality in the learning stakes and there may be little point in viewing them as distinctly different processes. It is the construction of elaborate mental schemas in LTM that provides the conscious mind, operating in working memory, with room to think when solving problems. Repetition and review are vehicles enabling knowledge to be stored in reliable retrievable units which, over time, accelerate mental growth through conceptual mastery and deeper understanding. As Willingham (2009) argued:

> As far as anybody knows, the only way to develop mental facility is to repeat the target process, again, again. (p. 87)

There is an elegant simplicity here; mastery of knowledge bases, good neural interconnectedness and plenty of varied retrievals of such knowledge, reduces the need to activate slow deliberate thinking processes—System 2 thinking. Hence, when solving known problems, the solutions are readily retrievable as memory algorithms or at least solid heuristics from LTM. That's the beauty of top-level expertise, and why persons possessing such capability typically get paid so much more than mainstream professional folk. I like the story about the expert chemical engineer who was called into a plant emergency where the on-site engineers could not identify why a reactor was not starting up, and where losses could run into many thousands of dollars a day if not rectified. The expert engineer walked around the plant, looked at various parts of the system, made certain adjustments to various parameters in the units, and within a couple of hours had the reactor working as it should. Later she billed the company $20,000. The company, not challenging the cost, given the alternative scenario, did ask the consultant engineer for a breakdown of the bill. The reply went something like this, "$1000 for the call out, $19,000 for what's in my head".

This heuristic has many implications for how we teach. Perhaps most apparent is the need to chunk up information into manageable learning structures to prevent cognitive overload on WM. Also, to take account of students' prior knowledge, as its level of completeness, integration and ease of access for retrieval will impact our pace and focus when teaching. For example, students with limited prior knowledge, when presented with new information in that area, may be especially vulnerable to cognitive overload. In this situation, they will struggle to process it meaningfully, feel confused, and fail to assimilate it meaningfully in LTM. Cognitive load has been distinguished in terms of two main interrelated components: intrinsic cognitive load and extraneous cognitive load (e.g., Van Merrienboer and Sweller 2005). Essentially intrinsic cognitive load is related to the task complexity itself and the ability of WM to deal with it. For the novice, a complex learning task will create cognitive overload, simply trying to make sense of it. Extraneous cognitive load refers to introducing information into the learning situation that is not relevant to the learning (e.g., unnecessary text, graphic or colour change) or being poorly organized. This can be significantly reduced by good instruction design. As the authors emphasize, "There is no substitute for evidence-based instructional design" (p. 173). In contrast to the novice learner, when teaching students who have a high level of knowledge and expertise in an area, we can present information much quicker and in more elaborated forms, as they already have highly developed mental schemata in that knowledge field. In terms of analogy, therefore, I can read (and usually make good sense of) several psychology journals in a day but cannot retrieve the television picture when my dog sits on the remote-control device and scrambles the channels.

Students need time to rehearse new information in their minds and consolidate to existing mental schemata, which is facilitated through application activities that generate appropriate types of thinking (e.g., analysis, comparison, inference & interpretation, and evaluation), as this facilitates understanding. The wise teacher will provide this structure for students and adjust the pace of instruction accordingly. Consolidation of learned material in LTM is further reinforced through providing systematic reviews stimulating the retrieval of key information from LTM and bringing it to conscious attention in WM, as illustrated in Fig. 2.4. Students and the teacher can then do a quality check on what has been learned, remediate lost elements, clarify overall understanding, as well as reinforce desired learning. This very act of conscious retrieval from LTM to WM fires related neural structures, which result in the secretion of myelin, an enzyme-based substance that forms an insulating sheath around the axon in a neuron. In basic terms, this further strengthens the learning bond in LTM. Talking to oneself, when memorizing for an exam, if it is about the 'right stuff', is far from madness; it is a good learning strategy.

Another aspect of how memory systems work, which has important implications for the design of learning and teaching practices, concerns how information is selected when presented to learners. It is well documented that apart from the

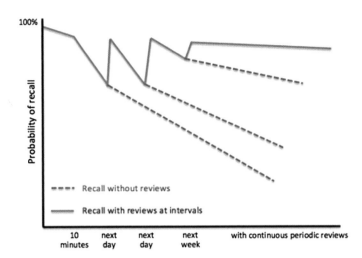

Fig. 2.4 Retention of information with and without reviews

limited capability of WM to deal with incoming information, the attentional and information processing is not uniform. The Serial Position Curve (Murdock 1962) demonstrated that when presented with a list of 16 items to memorize, people typically memorize more at the beginning and the end, tending to forget what was in the middle. These information acquisition biases have been labelled the Primacy effect (the tendency for the first items presented in a series to be remembered more easily as compared to most other items) and the Recency effect (the tendency for the most recently presented items to be remembered more easily as compared to most other items). Another important effect is what is referred to as the Von Restorff effect (the tendency to remember distinct or novel items more easily as compared to most other items), named after the psychologist who discovered it (See Fig. 2.5, incorporating the Von Restorff effect). Even a quick break in a session can represent a change in the stimulus situation and has benefits in attentional and memory processing—"a change is as good as a rest"; another of those old folk sayings that has acquired validity from cognitive neuroscience. Hence, from a practical teaching point of view, the early part of the lesson is where a key impact can be made, both in terms of teaching the main concepts and building rapport. The best motivational speakers know this well and exploit it to the limit. Similarly, the end of the lesson is also important as it facilitates retention of the key points in summary, as well as linkages with other resources and possibly a short advance organizer for the following lesson. Also, irrespective of what has happened in the lesson earlier (e.g., students did not do particularly well on a test), it can be used to finish on an upbeat and positive note, lifting the psychological state of the students. This is like ending an interview with a firm handshake and positive eye contact. Such non-verbal behaviours can significantly influence the perception, and subsequent behaviour, of others, albeit largely unconsciously. In this context, the creative teacher, through well designed changes in method and activity,

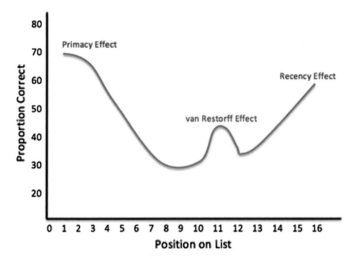

Fig. 2.5 Serial position curve, incorporating the Von Restorff effect

can trick the brain into paying much more attention than it would customarily give over a given time duration. Finally, the creative application of a Von Restorff effect, will put the 'icing on the cake', metaphorically speaking.

Many teachers have long recognized the importance of presenting information in manageable chunks and then structuring activities that give students time to make sense (digesting) of it through discussion and/or other forms of application. Over time they do periodic recap and review to increase the chances of effective transfer and retention in LTM, as well as remediate gaps in learning. We are developing a more precise science that underpins how this works and can now confidently predict that when utilized thoughtfully in practice, there is likely to be significant gains in student attainment, as well as better engagement. We can all remember the teachers who bored us. Several behaviours can contribute to boredom in the classroom; many are violations of memory processing. Teachers who consistently use practices that conflict with the workings of the human memory system will experience frustration with the gaps in many students learning, lack of attention and engagement, and possibly acts of indiscipline. The frustration and consequences will invariably be greater for the students themselves.

2.8 Core Principle 7: The Development of Expertise Requires Deliberate Practice

Most Saturday mornings I usually go into the gym at my apartment block and do around 30 min of weight training; apparently, this is a good exercise regime. Whilst doing my exercise programme, I occasionally look out of the window at people playing tennis. I have noted that several players never seem to get any better even after a few years. They play the same novice game every week. They are unable to serve with any consistency, do not adopt proper body positions when striking the ball and don't even seem to focus attention on the ball when they hit it, and I'm not a professional tennis coach.

The notion that learning inevitably improves over time and that experience is central to improvement is highly questionable. Yes, time on task is important and so is experience. However, it is more about what is done when on a task that really makes a big difference in the experience. For example, why is it that some people, who have many years of experience, still display limited competence, whereas relative newcomers achieve good competence in a comparatively short time? The conclusion of Berliner (1987) offers insight into such questions:

> …experience will probably only instruct those who have the motivation to excel in what they do and the metacognitive skills to learn from their experience…we believe that individuals with that kind of motivation to learn and in possession of a set of strategies for learning from experience are literally transformed by their experience. (p. 61)

It is certainly the case that motivation is a key factor in effective learning. However, it's also about being clear about learning goals and having the strategies and resources to achieve them. Furthermore, in skill development, one must put in a lot of actual *doing*, and much of this is what is typically called the practice. It seems obvious to assume that practice is important. I like the quote from the legendary golfer Gary Player, who once said, "The more I practice, the luckier I get." However, from research, we are becoming aware that it is not just practice per se that facilities improvement but, more importantly, how it is structured, and the way feedback is utilized. Colvin (2008) noted that exceptional performers were not necessarily the most talented in terms of their earlier biographies but had certain attributes and practices that distinguished their expertise over time. Of most importance was what is now referred to as *Deliberate Practice*. According to Colvin, Deliberate Practice is characterized by many interrelated key elements, which include:

- The activity (practice) is carefully designed to improve specific aspects of the performance, often with a teacher's help
- It requires much repetition
- Feedback on results is continually available
- It is highly demanding mentally (whether a physical or mental task)
- It isn't much fun (in the main; but maybe for some).

If we analyse these components of deliberate practice it becomes apparent what makes Deliberate Practice so important in the development of competence and eventually expertise.

Firstly, deliberate practice fully utilizes the learning affordances of Core Principle 1: *Learning goals, objectives and proficiency expectations are clearly visible to learners*. Hence, there needs to be a desired goal that is both challenging and achievable, with clear expectations of what is involved. It must also be highly contextualized to the individual who is to invest the necessary effort required to meeting the goal. If learners take on unrealistic goals, this invites unnecessary failure, frustration, and this can undermine confidence and belief in one's potential. Equally, if it is too easy to achieve, then we are dumbing down a person's potential and creating false confidence, who when faced with challenging tasks, as will inevitably happen at some future time, will not well prepared for this. Hence, this process takes both openness and good two-way feedback to frame the right goals for the right person at the right time.

Secondly, it requires repetition, and this is often where perhaps the difficult bit part of practice comes into play. Repetition is so important for building both clarity of understanding and skill acquisition at the neural level. To build a strong mental model in the brain at the neural level requires much rehearsal and retrieval activity in and out of LTM to WM. This is generally referred to as *Retrieval Practice*. In most basic terms learning at the level of the brain, involves the development of integrated neural structures. As Lang (2016) describes:

> Neurons form new connections with one another with every new experience we have: new sensations, new thoughts, new actions. As the neurons are connecting to one another in novel ways, growing and strengthening new connections, they are forming networks. The first-time neurons link up in a new way, that connection is a temporary or fleeting one; if that connection is used again (because we repeat the thought, or recreate an experience), the link strengthens. The more times the pathway is used, the stronger the connection. The more times the pathway is used, the stronger the connection. Neurons that fire together, goes the saying in this corner of the biological world, wire together.
>
> …we learn when our brains form new neural networks or modify existing ones as a result of our experiences; this means that quite literally, learning requires the continual formation of new connections between our neurons. (p. 95)

Retrieval practice then facilities this process of building stronger and better neural networks, which results in a deeper and deeper understanding of that area of learning. Hence, if you are doing plenty of retrieval practice as you read through the various sectors of this book, you will be linking together the content knowledge in the various areas, as well as the key concepts and principles, (hopefully in multiple ways) and, as a result, will be building understanding. This will be growing neurologically in your brain, and as you continue the retrieval practice, maintain interest and thinking about how this applies in different contexts and situations, the conceptual understanding will grow, getting deeper, until you have that total 'aha' feeling—often referred to as the Eureka Effect. Every time you do a retrieval practice activity from LTM into WM, these neural pathways will fire and when they go back to LTM the connections get stronger. There are biochemical processes involved in such activity, which the interested reader can research. From a pedagogical perspective, we don't need to know the neuroscience—but we do need to understand the cognitive behaviour and ensure that our teaching practice includes the necessary opportunities for retrieval practice. Hence, Brown et al. (2014), in summarizing this process concludes:

> …while the brain is not a muscle that gets stronger with exercise, the neural pathways that make up a body of learning do get stronger when the memory is retrieved, and the learning practised. Periodic practice arrests forgetting, strengthens retrieval routes, and is essential for hanging on to the knowledge you want to gain. (pp. 3–4)

I explicitly teach students how this works and why they should do it. In fact, I teach students all the core principles of learning, of course, customized to their context. If we want students to learn well, we should teach them how learning occurs and how they can make this part of their own self-regulation. If we are seeking to develop self-directed learners and we want them to become increasing agentic in their learning (e.g., be proactive, seek out resources and develop their own personalized learning strategies) this is surely a no-brainer—no pun intended. In practical terms, the more students do retrieval practice, the better they learn and retain that knowledge gained. Furthermore, as Brown et al. (2014) points out:

> *Testing doesn't need to be initiated by the instructor.* Students can practice retrieval anywhere; no quizzes in the classroom are necessary. (p. 44)

Another consideration in doing practice concerns how often you do it and for how long. For example, we may do our practice every hour, every-day, every week—or whatever; this is typically referred to as *Spaced Practice*. Also, there is the question of how long should we do our practice sessions? For example, we may do it for a designated amount of time or until we are exhausted or fed up—or both. Doing long practice sessions is usually referred to as *Massed Practice*. We can, of course, do both. In the movie *Desperado*, the main male actor, Antonio Banderas, who apart from other things, was a guitarist who in one scene was asked by a young boy in the street how to get good at playing this instrument. The actor replied with the statement "practice all day, every day". Well, that combines spaced practice and massed practice into continuous practice—is this the best option? The answer is, of course, no, it's not possible to do that—there is a limit physiologically and psychologically to the amount of massed practice one could put in, though this would vary depending on the activity and person.

However, research shows that both types of practice, when done effectively in context, can enhance learning, but Spaced Practice has been shown to be most beneficial for learning. As Lang (2016) summarized:

> Research has demonstrated the power of spaced learning. Carey wrote, "nothing in learning science comes close in terms of immediate, significant, and reliable improvements to learning". (p. 76)

Brown et al. (2014) explains how this works:

> It appears that embedding new learning in long-term memory requires a process of consolidation, in which memory traces (the brains representations of new learning) are strengthened, given meaning, and connected to prior knowledge – a process that unfolds over hours and may take several days. Rapid-fire practise leans on short term memory. Durable learning, however, requires time for mental rehearsal and other processes of consolidation. Hence, spaced practice works better. (p. 49)

In addition, when done effectively and is ongoing (as and when needed) is another core principle of learning and will be examined in some detail in the next section. For now, let's see feedback as a two-way communication process between the learner and others who can contribute to his/her learning in some useful way. This would certainly include the teacher/coach, but also peers and any other persons who can contribute input for an individual learner. In the feedback process, irrespective of who the others are, the learner must go through the retrieval practice process, whether it is pulling out knowledge from LTM or displaying skill and engaging in dialogue as to its development in terms of the desired goal. Essentially, this is a process of testing (not summative assessment), but formative assessment (assessment for learning). The purpose of retrieving knowledge from memory (or displaying one's skill in a visible performance) is that the learning becomes visible to both learner and others (especially for the expert teacher/coach). This has two important benefits for future learning. Firstly, it provides direct performance evidence on what the learner knows, doesn't know, and any fuzzy areas or misconceptions. It is then possible, through mediation between the teacher (or other), to identify where the learner is now in terms of meeting designated goal(s), what is needed to move the learner progressively forward, and how best to do this (e.g., enhance the learning strategy in specific ways). Remember, every time students do these retrieval practice activities, they are not only going through the testing processes with feedback but also further cementing the learning neurologically in the brain, which makes it easier and easier to recall in future testing activities.

The pedagogic implications of this are significant. For the purposes of teaching, the use of performance tasks that directly test the key concepts and skills involved will facilitate the learning process, especially with supportive feedback. Hence, in terms of the core principles covered so far, clear goals, knowing what students already know, focusing on key concepts, getting students to think, managing cognitive load, and doing deliberate practice is offering a potentially high impact instructional design framework. There's more to come, so we'll stay with the retrieval practice concept, and re-fire those neurons in spaced practice—seems like a well-reasoned approach.

Research also strongly supports the use of what is referred to as Mastery Learning. For example, Bloom (1985) in evaluating optimal conditions for learning, summarized its impact on learning:

> One example of such conditions is *mastery learning* where the students are helped to master each learning unit before proceeding to a more advanced learning task. In general, the average student taught under mastery-learning procedures achieves at a level of above 85% of students taught under conventional instructional conditions. (p. 4)

From an evidence-based approach, this makes perfect sense. In terms of my math learning (or not learning) experience, once I got lost, which was almost immediately, nothing else made sense. When taught differently (better), with step by step learning, in manageable chunks, with practice and feedback, and patience on the teacher's part, my grade went from 9 (highest failing grade) to 3 (average passing grade). I did not achieve expert mastery for '0' level, but it was good enough for university entry at that time. As Brown et al. (2014) points out:

> Mastery in any field, from cooking to chess to brain surgery, is a gradual accretion of knowledge, conceptual understanding, judgement, and skill. (p. 18)

It is important that there are clear and realistic improvement targets for the particular learner. This involves stretching the individual beyond an existing performance level to a recognizably improved level in some aspect but a level that is achievable with effort and coaching from a teacher. As outlined in Core Principle 1, it is important to have as much clarity—visibility—of the learning goal, objectives and proficiency level as possible. In this way, motivation is maintained as the learner will have a perceived experience of a higher mastery in at least some aspect of the performance, which further reinforces the belief and sustains effort in continuing this learning strategy. To reiterate the point, "nothing breeds success like success". It is often noted in professional sport that when a player finally wins that elusive major tournament, more seem to quickly follow. Andy Murray winning the men's tennis tournament at the Olympics, the US Open and Wimbledon, is perhaps an illustrative recent example. Prior to that, he had failed to win a major tournament, losing in 4 finals.

Of key importance is the role played by expert teachers in helping the learner identify what specific aspects of the performance to improve, structure the practice programme accordingly and provide ongoing quality feedback to maintain focus on the skill development. Again, to use the Andy Murray example, the appointment of Ivan Lendl in this role may have been more than coincidental in his attainment of two major titles within one year. Lendl himself had gone through the experience of losing his first 4 finals in major tournaments but eventually went on to win 8 singles titles in such events. Certainly, he had learned something important and this may have helped in coaching Andy Murray. It seems that even the very best in the world still desire and need an expert teacher. It is necessary to emphasize that while the deliberate practice is fundamental to effective and efficient learning, it is not a shortcut to expertise or even competence (however defined).

What is particularly interesting is that in the process of developing expertise, not only is there an enhancement in understanding and skill, but significant changes in neurology and sensory acuity relating to the field of expertise. Many years of intensive deliberate practice changes the body and the brain, enabling great performers to perceive more, to know more and to remember more than most people. Colvin particularly noted the following key attributes of great performers:

> They all possess large, highly developed, intricate mental models of their domains. (p. 122)
>
> …observe themselves closely… monitor what is happening in their own minds and ask how it's going. Researchers call this metacognition …top performers do this more systematically than others do; it's an established part of their routine. (p. 118)

This enables them to:

- add and make sense of new knowledge more quickly and in more qualitatively useful ways
- to distinguish relevant information from irrelevant information
- predict what will happen next in a domain-specific situation.

Perhaps, what is most significant is the relative ease in terms of cognitive load and strain that they must expand in doing most tasks in their field. As Kahneman (2012) explained:

> As you become skilled in a task, its demands for energy diminishes. Studies of the brain have shown that the pattern of activity associated with action changes as skill increases, with fewer brain regions involved. (p. 35)

Expertise then enables a better understanding of a situation and facilitates a heightened perception of what is most relevant for the task at hand. This allows the expert to do many things quickly and automatically, releasing time to be more situationally responsive and potentially creative. In the context of teaching, Turner-Bisset (2001) identified such capabilities in expert teachers:

> Expert teachers are able to read and process the complex mass of information which any classroom provides, much more rapidly and meaningfully.
>
> …expert teachers use a repertoire of strategies, selecting the most appropriate for use in a particular context and adapting it if necessary for a group of learners. (p. 69)

Hattie (2012) from extensive research supports this heightened capability of expert teachers as well as providing insight into how they are likely to be more creative:

> Experts possess knowledge that is more integrated, in that they combine the introduction of new subject knowledge with students' prior knowledge; they can relate current lesson content to other subjects in the curriculum; and make lessons uniquely their own by changing, combining, and adding to the lessons according to their students' needs and their own teaching goals. (p. 28)

This heuristic focuses attention on the important role of deliberate practice in skill development and attainment. From an evidence-based perspective, we are now able to be much more precise and specific in terms of what types of practice and how best to structure and manage practice to enhance attainment. The saying that "practice makes perfect" is not quite right, though well-intended. Simply getting students to practice and spend more time on task may have limited value in optimizing competence and expertise without the systematic structuring of the practice activity, calibrated to the learner's proficiency level, and with expert feedback. Practice on its own may simply lead to consistent proficiency at not doing an activity well, as Berliner noted above, and I observe from the gym window. It is through deliberate practice over time that is most likely to lead to higher proficiency levels and eventually expertise. However, deliberate practice is very much intertwined with the building of dense and well-integrated mental schemata in LTM and the ability to use metacognitive capabilities at heightened levels. As emphasized earlier, each Core Principle of Learning, while focusing on a specific aspect of the learning process, is ultimately part of a dynamic and synergistic system in which specific areas of learning capability become mutually supporting in enhancing human learning (e.g., attainment, engagement and well-being).

In applying this core principle of learning in practical teaching it is important to ensure that the process of using deliberate practice is adhered to as best as is practically feasible. Invariably, in working with large classes it is harder to be as precise in diagnosis, task structuring and providing the time for ongoing feedback, as in the case of purely individualized coaching. However, by making the process of deliberate practise visible and meaningful to students, it is possible with some thoughtful application of collaborative learning and peer coaching—creative teaching—provide better opportunities for enhancing learning and attainment in this area.

2.9 Core Principle 8: Assessment Practices Are Integrated into the Learning Design to Promote Desired Learning Outcomes and Provide Quality Feedback

At school, I don't recall the word 'assessment' being used, and certainly not 'learning outcomes' or 'feedback'. We had to sit end of year exams and we were given homework each week, which was marked by teachers. On receiving homework back, we got a graded mark often with a '+' or '−' sign next to it, and a short comment such as, "fairly good", "could do better", etc. I also never recall giving this much thought in terms of what I might have done well and what I had not done well, and certainly not what I needed to do in order to improve and how. It was done and out of the way and that was that. I attach no blame to the teachers as that was the assessment practice in that context. The assessment was largely seen in terms of summative grading and not as a key facilitating aspect of the learning process. The question, in the present context, is what do we now know about assessment practices that are evidence-based in terms of providing an important heuristic for significantly improving student learning and attainment?

Firstly, it is now clearly recognized that assessment is not simply a means to measure learning that has already occurred but a major facilitator in the learning process itself. As Boud (1988) illustrated:

> There have been a number of notable studies over the years which have demonstrated that assessment methods and requirements probably have a greater influence on how and what students learn than any other single factor. This influence may well be of greater significance than the impact of teaching or learning materials. (p. 35)

In my experience, whether teaching pupils in the mainstream school context, or on Masters' degree courses, learners typically focus on what is assessed. I have taught many students on such degree programmes who have been very explicit about what their main priority is, and that was passing and getting a good grade. To do this they want to know what to learn and how to apply it to meet these goals. I am not saying there is no intrinsic motivation in their overall approach, but assessment largely drives the learning process. For higher education programmes, there is an emphasis on complex understanding and application, which inevitably pushes students towards engaging in more complex levels of thinking and problem-solving. However, this is not always seen as a pleasurable activity, even for Masters' degree students. Many like it when you model the answers for them, and why would they not, as many do the programme after a long day's work and are already suffering from cognitive strain. Similarly, in school, if an assessment is mainly focused on memorizing large bodies of factual content knowledge, then that's what most motivated students will do. Teachers talking about learning for passion and the importance of becoming self-directed lifelong learners will mean little when the marks on test papers suggest otherwise.

Assessment serves many purposes for different stakeholders (e.g., selection, maintaining standards, identifying and diagnosing learning difficulties, and enhancing teaching). Most significant, in this context, is the important role that formative assessment (e.g. where learning is focused on supporting the learning process) plays in influencing student attainment and engagement, especially through the process of ongoing two-way feedback. This contrasts with summative assessment (e.g., where a terminal assessment decision is made and the learner either passes or fails or is graded accordingly). The high impact of feedback on attainment (e.g., the average effect size of 0.79, which is twice the average effect of all other schooling effects) is well documented by Hattie (2009). However, it is only relatively recently that this has been subjected to detailed scrutiny in terms of its impact and how it works on specific aspects of the learning process.

There are many interrelated aspects that contribute to the high impact potential of feedback on learning. Nicol and MacFarlane-Dick (2006), in synthesizing the research literature suggest the following seven principles:

Good feedback practice:

1. helps clarify what good performance is (goals, criteria, expected standards);
2. facilitates the development of self-assessment (reflection) in learning;
3. delivers high-quality information to students about their learning;
4. encourages teacher and peer dialogue around learning;
5. encourages positive motivational beliefs and self-esteem;
6. provides opportunities to close the gap between current and desired performance;
7. provides information to teachers that can be used to shape teaching. (p. 203)

As prior learning (Core Principle 2: *Learners' prior knowledge is activated and connected to new learning*) is always the entry point for new learning (and feedback is new learning) it must find some anchor point in prior learning to connect meaningfully to it, in order to result in some enhancement of understanding in that given area of knowledge/ skill-building. For example, if students are unclear about what they are supposed to be learning (e.g., the goals, criteria, expectations) even good feedback may not make much sense. Indeed, research (e.g., Hounsell 1997) has found that teachers and students often have quite different conceptions about what is involved in meeting the goals and what the criteria mean in specific terms. The significance of this is, as Nicol and MacFarlane-Dick (2006) explain:

> Weak and incorrect conception of goals not only influence what students do but also the value of external feedback information. If students do not share (at least in part) their teacher's conceptions of assessment goals (and criteria and standards), then the feedback information they receive is unlikely to 'connect'. (p. 206)

Hence, good feedback is very much an ongoing dialogue between teacher and learners (as well as between learners) to identify gaps in knowledge, understanding and skills, as well as directing the necessary action to resolve these gaps. As Hattie (2009) states, effective feedback must:

…reduce discrepancies between current understandings and performance and a learning intention or goal. (p. 175)

It's not surprising that quality feedback has such high impact in terms of effect size on student attainment, as it connects to so many aspects of the learning process. However, to maximize the positive impact of feedback on attainment a few conditions need to be effectively met. Sadler (1989) summarized these as follows:

- What good performance is (i.e. the learner must possess a concept for the goal or standard being aimed for)
- How current performance relates to good performance (for this, students must be able to compare current and good performance)
- How to act to close the gap between current and good performance.

A useful approach for helping students to identify what good performance is in relation to a goal, what assessment criteria are and how they work, and to make comparisons between present competence and the competence needed (e.g., knowledge bases, skills and levels of performance) is to provide various exemplars of performance at these different levels. As we know, providing students with examples (and non-examples) is an effective instructional method. With deliberate practice, including plenty of testing, such capability with emerge—given student motivation and volition. In the most basic terms, students must understand the nature and qualities of good work if they are to create it themselves. Specifically, they need to know:

- The meaning of key tasks language (e.g., what does 'describe' and 'evaluate' mean)?
- The meaning of assessment criteria (e.g., "what does give evidence", "show your working" mean)?
- How actual tasks and criteria can be demonstrated in practice (e.g., what are acceptable evidence formats)?

The manner and types of questions asked during feedback sessions are also important. A friendly supportive mediating approach is essential to create a level of rapport in which learners feel comfortable in providing feedback to the teacher. Once established, teachers can then ask students focused questions in order to ascertain what they know and understand, identify specific gaps in knowledge and understanding, as well as misconceptions, thus enabling learning to become more visible to both.

Furthermore, effective teachers, just as they adjust their communication style to different student personalities, also adjust their provision of feedback accordingly based on students' need in different contexts. For example, Hattie and Yates (2014) suggest that novices require more specific task-related corrective feedback, to be gradually replaced with more process feedback as they become increasingly proficient and self-regulated in their learning. What this means is that initially, feedback will focus on detecting errors in what students are doing on a task and help to reduce and eventually eliminate these errors. Such feedback will include showing students what went wrong, examples of correct performance and ways to improve on these types of learning tasks. Process feedback is more focused on how the students are tackling the tasks given, such as their thinking (e.g., analysing, comparing, making inferences & interpretations, evaluating) and the learning strategies they are using. In providing feedback it is often the case that both aspects are needed, and this is where the teacher's judgement and skilful action are most impactful. As students become increasingly proficient, feedback is usually more focused on their abilities to monitor and evaluate their own learning, both at cognitive and affective levels (e.g., metacognition). Questions of how much feedback and the frequency of feedback, as with all aspects of differentiated instruction, will depend on the situation and learners' readiness. As Hattie (2012) summarized:

> The key is the focus on decisions that teachers and students make during the lesson, so most of all the aim is to inform the teacher of student judgements about the key decisions: 'Should I relearn…Practice again…To what?' and so on. (p. 143)

The strategic use of ongoing formative assessment is an essential part of the overall assessment strategy and, as Perkins (1992) suggests, once considered thoughtfully:

> Teaching, learning, and assessment merge into one seamless enterprise. (p. 176)

Furthermore, feedback is not something that occurs only between the teacher and individual students but can, and should be, an ongoing collaborative process with students as active participants. Hence, any activity that tests what students know/dont know, etc., is retrieval practice which, as detailed prior, is essential for memory consolidation and building understanding—with good thinking of course. This can be done through quizzes, conversations, or any performance task that authentically enables students to elicit/display the knowledge and skills involved.

A method that I have found particularly useful is that of peer assessment. As Petty (2009) summarizes the procedure:

1. Students come to understand the nature of good work more deeply, as they must use this understanding to judge peers' work. This helps them understand their goals as learners, for example how marks are gained and lost. These goals are learned from concrete to abstract; this is the most powerful way to learn.
2. They learn other ways of approaching a task than the approach they used.
3. They become more reflective about their own learning and gain understanding by discussing disagreements. For example, if students realize they did one calculation wrong because they confused a sine with a tangent that is very helpful.
4. Students can do more work than you can mark.
5. Students tend to take pride in work that will be peer-assessed: they are more likely to complete it and to write more neatly than if you assess it.
6. Students accept criticism from each other that they would ignore if given by you! For example, 'Your writing is really hard to read.'
7. Students greatly enjoy this method, and both 'helpers' and 'helped' learn if they support each other constructively. (The standard of discussion is commonly higher than you expect!)
8. It helps to develop the skills required for self-assessment. (p. 263)

Feedback from students is invaluable in helping teachers appraise the effectiveness of their own teaching strategies. Unfortunately, many fail to take advantage of this easy-to-use approach for monitoring the effectiveness of their teaching on an ongoing basis and are often dismayed and surprised when they receive negative feedback at the end of a course programme. Teachers who are in regular dialogue with their students concerning learning and collaboratively finding ways to enhance attainment and engagement are rarely surprised by the findings of programme evaluation exercises, and their feedback is likely to be very positive. The very act of seeking feedback from students concerning what aspects of the instructional process are most useful (and least useful) in supporting their learning, is positively impactful in two main ways. Firstly, from a technical point of view, it will enable the teacher to identify what is working well, what is not working well, etc., and subsequently, make appropriate modifications in instructional strategy. Secondly, this is likely to subconsciously impact affective aspects of the relationship, in terms of communicating care and concern for their learning. When positive feelings are communicated in this area, students are more likely to be motivated and maintain the necessary volition and belief that, with the teacher's help, they will be successful in meeting the learning goals. The likely outcome of this scenario is that students will experience both self-efficacy and mastery in their learning, which are mutually positively reinforcing in the learning stakes.

Obtaining feedback on one's teaching is not difficult or time-consuming. It can be done very informally as part of the everyday dialogue of instruction. For me, it is a routine practice to make explicit and remind students (yes, they need reminding) that they must let me know if they are finding learning difficult or not understanding key concepts, etc. With new groups, I usually initiate this with some humour by referring to my East London accent, which does not use the letter H, and they may need to check what I'm saying occasionally and 'pull me in a bit' if I am drifting into local East London diction. Equally useful and a more structured easy to use method for obtaining key student feedback is what is often referred to as a 'One-Minute Paper' (see Table 2.3). This is a simple feedback questionnaire of only two question areas, one identifying a key positive aspect of the lesson and the other one identifying a key negative aspect of the lesson. It can be framed in various ways, as well as modified in terms of focus or terminology. Essentially it provides a quick communication tool to get across your intention to consider their experience and identify what seems to be working well and not working well, from their perspective. This enables a better diagnosis of what the areas of difficulty are and how best to situate the instructional

Table 2.3 Example of a one-minute paper

Example of questions for a one-minute paper
What was the most important learning point for you from this lesson?   • Can also use concept, idea, etc.
What is still not clear to you from this lesson?   • Can also use "muddy", confusing, etc.

approach for addressing them. Good feedback, when used effectively is another of those "Russian Dolls" (Hattie 2009) and it supports learning, both for students and the teacher.

The importance of this heuristic is fundamental to the whole instructional process as our goal is to help develop in our students the capability to more agentic in how they learn so that they can increasingly (and this does take considerable time) to become self-directed learners; that is to be able to plan, monitor and evaluate their learning—and do this well. As Hattie (2012) concluded:

> …all students should be educated in ways that develop their capability to assess their learning. (p. 141)

2.10 Core Principle 9: A Psychological Climate Is Created Which Is Both Success-Orientated and Fun

In Visible Learning (Hattie 2009), the importance of the climate of the classroom was noted as among the more critical factors in promoting learning, emphasizing that teacher-student relationships were the major determiner of such climates—having an effect size of 0.73.

We are all very clear on what constitutes a physical climate, and its various features. It was a typical everyday conversation in England, especially when in a lift with a stranger. How many times have I heard the comment, "Looks like rain shortly"? One of my reasons for leaving the UK was quite simply the climate. I did not like the long winter months, which seemed to last forever. I prefer the perpetual summer weather in Singapore, and what an easy job weather forecasting is here: "26–33 centigrade with some chance of rain in the afternoon" is a 90% correct call on most days. In the UK, as I remember it some 20 years back, weather forecasting was a combination of thoughtful roulette and serendipity, at best.

Now defining a psychological climate is a bit like defining thinking, as we can't see, touch or smell it. However, when it is very good or very bad, we can certainly *feel* it. People typically use terms like, "The atmosphere is terrible in there", or "Everything's cool here". Essentially, it's about the nature and types of interactions that are going on—or not going on—between people in each social and geographical context (in educational contexts it's typically a classroom) and their impact on perception, feeling and subsequent behaviour. The ability to create and facilitate a positive psychological climate in a range of informal interpersonal situations is a great skill set to have. If you have such capability, it's likely that you will always be high on the invite list for socially orientated parties, as you have the skill of creating lively conversational content which helps folks to relax and feel comfortable. As classrooms are not fundamentally different from other social interaction situations in that there are human actors (teachers and students) involved in interpersonal communication over time for a purpose (e.g., teaching and learning), a psychological climate will inevitably result. Furthermore, there is no doubt that certain types of psychological climates are much more conducive to attainment, engagement and well-being than others, which may have adverse effects. Research suggests that several key factors are very important for promoting a positive psychological climate. First and foremost, as Hattie and Yates (2014) summarized, this entails the teacher exhibiting attributes that:

> …promote positive and open human communication. Students value being treated with (a) fairness, (b) dignity, and (c) individual respect. These threefold aspects have emerged strongly in all studies in which students are interviewed and surveyed as to what they expect of their teachers. (p. 26)

Similarly, Ornstein and Behar (1995), from research, concluded that:

> . …the most effective teachers endow their students with a "you can do it" attitude, with good feelings about themselves, which are indirectly and eventually related to cognitive achievement. (p. 86)

Very much in the context of this Core Principle, Jensen (1996) found that:

> Learners in positive, joyful environments are likely to experience better learning, memory and feelings of self-esteem. (p. 98)

Finally, in this context, Sullo (2007) argues:

> Every time we learn something new, we are having fun…It is our playfulness and our sense of discovery that allows us to learn as much as we can…A joyless classroom never inspires students to do high-quality academic work on a regular basis. (p. 9)

However the really important questions concern what are the specific things that teachers can do, and how best to do this, in order to create and sustain a psychological climate that results in the students perceiving and feeling that they are being treated with 'fairness', 'dignity' 'individual respect', developing a 'you can do it attitude' and experiencing some sense of joy in participating in the classroom learning activities?

It is easy, though somewhat limited, to address these questions in terms of intent or generalizations. For example, we might say, "show respect", "Be enthusiastic in how you teach" or "Display passion about learning". This is in many ways like going on a first date and having little idea on what to say or do and being told by a friend to "Be interesting". Such statements are, in both the above contexts, valid and will make sense to both the cognitive neuroscientist and the layperson alike—but there is something significantly missing. It is interestingly and annoyingly (for me anyway) captured in the saying "Everything is easy when you can do it." Being interesting certainly was not the case for me on my first ever proper date with a girl as a seventeen-year-old. Getting ready to meet Geraldine (that was her real name—it will give her a chuckle if she ever reads this book) at a local cinema on a Saturday night, I suddenly posed myself the essential question, "What do I talk to her about?" Instantaneously, I became anxious, which quickly escalated to panic (we have all been there, and we know what this does for good thinking and confidence). In delving into my LTM system it was not long before I realized that all I ever talked about was football, boxing and fishing with my friends, who were all boys. I had no idea at all on what to talk to a girl about, a real lack of prior knowledge containing, in retrospect, only misconceptions. The inevitable happened and the date was a disaster. I had nothing to say, was visibly uncomfortable all night, and this clearly contributed to her feeling equally uncomfortable. At the end of the film, the encounter quickly ended with a statement from me like, "How do you get home?" I had a reply something like, "I get the bus from over there." Geraldine never contacted me again, and that's not too difficult to explain. That was my first date and my last for a while; I was afraid to go through that again. If there was an 'O' level in conversational literacy with females, another grade 9 was an absolute certainty for yours truly, at that time. My Jack Russell dog would have fared better, and you will know why shortly. Despite my intention to be interesting and build rapport, I had no knowledge in LTM on how to do this. As Molden (2001) makes clear:

> It is our behaviour that directly connects to results, even though our thinking may be responsible for generating behaviour. (p. 59)

As we have explored in detail in previous core principles of learning, limited prior knowledge (in my case at that time, very limited) resulted in poor thinking, and the evitable happened—poor results. Did that in any way enhance my self-efficacy in terms of my ability to talk to girls? Not in the least, especially when Geraldine's friend later told me that she thought I was a 'moron'—a word not used much now, but prevalent in the 1960s, as a negative term for someone who appears stupid. Mastery learning for talking to girls?—not me, sadly, at that time. How did I learn to improve my skills in talking to girls—read on?

At university, in the first year of my psychology undergraduate programme, I learned something useful from a fellow student. I noted that he had an attractive girlfriend and he wasn't Chris Hemsworth. I once asked him about this, a kind of "How do you do this?" type of question. His reply was initially strange, "You need a nice-looking friendly dog and walk around the local park." This made no sense to me, until he explained further, "If you do this, girls will notice the dog and want to pet it." I was still no clearer at the time, but you will have probably worked this out by now. As my fellow classmate pointed out, you talk about the dog, mention that you are going for a coffee and ask her if she would like to have a drink with you. It's just then a matter of generating mutually interesting content for conversation. You might be ready to ask, "How do you do this?", and that was my immediate question to him. Summarizing his response in more technical terms, which all seems so easy now, it's about generating content that the other person is interested in talking about, then showing that you are interested in the responses made (whatever this entails), which is done initially by asking the person what is of interest to him or her. Invariably, as we know, highly impactful interpersonal communication is not just about the verbal content, but also (and probably more importantly) the tone and pitch of voice and the accompanying body language components (e.g., posture, eye contact, and gestures). Furthermore, these all need to be appropriately calibrated to create a *total communication experience*.

When one is confident, and this typically comes from one's own self-efficacy and perceived mastery, the communication package comes nicely into place and flows. We should not be surprised as this is simply the result of good learning for these skill sets. Good understanding plus deliberate practice over time will get one to this desirable state. The converse is equally true. In most basic terms, to be effective at something, having intent is only an initial motivator, you must know how to do it well, and be able to do it at the behavioural level. Ultimately perception and judgements about other people, accurate or otherwise, is the product of their behaviour, and of course, our pre-existing beliefs.

In several teaching situations, I have seen novice teachers tremble at the front of a classroom, even run out in fear and despair when faced with challenging students or sometimes from forgetting the details of their teaching plan. Quite simply, they don't know what to do next and lack the strategies in their long-term memories that might be effective in such situations. In contrast, highly competent and creative professionals when confronted with a challenging group of students or even noticing boredom developing on some of the students' faces, while never complacent, can typically and smoothly change the teaching strategy in situ (re-create the pedagogy situationally). In most cases, such action results in regaining attention, settling the group down and changing the psychological climate to one that is more positive, and task-focused. To a novice teacher or outsider, this may seem almost like magic, as creativity in any domain often feels a bit like that. However, as for most things (including magic), once things are made explicit at the behavioural and cognitive level, it all seems rather obvious and logical.

Of course, understanding is not competence, and deliberate practice is needed in skill acquisition, but it certainly helps if one knows very clearly what is involved in the learning process. What I have been describing may seem somewhat behaviouristic and contrived, and that is partly true. However, customer service professionals don't learn how to speak, smile and use their voice in certain specific ways just to fill up training hours on their staff development plans. As Mlodinow (2012) summarized:

> The gestures we make, the position in which we hold our bodies, the expressions we wear on our faces, and the nonverbal qualities of our speech, all contribute to how others see us. (p. 110)

He goes as far as to argue that:

> The pitch, timbre, volume of your voice, the speed with which you speak, and even the ways you *modulate* pitch and volume, are highly influential factors in how convincing you are, and how people judge your state of mind and your character. (p. 132)

It's therefore not surprising that politicians and other high-profile media people employ communication specialists and psychologists to create certain positive appearances to influence the public at large. They do this because it works in large part with many people, and there is an underlying set of reasons why it works. For many years, I mentored and coached 'underperforming' teachers. These were academic faculty who received below 3.25 on a rating scale (where a score of 5 was 'very good' and a score of 1 was 'very poor') from student feedback for 2 semesters on the formal end of semester online questionnaire. Over the years this highlighted how, in a communication encounter, the relationship between a communicator's intention and the perception and meaning by others can be so incongruent. Many of these teachers also had very negative qualitative comments relating to such things as "shows little interest", "no care and concern" etc. In conversation with them, some were very disturbed by such student responses, and could not explain on what basis and how they might have been perceived in such negative light. They seemed unaware that such perceptions originate from specific behavioural aspects of personal presentation.

Essentially, the psychological climate is largely shaped based on how the teacher behaves on an everyday basis with the student group. Hattie and Yates (2014) summarized the specific behaviours that are positive in this respect; they are noteworthy, but obvious when made explicit:

> The key aspects, as described by a significant body of research involve the teacher's positive open gestures when dealing with the class, physically moving around the room, relaxed body orientation, frequent use of smiles, direct eye contact, and using a variety of friendly and encouraging vocal tones, especially when dealing with an individual student. (p. 28)

They go on to point out:

> The human brain is hard-wired to instantly apprehend emotional states in other people …while some cultural differences are found …The notion that humans everywhere share a common basis in being able to recognize emotions in others embodies considerable truth. (p. 266)

Mlodinow (2012, p. 118) quotes research by Ekman and Friesen (1971) who showed people in an isolated Neolithic culture in New Guinea pictures of American faces displaying a range of typical emotions. These primitive people had never been exposed to outside cultures, used no written language, were still using stone implements, and very few had seen a photograph let alone television or films. However, when they were shown American faces of basic emotions, they were as able, as people from the twenty-one literature countries who participated in the research, at recognizing happiness, fear, anger, disgust, sadness and surprise in the faces of the emoting Americans.

Certainly, from my experience of facilitating many workshops in a wide range of cultural contexts, I would also make the case for there being much similarity in terms of people's perception and comprehension of what constitutes good human conduct, a positive psychological climate, as well as the way they learn. Several years ago, I was attending a conference in which one of the keynote speakers was emphasizing how people from different cultures learned very differently, and that we should be thinking of culturally relative pedagogies. In listening, I was reflecting on my own experiences and feeling a bit confused and somewhat annoyed. Yes, of course, there are cultural differences, and pedagogy must consider relevant culturally determined situated factors for obvious reasons. However, in large part, the main specific cultural factors relate more to specific social norms and custom, rather than pedagogic or fundamental interpersonal communication practices. For example, I am mindful of touch, even handshakes in certain cultural contexts, as well as the humour I use. I also notice that in different cultural and ethnic contexts, one must modify the level of informality accordingly. For example, I tend to be more informal quicker in the Philippines than other countries, as participants seem to respond well to this. In certain countries, I tend to retain formality longer as I feel that the early display of humour may be detrimental to a perception of high professional credibility.

However, my experience is that irrespective of cultural context, learners will become more informal and appreciate some humour, once they feel comfortable and perceive high credibility in terms of what is on offer in the learning stakes. Culture is impactful, but it may be less so than personality configurations. In terms of how people learn, I find little difference, and that's because we share the same brain structure and we learn structurally in the same way. In most basic terms, learners must acquire knowledge through memory processing, make meaning of it (build understanding) through thinking, and acquiring skills by doing. In this context, there is motivation and beliefs that will come into play, but the essential principles of human conduct and learning seem largely universal, based on my experience.

I am convinced that highly competent and creative teachers will be positively impactful anywhere, but not with everyone all the time—that's impossible. Equally, very poor teachers will be similarly experienced in negative ways, wherever, in most cases. What is often of noticeable difference is how learners across cultures and contexts respond to the variety of teachers they experience. The best are generally always appreciated. However, how the worse teachers fare may vary significantly depending on cultural contexts. In some cultures, it seems that few learners will disrupt or react negatively even in the face of poor teaching, as there is a deep respect for the profession of teaching. They probably remain just internally bored or upset, depending on whether the teacher is just incompetent technically or socially, or both. The latter is a sorry state to experience. In summary, Sale and Mukerji (2006) were delighted to report:

> …in our experiences of co-facilitation over several years, we were initially surprised but ultimately delighted to find that there appears to be several generic principles and practices that facilitate rapport and effective learning irrespective of cultural and ethnic contexts, (abstract)

There is little doubt that students in a success-orientated psychological climate are more likely to develop a mindset that with effort on their part and with good teaching, they can be successful in their learning and in a sense 'grow their intelligence'—so to speak. The work of Dweck (2006) has been extensive in this area and will be explained and applied further in later chapters. Most significant is the impact of prior knowledge and beliefs on learning, self-efficacy and the ability to frame and enact the appropriate behaviours in real-life contexts—as my dating story above aptly (though sadly for me at the time) illustrates.

Fun or humour was certainly not a significant feature of my school experience, well not in classroom time. It seemed that learning was a very serious business and anything resembling a joke was a prelude to classroom disruption. As a Cockney from East London, I have always felt that humour was one of the most important aspects of human experience, and this is now supported through a wide range of research (e.g., Garner 2006; Lei et al. 2010). Most significantly, the world-famous psychologist, De Bono (2003), frequently refers to humour as "…by far the most significant activity of the human brain" (p. 12). Humour for de Bono is very much related to creativity as it involves the disruption of the brain's natural tendency to self-organize based on already existing neural pathways, which will typically restricts creative thinking or *Lateral Thinking* in his terminology. As he points out:

> Humour not only indicates the nature of the system but also shows how perceptions set up in one way can suddenly be reconfigured in another way. This is the essence of creativity. (p. 12)

Humour makes us feel better, and this has a positive effect on our psychological state. Of course, humour must be used thoughtfully and in context. However, far from limiting the learning experience, humour is now seen to have a wide range of positive impacts on aspects of the learning process, such as:

- Refreshing the brain
- Creating mental images that retain learning
- Reinforcing the desired behaviour and making classroom management easier
- Developing positive attitudes
- Promoting creativity
- Contributing to the enjoyment of teaching.

Furthermore, humour seems to have a role in learning more generally. Earleywine (2011) summarized:

> Funny instructors get higher ratings perhaps because humour affects immediacy – the sense that an instructor is present and attentive with students…
>
> …a full semester of instruction that includes relevant jokes that illustrate key concepts lead to better scores in final exams. (p. 138)

The use of humour in terms of creatively enhancing the learning experience, student attainment and well-being is explored in detail in Chap. 5.

This core principle of learning is both fundamental to and generic across all aspects of teaching, and in all interpersonal communication contexts. It is truly a generic existential competence that should be cultivated as it not only enhances learning, in all the ways outlined here, but also well-being. It is fundamental to building good relationships and positive feelings. Gregory and Kaufeldt (2015) referring to O'Doherty's work (2004) points out that the brain naturally seeks dopamine releasing activities that we feel good in, and positive relationships, humour and feelings of mastery certainly do this. The pedagogic implications are that:

> If students receive a dopamine release, accompanied by positive enthusiastic feeling, they are more likely to persevere through the next challenge and continue to be engaged. (p. 40)

When I first arrived in Singapore some 24 years prior, there was a Smile Campaign, and since then there have been campaigns for Graciousness and, more recently, Kindness. Cynics may say this is 'social engineering'. Well, any act of socialization is social engineering, and my response to such folk is that it's much nicer to have smiling, gracious and kind people, than the opposite. In terms of moral values, there is much that can be learned from neuroscience as Harris (2010) Suggests. People generally feel better in a psychological climate that encourages kindness and graciousness, and smiling is both uplifting and builds rapport. In the context of education, the teacher's interactions with students will largely shape the psychological climate of the classroom, and as Rogers (1998) described:

> …the facilitation of significant learning rests upon certain attitudinal qualities that exist in the personal relationship between the facilitator and the learner. (p. 121)

Many of the important components that underpin the shaping of this relationship have been outlined and illustrated in this chapter and some key areas will be developed further in subsequent chapters. Most significantly as a teacher, shaping the psychological climate is in large part your responsibility, though it can be challenging in many situations. However, as Hattie and Yates (2014) argued:

> As their teacher, you are an inevitable coach in interpersonal mannerisms. Hence a deep understanding of how these social processes operate will prove of inherent value in your professional work. (p. 269)

2.11 Using the Core Principles Thoughtfully: The Fly-Fishing Analogy

For the uninitiated, fly fishing involves a sophisticated fishing technique in which an artificial fly is cast to catch trout. However, whether the fisherperson catches trout, involves much more than this. Choosing the strategy, type of fly, identifying the species of trout in the location, interpreting the impact of weather conditions are some of the critical considerations in catching trout. The expert fisherperson negotiates these almost intuitively and catches fish regularly. Suffice to say, as a novice fly-fisherman, I did not catch many trout and never reached any great heights of expertise.

Fly fishing is a useful analogy when applying the core principles of learning to the context of teaching in that both involve a solid knowledge base relating to the design and conduct of the respective activities. Similarly, they are also mediated by the situated context in which they are enacted in that both the fly fisherperson and the teacher must deal with the here and now environmental situation. For the fly-fisherperson, there is a need to carefully consider such factors as the nature of the water locality (e.g., river, lake or sea), type of trout inhabitants in the locality, the season of the year and prevailing weather conditions. For the teacher, key considerations include the nature and composition of the student group (e.g., prior knowledge and competence levels, motivational status), classroom resources and time of the day. Based on their knowledge and framing of the situated context, both fly fisher-persons and teachers select methods and resources, and create strategies to try to produce good results—whether defined in terms of 'trout caught' or 'students taught'.

In teaching, while the core principles of learning are enduring heuristics in the design of the learning experience and the conduct of teaching, their relative importance as focal points in the design and teaching process is typically mediated by such situated factors. For example, if I am aware that a learning group has many students who have a generally low intrinsic motivational level for the subject, I will give more thought concerning how best to incorporate appropriate motivational strategies and work on creating a positive psychological climate as the central consideration. In this situation, I may 'sacrifice' cognitive considerations for better motivational or affective outcomes, at least in the short term. However, I would maintain a strong focus on avoiding cognitive overload and developing some mastery of key skills as priority pedagogic features. In contrast, when teaching fee-paying students on higher degree programmes motivating them may not be such a central concern, though they typically appreciate it anyway. This thoughtful and situated application of the core principles of learning has been well captured by Darling-Hammond and Bransford (2005):

> ...teachers not only need to understand the basic principles of learning but must also know how to use them judiciously to meet diverse learning goals in contexts where students differ in their needs. (p. 78)

2.12 Instructional Design from an Evidence-Based Approach

The cognitive scientific principles (Core Principles of Learning) presented and illustrated in this chapter are not meant to be exhaustive or summative—as noted in the introduction. However, they do constitute powerful universal heuristics in the design and facilitation of learning in all contexts (e.g., face-to-face teaching, blended learning, fully online). What this means is that teaching can be designed, conducted, and evaluated from a sound pedagogic base. In other words, teaching can be subjected to a systematic evidence-based pedagogic analysis which will increase both diagnoses of learning events in terms of their effectiveness and efficiency, and well as designing learning events with a high probability of successful outcomes. This will significantly enhance all teaching and training professional's ability to conduct a more rigorous and useful process of reflective practice; whereby they can not only identify what has worked well, or not so well, but also the underlying psychological principles that have led to such outcomes. Evidence-Based Reflective Practice is fully explained and illustrated in Chap. 9.

Whatever frameworks or models of teaching are used, teachers must inevitably design learning events, and it makes better sense to be as evidence-based as possible in relation to the situated context as documented above. Kilbane and Milman (2014) argued that teachers should be educational designers, and define this in these terms:

> An educational designer is a teacher who approaches instructional planning with purpose, uses knowledge of specialized systematic processes to identify and frame instructional challenges related to learners and content, and competently addresses these challenges through the skillful application of a broad repertoire of instructional models, strategies, and technologies. Educational designers approach the work of teaching with a new mindset, a broadened skill set, and a high-quality tool set – all of which assist them in developing instruction that

responds to their learner's needs. The new mindset enables teachers to approach their work as empowered problem solvers who are aware of their ability to direct important dimensions of practice. The expanded skill set allows them mastery over systematic approaches to instructional planning and assessment processes. The high-quality toolset consists of a collection of models, strategies, and technologies for teaching that can make learning more efficient, effective, and engaging. (xxi)

Evidence-Based Teaching provides the framework for teachers to become experts in instructional design, as well as how best to facilitate the learning process in the situated contexts of practice. Different teaching models, strategies and tools (e.g., signature pedagogies, high effect methods, e-tools) can be creatively blended to produce effective, efficient and engaging instructional approaches.

The educational literature is awash with books, journals, internet resources on teaching and learning, just as there are on childrearing and other significant life-related activities. While there is much of merit in many sources, there is probably as much that abets rather than informs understanding and practice. Also, how does one find the time to extract what is most useful as time and cognition are both limited? For example, while I follow new research from neuroscience, and have an interest in how the brain behaves in response to various environmental stimuli (especially in the context of teaching) there is, for practising teachers anyway, little need to plough through the complex literature on neuroscience for validation of good instructional design. As Dougherty and Robey (2018), from an extensive review of the literature, concluded:

> …the gap between neuroscience and education cannot be bridged without the intermediary stepping-stone of cognitive theory. But more to the point, neuroscience is not even needed… (p. 403)

> What is needed is whether training leads to changes in core cognitive processes (e.g., WM or IQ) necessary for education success and whether improvements in these abilities generalize to authentic education settings. The theory of change requires specification at the cognitive level but not at the neurological level (p. 3).

> Similarly, Brown et al. (2014) pointed out that while neuroscience is enhancing our understanding of the brain mechanisms that underlie learning it is still a long way from being able to tell us, in specific empirical terms "about how to improve education". (p. 8)

2.13   Summary

This chapter has outlined and illustrated key heuristics—Core Principles of Learning—for planning and conducting the practices of teaching. They are underpinned by current and established knowledge relating to human learning and research on what methods are most effective. The extent to which cognitive scientific principles (e.g., Core Principles of Learning) can be said to constitute an essential *Pedagogy Literacy* for the planning and facilitation of learning may rest on how other literacies are framed and on what basis. The term literacy has been typically used in the context of language acquisition and use. For example, persons who cannot read, speak or write effectively are sometimes referred to as 'lacking in literacy'. When such competencies are severely lacking, the term illiteracy is often used. How lacking one must be in these areas to meet the criteria of illiterate is a value judgement to some extent and reflects the proficiency standards used. Whatever the standard, I certainly meet such labelling in terms of my fluency in foreign languages. As a Brit I am somewhat ashamed, in my travels, to have to explain that the only language I have any acceptable literacy in is English. More recently, the term literacy has been applied to a wide range of domain areas (e.g. computer literacy, media literacy, and political literacy). This is similar in many ways to the proliferation of different bits of intelligence (e.g., emotional intelligence, social intelligence, and cultural intelligence). Whether different literacies or intelligences merit such grand description is open to debate, but there are clearly valued areas of human capability implicit in these designations. In the present context, Pedagogic Literacy would meet such criteria.

Indeed, once teachers have strong pedagogic literacy as well as the technical knowledge and skills to use a range of instructional methods thoughtfully and skilfully, they can evaluate the impact of their teaching on student learning and attainment from an evidence-based approach. It is then possible to achieve what Hattie (2009) emphasized as fundamental to improvement:

> The ultimate requirement is for teachers to develop the skill of evaluating the effect that they have on their students. (p. 36)

References

Apter MJ (ed) (2001) Motivation styles in everyday life: a guide to reversal theory. American Psychological Association. ISBN: 1-55793-739-4, Washington, D.C.

Ausubel DP et al (1978) Educational psychology: a cognitive view. Holt, Rhinehart and Winston, New York

Bandler R, Grinder J (1990) Frogs into princes: the introduction to neuro-linguistic programming. Eden Grove Editions, Middlesex

Berliner DC (1987) Ways of thinking about students and classrooms by more and less experienced teachers. In: Calderhead J (eds) Exploring teachers' thinking. Cassell, London

Bloom BS (ed) (1985) Developing talent in young people. Ballantine Books, New York

Boud D (ed) (1988) Developing student autonomy in learning. Kogan Page, London

Brown PC, Roediger III HL, McDaniel MA (2014) Make it stick: the science of successful learning. Harvard University Press, Massachusetts

Bruner JS (1966) The process of education. Harvard University Press, Cambridge

Clark RC, Lyons C (2005) Graphics for learning: proven guidelines for planning, designing, and evaluating visuals in training materials. Pfeiffer, San Francisco

Colvin G (2008) Talent is overrated: what really Separates world-class performers from everybody else. Penguin, London

Csikszentmihalyi M (1990) Flow: the psychology of optimal experience. Harper Row, New York

Darling-Hammond L, Bransford J (2005) Preparing teachers for a changing world: what teachers should learn and be able to do. Jossey-Bass, San-Francisco

De Bono E (2003) Serious creativity 2. Allscript Establishment, Singapore

Dilts R et al (1980) Neuro linguistic programming Vol. 1: the study of the structure of subjective experience. Meta Publications, California

Dougherty MR, Robey A (2018) Neuroscience & education: a bridge astray. Association for Psychological Science. Curr Dir Psychol Sci, 1–6. https://www.researchgate.net/publication/326143095_Neuroscience_Education_A_bridge_astray

Dweck CS (2006) Mindset: the new psychology of success. Ballantine, New York

Earleywine M (2011) Humour 101. Springer, New York

Ekmann P, Friesen WV (1971) Constants across cultures in the face and emotion. J Pers Soc Psychol 17(2):124–129

Flavell JH (1979) Metacognition and cognitive monitoring: a new area of cognitive-developmental inquiry. Am Psychol 34(10)

Garner LA (2006) Humor in pedagogy: how haha can lead to aha! Coll Teach. Last accessed 7 Nov 2019

Gregory G, Kaufeldt M (2015) The motivated brain: improving student attention, engagement, and perseverance. ASCD, Alexandria, USA

Goulston M (2009) Just listen: discover the secret to getting through to absolutely anyone. AMACON, New York

Halpern DF (1998) Teaching critical thinking for transfer across domains: dispositions, skills, structure training, and metacognitive monitoring. Am Psychol 53(4):449–455

Hart LA (1983) Human brain and human learning. Longman, New York

Harris S (2010) The moral landscape: how science can determine human values. Free Press, New York

Hattie J (2009) Visible learning. Routledge, New York

Hattie J (2012) Visible learning for teachers: maximizing impact on learning. Routledge, London

Hattie J, Yates GCR (2014) Visible learning and the science of how we learn. Routledge, New York

Hounsell D (1997) Understanding teaching and teaching for understanding. In: Marton F, Hounsell D, Entwistle N (eds) The experience of learning: implications for teaching and studying in higher education. Scottish Academic Press, Edinburgh

Jensen E (1996) Brain based learning. Turning Point Publishing, Del Mar, CA

Kahneman D (2012) Thinking fast and slow. Penguin Books, London

Keen A (2007) The cult of the amateur: how today's internet is killing our culture and assaulting our economy. Nicholas Brealey Publishing, London

Kilbane CR, Milman NB (2014) Teaching models: instruction for 21st century learners. Pearson, London, England

Kircher PA et al (2006) Educ Psychol 41(2):75–86

Knowles MS (1973) The adult learner: a neglected species. Gulf Publishing Company, Houston

Koffka K (1915) Zur Grundlegung der Wahrnehmungspsychologie: Eine Auseinandersetzung mit V. Benussi. Beiträge zur Psychologic der Gestalt, No. 3. Zeitschrift fur Psychologie 73:11–90

Kohler W (1929) Gestalt psychology. Liveright, London

Lang JM (2016) Everyday lessons from the science of learning. Jossey-Bass, San Francisco, CA

Lei SA, Cohen JL, Russler KM (2010) Humor on learning in the college classroom: evaluating benefits and drawbacks from instructor's perspectives. J Instr Psychol

Marcus G (2009) Kluge: the haphazard evolution of the human mind. Mariner Books, New York

Martinez ME (2006) What is metacognition? Phi Delta Kappan, 696–699

Marzano RJ et al (1988) Dimensions of thinking: a framework for curriculum and instruction. ASCD, Alexandria, VA

Marzano RJ et al (2007) Designing and teaching learning goals and objectives: classroom strategies that work. Marzano Research Laboratory, Colorado

Mayer RE, Alexander PA (2011) Handbook of research on learning and instruction. Routledge, London

McTighe J, Wiggins G (1998) Understanding by design. ASCD, Alexandria, VA

Miller GA (1956) The magical number seven, plus or minus two: some limits on our capacity for processing information. Psychol Rev 63(2):81–97

Miller MD (2016) Minds online: teaching effectively with technology. Harvard University Press, Massachusetts

Mlodinow L (2012) Subliminal: how your unconscious mind rules your behaviour. Vintage Books, New York

Molden D (2001) NLP business masterclass. FT-Press, New Jersey

Mortimore P (ed) (1999) Understanding pedagogy & its impact on learning. Paul Chapman, London

Murdock BB (1962) The serial position effect of free recall. J Exp Psychol 64(5):482–488

Nicol DJ, MacFarlane-Dick D (2006) Formative assessment and self-regulated learning: a model and seven principles of good feedback practice. Stud High Educ 31(2):199–218

Nuthall GA (2005) The cultural myths and realities of classroom teaching and learning: a personal journey. Teachers Coll Rec 107(5):895–934

O'Doherty J et al (2004) Dissociable roles of ventral and dorsal striatum in instrumental conditioning. Science 304:452–454

Ormrod JE (2011) Our minds, our memories: enhancing thinking and learning at all ages. Pearson, London

Ornstein AC, Behar LS (1995) Contemporary Issues in Curriculum. Allyn & Bacon, Massachusetts

Paul RW (1993) Critical thinking. Foundation for Critical Thinking, Santa Rosa, CA

Paul R, Niewoehner R, Elder L (2006) The thinker's guide to engineering reasoning. Foundation for Critical Thinking, Tomales, CA

Perkins DN (1985) 'What creative thinking is'. Excerpt from Perkins DN. Creativity by design. Educ Leadersh 42(1):18–24

Perkins DN (1992) Smart schools. The Free Press, London

Perkins DN et al (1995) Software goes to school: teaching for understanding with new technologies. Oxford University Press, Oxford

Petty G (2009) Evidence-based teaching: a practical approach. Nelson Thornes, Cheltenham

Piaget J (2001) The psychology of intelligence. Routledge, New York

Pinker S (2003) The blank slate: the modern denial of human nature. Penguin, London

Pugh KJ, Bergin DA (2006) Motivational influences on transfer. Educ Psychol 41(3):147–160

Ramsden P (1992) Learning to teach in higher education. Routledge, London

Resnick LB (1989) Assessing the thinking curriculum: new tools for educational reform. In: Gifford BR, O'Connor MC (eds) Future assessments: changing views of aptitude, achievement, and instruction. Kluwer Academic Publishers, Boston

Rogers A (1998) Teaching adults. Open University Press, Buckingham

Sale D (2014) The challenge of reframing engineering education. Springer, New York

Sale D (2015) Creative teaching: an evidence-based approach. Springer, New York

Sale D, Mukerji S (2006) Establishing rapport in facilitating workshops in different cultural and ethnic contexts: key principles of effective human conduct and engagement. In: Paper presented at ERAS conference, Singapore

Sadler DR (1989) Formative assessment and the design of instructional systems. Instr Sci 18:119–144

Schank R (1997) Virtual learning. McGraw-Hill, London

Schank R (2011) Teaching minds: how cognitive science can save our schools. Teacher College Press, New York

Shulman L (1991) Pedagogic ways of knowing. Institute of Education, London

Stigler JW, Hiebert J (1999) The teaching gap. Free Press.

Sullo B (2007) Activating the DESIRE to learn. ASCD, Alexandria, VA

Swartz RJ, Parks S (1994) Infusing critical and creative thinking into content instruction. Critical Thinking Press & Software, Pacific Grove, CA

Sylwester R (ed) (1998) Student brains, school issues. Skylight Training and Publishing Inc, Arlington Heights, IL

The Free Dictionary (2019). www.thefreedictionary.com. Last accessed on 8 Nov 2019

Turner-Bisset R (2001) Expert teaching: knowledge and pedagogy to head the profession. David Fulton, London

Van Merrienboer JJG, Sweller J (2005) Cognitive load theory and complex learning: recent development and future directions. Educ Psychol Rev 17(2)

Wagner T (2010) The global achievement gap. Basic Books, New York

Willingham DT (2009) Why don't students Like school: a cognitive scientist answers questions about How the Mind works and what it means for the classroom. Jossey-Bass, San Francisco

Wlodkowski RJ (2008) Enhancing adult motivation to learn: a comprehensive guide for teaching all adults. Jossey-Bass, San Francisco

Chapter 3
Metacognitive Capability: The Superordinate Competence for the Twenty-First Century

Abstract This chapter argues that *Metacognitive Capability* is the Superordinate twenty-first century competency. While there is much reference to the 4 C's of critical thinking, creativity/creative thinking, communication and collaboration, these are essentially no more than extended versions of first century skills. However, in a world of exponential knowledge, but with a brain perfectly equipped for the world of primitive life in centuries past, we are increasingly suffering from cognitive overload and strain. No matter what we do pedagogically, we cannot solve the knowledge explosion with a stone-age brain. We must develop the capability to think, learn and self-regulate better. *Metacognitive Capability*—as framed here—is the unique human quality that can best help us deal with the challenges of modern life, as well as maintain positive well-being. The chapter extensively unpacks the components of metacognition, how they work in terms of psychological functioning and, most importantly, how to use pedagogic strategies that can develop this capability to optimal levels.

3.1 Introduction

Akturk and Sahin (2011), in reviewing the literature on metacognition, argued that:

> Metacognition is a structure that is referred to as fuzzy by many scholars and has very diverse meanings. Much research has been conducted for more than 30 years to access the inner side of this structure, which is hard to grasp. (p. 3731)

Having conducted numerous workshops with adults, I am still surprised that many people have little or no knowledge relating to metacognition; some are even totally unfamiliar with the term. Just using the word metacognition often results in glazed eyes as the initial hearing of this unfamiliar term seems to evoke feelings akin to 'not more psychobabble'. Brown's (1987) framing of metacognition, while making sense to me having surveyed the literature extensively, may also do little to enthuse the lay reader to explore it further:

> Metacognition is not only a monster of obscure parentage but a many-headed monster at that. (p. 105)

© Springer Nature Singapore Pte Ltd. 2020
D. Sale, *Creative Teachers*, Cognitive Science and Technology,
https://doi.org/10.1007/978-981-15-3469-0_3

However, at the same time, metacognition is being heralded as one of the main twenty-first century competencies. For example, Lai (2012) noted that while many frameworks mapping such skills have been developed, underpinning these frameworks, and related to student attainment, positive learning and career development, 5 key research-based competencies have been identified: Critical Thinking; Creativity; Collaboration; Motivation; Metacognition.

Certainly, voices advocating the importance of metacognitive activity within educational contexts have resulted in placing metacognition high on educational research agendas. For example, Martinez (2006) argues:

> Metacognitive ability is central to conceptions of what it means to be educated. The world is becoming more complex, more information-rich, more-full of options, and more demanding of fresh thinking. With these changes, the importance of metacognitive ability as an educational outcome can only grow. (p. 699)

Similarly, Noushad (2008) concluded:

> ...metacognitive strategies are essential for the twenty-first century because they will enable students to successfully cope with new situations and the challenges of lifelong learning. (p. 16)

In this chapter, I will get to grips with this 'many-headed monster of obscure parentage', as it is an issue of pedagogic necessity. While accepting that there is much by way of conceptual confusion in the literature—yep, it's fuzzy—metacognition is not just a twenty-first century competence, it is *the* superordinate competence of the twenty-first century. The ability to be highly effective in this competence, I frame as *Metacognitive Capability* (MC). While this focus on MC is in the field of education and training, it is of relevance to all professional and human activity, as it is central to how humans can learn better, be more self-directed in their learning and careers as well as experience higher levels of personal well-being.

Learning to be competent in any area of human performance always requires a combination of knowledge, skills, and attitudes (however defined). Also, there are universal principles of learning that underpin such processes, and MC is no exception. To develop competence and expertise in MC, it is necessary to fully understand what it is, the benefits for learning, well-being and, over time, self-directed lifelong learning. Invariably, one must then invest the motivational and volitional strategies, as well as the spaced and deliberate practice, to achieve such competence.

3.2 Making Sense of Metacognition: *Unpacking the Metacognition Monster*

Let's start with direct experience as an anchor point. Whether we like it or not, we inevitably reflect on our experiences and actions in living in the world. We also typically do this in relation to the actions of others, making judgements of worth, often of right or wrong, based on our beliefs and perception. There is variation in terms of the extent, form and nature of how humans do this, and it is influenced both by hereditary and experiential factors. As you are reading this text you may choose or simply drift into reflecting on what you already know, or don't know about metacognition. Hence, you are thinking about it and you'll probably be doing some analysis, comparing and contrasting, making inferences and interpretations, even some evaluation on what you know and what you are reading. These are the key cognitive heuristics of critical thinking which help to build understanding—when done well with the right content knowledge, as we saw in Chap. 2.

This capability to be able to consciously think about our actions, be aware of the potential consequences of action, and the likelihood of experiencing a range of feelings from heightened pleasure to extreme dread, seems to define mankind as distinct from other animal species. Over the years, my family has acquired a few domestic cats as pets, which they see as lovely friendly creatures who are so affectionate—after all they are pussycats, right? For me, I am ok with cats, but the occasional gecko or cockroach, not to mention mouse, that happened to venture into our apartment would not share such positive experience of *the pussycats*. Of note, the cats never show any visible signs of remorse for torturing these small creatures. Fromm (1987) referred to man "as a freak of nature, being within nature and yet transcending it" which makes humans beset with existential paradoxes. Humans, in contrast to animals, must consciously live with the consequences of their actions, as well as deal with knowing that suffering and finality to life (in the existential sense) is probably inevitable. As Ursula, the sea witch, said to Ariel, the little mermaid, in the Disney film The Little Mermaid (1989), when she bargained with Ariel an exchange of her beautiful voice for a pair of legs to be fully human, "...life is full of tough choices innit?".

It is this distinctive human capability to be able to self-reflect, think, plan, monitor and evaluate our thinking and actions that underpin metacognition. Central to self-directed learning is the capability to use metacognition to self-regulate one's thinking, affect and behaviour in meeting learning goals. There is no significant evidence that other creatures possess such a capability, though of course they experience pleasure and pain and other emotions that seem to have similar correlates to those in humans.

Flavell originally coined the term metacognition in the late 1970s to mean "cognition in relationship to cognitive phenomena," or more simply "thinking about thinking" (Flavell 1979, p. 906). He went on to suggest that while metacognition mainly focuses on knowledge and cognition about cognition, the concept could be broadened to anything psychological, rather than just anything cognitive. It could be related to include executive processes, formal operations, consciousness, social cognition, self-efficacy, self-regulation, reflective self-awareness, and the concept of psychological self or psychological subject (Flavell 1987).

Flavell's suggestion that metacognition could be more than just thinking about thinking, but "broadened to anything psychological" opened-up the 'many headed monster' analogy. Recent research tends to confirm both Flavell and Brown's analysis. The most current framing of metacognition, especially from neuropsychology, goes beyond monitoring and evaluating cognition, but also regulating affective and motivational aspects of being—the whole person—such as beliefs and emotions (e.g.,

Dweck and Legget 1989). There is now an integration of metacognitive and motivational approaches to explaining the development of student's success (or otherwise) in school learning. Of note, Martinez (2006) argues that metacognition entails the management of affective states, and that metacognitive strategies can improve persistence in the face of challenging tasks. Paris and Winograd (1990) support this view as they see affect as an inevitable element of metacognition because as students monitor and appraise their cognition, they will become more aware of strengths and weaknesses and can take the necessary action to enhance learning capability and well-being.

This makes perfect sense as we know cognition (thinking) does not exist as an independent entity in human decision—making and behaviour. Apart from hereditary factors and personality configurations, beliefs, emotions, and motivation all play out as a dynamic system in determining perception and behaviour. Also, there is increasing evidence that much of this activity (which may give us the illusion of conscious self-control) is operating sub/unconsciously (e.g., Mlodinow 2012).

Education and many branches of psychology abound with perspectives, theories and models, and this poses problems for even the theoreticians and researchers, let alone practitioners in the field of teaching and training, as well as the wider lay public. It is, therefore, necessary to firstly clarify key conceptual issues in the field of research on metacognition. For example, the term metacognition is often used simultaneously or alternately with other frameworks relating to mental processing and self-regulation. The major frameworks include:

- Self-Directed Learning
- Self-Regulated Learning
- Meta-Learning.

Self-Directed Learning (SDL) is becoming somewhat of a buzzword in the educational landscape. Certainly, in the rapidly changing and volatile world of today, the aim of developing persons, in-school environments or elsewhere, who are self-directed lifelong learners would seem a pertinent educational aim. There is nothing new about the need for self-directed and lifelong learning. It's just that it is more essential in the modern context than yesteryear. A useful reference point in framing SDL is the enduring definition of Knowles (1975):

> Self-Directed Learning describes a process in which individuals take the initiative, with or without the help of others, in diagnosing their learning needs, formulating learning goals, identifying human and material resources for learning, choosing and implementing appropriate learning strategies, and evaluating learning outcomes. (p. 18)

More recently, Ambrose et al. (2010) referred to SDL as a key research-based principle of learning:

> To become self-directed learners, students must learn to assess the demands of the task, evaluate their knowledge and skills, plan their approach, monitor their progress, and adjust their strategies as needed. (p. 191)

Based on a Delphi study of experts' framing of SDL; it involves many interrelated skills, dispositions/habits of mind and beliefs about self and learning:

> A highly self-directed learner...is one who exhibits initiative, independence, and persistence in learning; one who accepts responsibility for his or her own learning and views problems as challenges, not obstacles; one who is capable of self-discipline and has a high degree of curiosity; one who has a strong desire to learn or change and is self-confident; one who is able to use basic study skills, organize his or her time, set an appropriate pace for learning, and develop a plan for completing work; one who enjoys learning and has a tendency to be goal-oriented. (Guglielmino 1978, p. 73)

Certainly, to be a self-directed learner, a key underpinning competence is the ability to think well—*Good Thinking*—as outlined in Chap. 2.

Self-Regulated Learning (SRL), which also figures extensively in the literature has been referred to by Schunk and Zimmerman (2012) as:

> ...the process by which learners personally activate and sustain cognition, affects and behaviours that are systematically oriented toward the attainment of learning goals. (vii)

In practice, SDL and SRL have often been used interchangeably in the literature. However, Saks and Leijen (2013), in comparing and contrasting SDL and SRL, suggest that while both incorporate task definition, goal setting, planning and enacting strategies, monitoring and evaluation—as well as metacognition and Intrinsic motivation, they differ in certain important ways:

1. SDL originates from Adult Education; whereas SRL originates from Cognitive Psychology
2. SDL is practiced mainly outside school environments; whereas SRL is practiced more in school environments
3. SDL involves the learner more in the design of the learning environment and its trajectory; whereas in SRL, outcomes and tasks are usually set by the teacher
4. SDL is a broader macro-level construct than SRL.

Another term, that of *Meta-Learning*, originally framed by Maudsley (1979), has now been re-surfaced by the Center for Curriculum Redesign, (e.g., Fadel and Trilling 2015) who define it in terms of two main components:

- Metacognition—the process of thinking about thinking.
- Growth Mindset—the inner belief that abilities can be developed through hard work.

Meta-Learning has a 'catchy' tone to it, but these components—metacognition and growth mindset—are already embedded in more recent framings of metacognition.

From the above comparisons, it is not surprising that there has been confusion relating to what is metacognition and how it relates to other frameworks such as SDL and SRL. Indeed, the confusion is enhanced further as a result of some researchers considering self-regulation to be a subordinate component of metacognition (e.g., Kluwe 1987), whereas others regard self-regulation as a concept superordinate to metacognition (e.g., Winne 1995).

Metacognition may indeed be that many-headed-monster of obscure parentage. The latter aspect is more the domain of evolutionary theory and may be of academic interest only; the many heads are of major pedagogical, social and global interest.

What Makes Metacognitive Capability (MC) *the* Superordinate 21st Competency?

The nature and focus of learning, concerns of human conduct, improving intelligent behaviour (at individual, organizational, societal, and global levels) may reside in better MC, which I see as capturing the range of self-regulatory activities and process that are now subsumed under more recent framings of metacognition. In most basic terms, MC represents our best existential opportunity to become better learners, experience better well-being and become more self-directed in our personal and professional life directions. Such framing also encompasses various notions of lifelong learning.

Invariably, not everybody will develop such capabilities, as there are many foibles and paradoxes to the human condition, and our present stage of evolution may well just be a Kluge (an ill-assorted collection of parts assembled to fulfil a particular purpose) as Marcus (2008) suggests. Even a few nights review of world news depicting various genres of human tragedies occurring worldwide, more than suggests a significant absence of good thinking and sensibility occurring pretty much everywhere. We are living in a VUCA world (*volatile, uncertain,complex,and ambiguous*); the term seems to have migrated from U.S. military to global speak about modern society and life in general and is especially trendy in managerial dialogues when dealing with difficult present realities. However, we are having to deal with this scenario with a brain that is still largely that of our stone-age ancestors, which appears not to be naturally intelligently designed for notions of global wisdom and betterment in this VUCA world.

Much is written about the so-called twenty-first century competencies. Skills such as teamwork and communication, problem solving and creativity, are typically mentioned in various (often conflating) terminologies across the literature, and of course, these are essential. However, didn't our cave-dwelling ancestors also need such skills to stop hostile intruders, tigers, or whatever other 'nasties' were around to kill, maim or rob them? I tend to agree with Schank (2011) who argued:

> Twenty-first-century skills are no different from 1st-century skills. (p. 207)

Research suggests that human brains were pretty much the same (morphologically) some 50,000 years ago—indeed recent data shows that even 300,000 years ago, the brain size in H. sapiens already fell within the range of present-day humans (Neubauer et al. 2018). It is arguable, therefore, that such folk (let's be gentle on the language occasionally) had similar cognitive abilities and motivational dispositions—just played out in different contexts, with different resources, and different contingencies. I note that the Paleo diet is now becoming highly popular with certain sectors of the healthy-eating community, so there is probably still much in common between Homo sapiens today and those from the aeons of yesteryear.

One thing that certainly differentiates us from our ancestors is the rate of knowledge production. It's a cliché to refer to the exponential knowledge explosion of modern times, but the problem of constant cognitive overload is a problem for many of us. Before information can be stored in our long-term memory, it must first pass through our working memory, which has very limited immediate capability when processing new information, as we explored in some detail in Chap. 2, *Core Principle 6: Learning design takes into account the working of memory systems.* This limitation is further hampered in terms of learning effectiveness and efficiency by our brain's inherent systemic biases and tendency to be lazy (e.g., Kahneman 2012), and it not being particularly well designed for thinking (e.g., Willingham 2009), and our limited amounts of willpower (e.g., Baumeister and Tierney 2012). Dealing with them, at least in terms of mitigating their negative impacts on learning and well-being, will be imperative in future curriculum planning and instruction.

In terms of the twenty-first century underpinning research-based competencies identified by Lai (2012), I suggest that they are interrelated in complex ways. For example, metacognition is inevitably related to both critical and creative thinking and is the executive function monitoring and evaluating them—all being the main cognitive components of 'Good Thinking' as framed in Chap. 2. Also, motivation and metacognition are reciprocally related and synergistic and, finally, providing students with opportunities to work together may stimulate students' motivation and thinking.

MC in this context can be validly and usefully framed as the superordinate competence that encompasses a range of sub-competencies and skill sets (including those of self-regulation) that facilitates the development of SDL. MC *fuels* the process of becoming a self-directed learner in that it can develop and facilitate many functions that are highly beneficial for effective learning and well-being—helping learners to:

- Set key goals for learning (e.g., short term, long-term, and appropriate challenge) and deciding what needs to be learned for what purpose
- Know how to learn and plan a successful learning strategy (e.g., what, how, when and where)
- Use specific metacognitive, cognitive, and motivational strategies to achieve the learning goals
- Maintain positive beliefs and managing emotions to remain calm under pressure
- Persist, exercise volition to stay on track in the face of challenges and/or setbacks
- Monitor and review one's progress and modify/change aspects of strategy based on feedback (if necessary).

Research has shown that the effective use of metacognitive strategies has a high impact on student attainment (e.g., Effect Size of 0.69, Hattie 2009). Similarly, Dignath et al. (2008), quoted in Lai (2012), meta-analysed 48 studies investigating the effect of training in self-regulation on learning and the use of strategies among students from grades 1–6, as shown in Table 3.1.

The most effective metacognitive strategies included the combination of planning and monitoring (mean effect size = 1.50) and the combination of planning and evaluation (mean effect-size = 1.46), both of which were more successful than teaching any of the skills in isolation or teaching a combination of all three metacognitive skills (planning, monitoring, and evaluation). Training approaches that combined metacognitive components with other aspects of self-regulation, such as cognitive

Table 3.1 Summary of Dignath et al. (2008), meta-analysis

Type of treatment	Mean effect size
Any self-regulation training (metacognitive, cognitive, and motivational)	0.73
Metacognitive and motivational strategies training (all strategies)	0.97
Metacognitive and cognitive strategies training (all strategies)	0.81
Metacognitive strategies training (all strategies)	0.54
Metacognitive strategy training in planning and monitoring	1.50
Metacognitive strategy training in planning and evaluation	1.46
Training on metacognitive reflection—knowledge about and value of strategies	0.95
Cognitive strategies training (all strategies)	0.58
Cognitive strategy training in elaboration	1.19
Cognitive strategy training in elaboration, organization, problem solving	0.94

or motivational strategies were also successful, with effect sizes of 0.81 and 0.97, respectively (pp. 22–23).

Also, metacognition impacts other aspects of human psychological functioning, and their behavioural consequences, simply as a result of its executive function. For example, Critical Thinking—also one of the so-called twenty-first century competencies, is very much under the 'supervision' (for better or for worse) of metacognition. As Mango (2010) summarized:

> Higher-order thinking (like critical thinking) requires executive control and executive processes (that comes in the form of metacognition). Note, specifically, metacognition helps in developing critical thinking, because it is likely that critical thinking requires a form of meta-level operation. (p. 149)

Akturk and Sahin (2011) conclude:

> …teaching students how to use metacognitive strategies increases academic achievement (Biggs 1988). Students with advanced metacognitive skills are those who are aware of what they have learned and what they do not know. Generally, students with advanced metacognitive skills monitor their own learning, express their opinions about the information, update their knowledge and develop and implement new learning strategies to learn more. In comparison to other students, students using their metacognitive skills effectively are those who are more aware of their strengths and weaknesses and strive to improve their learning skills further (Bransford et al. 1999). According to Jones et al. (1995), the further students' awareness of metacognition is improved, the more students' effectiveness is increased. (p. 3735)

Similarly, Pintrich and De Groot (1990) have shown that the use of cognitive and self-regulatory strategies is an important component of student learning. For example, students who use cognitive strategies like elaboration (e.g., summarizing, paraphrasing) and organization (e.g., mind maps, concept maps) engage the content at a deeper level of processing and are more likely to be able to recall the information, understand it, and use it purposefully at a later time.

Fig. 3.1 Metacognition as
the superordinate 21st
competence

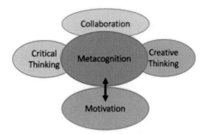

The view that metacognition can be developed through carefully planned instruction is supported in the literature. For example, drawing from the research of Ericsson et al. (1993), Schraw (1998) argues that:

> Well organized instruction or the use of effective learning strategies may in large part compensate for differences in IQ. In many cases, sustained practice and teacher modelling lead to the acquisition of relevant task-specific knowledge as well as general metacognitive knowledge that is either independent or moderately correlated with traditional IQ scores. (p. 117)

In summary, developing MC is a high leverage evidence-based pedagogic innovation for enhancing self-directed learning. Such competence encompasses both knowledge on how to manage one's thinking (e.g., critical and creative thinking), motivation (e.g., beliefs, emotions) and behaviour (e.g., collaboration with others), and the skills necessary to implement, monitor and evaluate strategy effectiveness (and efficiency) in different learning situations. It offers our best route to helping students to become self-directed learners, able to navigate the increasing complexities and existential challenges of the modern world, as well as find purpose and meaning in their lives. Figure 3.1 depicts the framing of Metacognition as the Superordinate 21st Competence.

3.3 Metacognition and Motivation: Two Bedfellows for Effective Learning

While the case is made for MC being the superordinate twenty-first century competence, based on its significance for effective learning and well-being, it is also widely accepted that metacognitive knowledge is insufficient to promote student achievement. Students must also be motivated to use their metacognitive knowledge and skills. In the context of metacognition, motivation is typically framed in terms of the "beliefs and attitudes that affect the use and development of cognitive and metacognitive skills" (Schraw et al. 2006, p. 112). According to them, motivation has two primary components: (1) self-efficacy, which is confidence in one's own ability to perform a specific task and, (2) epistemological beliefs, which are beliefs about the origin and nature of knowledge.

These two components of motivation are essentially concerned with beliefs, which affect perception, feelings and behaviour, though have their origins in prior perceptions and experiences. They work as an internal processing system for decision-making when people are confronted with new tasks. For example, Bandura (1997) referred to self-efficacy as:

…beliefs in one's capabilities to organize and execute the courses of action required to produce given attainments. (p. 3)

Without a belief in one's capability to execute a task successfully there may be a reluctance to embark on any activity related to tackling it in the first place, and as Schunk and Zimmerman (2012) points out:

the self-efficacy beliefs that students hold when they approach new tasks and activities serve as filters through which new information is processed (p. 18)

…unless people believe that their actions can produce the outcomes they desire, they have little incentive to act or to persevere in the face of difficulties. (p. 113)

Research shows that self-perceived competence is a key motivator for engagement. People with a strong sense of self-efficacy approach difficult tasks as challenges to be mastered rather than as threats to be avoided. They have a greater intrinsic interest and deep engrossment in activities, set themselves challenging goals as well as maintaining a strong commitment to them. High self-efficacy helps create feelings of serenity in approaching difficult tasks and activities. Consequently, self-efficacy beliefs powerfully influence the level of accomplishment one ultimately achieves (e.g., Fazey and Fazey 2001, p. 113).

Similarly, Weiner (1992), summarized how students attribute their outcomes (e.g., successes, failures) to such factors as ability, effort, task difficulty, and luck, have a significant impact on how they orientate themselves to academic learning. Most significant, in the present literature in recent years, has been the research by Dweck (2006) on the impact of a Growth Mind-set (as compared to a fixed Mind-set) on student learning and attainment:

…growth mindset is based on the belief that your basic qualities are things you can cultivate through your efforts. Although people may differ in every which way – in their initial talents and aptitudes, or temperaments – everyone can change and grow through application and experience. (p. 7)

Dweck's extensive research on students' beliefs (mind-sets) relating to intelligence has profound implications in terms of motivation, how students subsequently approach their learning and for how teachers teach (Table 3.2). In summary, she contrasted two fundamentally different mind-sets, relating to how students approach learning, a *Fixed Mindset* and a *Growth Mindset*. Students who possessed a fixed mindset tended to see intelligence as a stable genetic quotient and, consequently, you are either smart or you are not. In contrast, students who possessed a growth mind-set saw intelligence as a more fluid entity, reflecting effort and hard work, and a capability that can be developed and enhanced.

In most basic terms, our thinking and consequent behaviour are largely based on the 'pictures in our heads', which have their roots in our deep-seated belief systems,

Table 3.2   Comparison of fixed and growth mind-sets

Fixed mindset (intelligence is static)	Growth mindset (intelligence can be developed)
Leads to a desire to look smart and therefore a tendency to:	**Leads to a desire to learn and therefore a tendency to:**
• Avoid challenges	• Embrace challenges
• Get defensive and give up when faced with obstacles	• Persist in the face of setbacks
• See effort as something less able people need, and not for the smart	• See effort as the path to mastery
• Ignore useful negative feedback	• Learn from criticism
• Feel threatened by the success of others	• Find lessons and inspiration in the success of others
As a result, they may plateau earlier and achieve less than their full potential	**As a result, they reach ever-higher levels of achievement**

and if they are poor pictures, the consequences may turn out just that way also. Fortunately, they are changeable based on new experiences which is hardly surprising, if we think back to what we believed to be true as children. Do you still believe in the 'tooth fairy', Santa Claus, or the bogeyman under the bed? As Adler (1996) cleverly noted:

> We forget that beliefs are no more than perceptions, usually with a limited sell-by date, yet we act as though they were concrete realities. (p. 145)

It is not difficult to understand how beliefs profoundly affect the way people approach their learning and the subsequent impact on attainment and engagement levels. Beliefs act as major neurological filters that determine how we perceive external reality (Fig. 3.2). In this way, they provide the inner maps we use to make sense of the world around us. When we have a belief about something in our world, we act as though it is true. It is what is in our Inner Personal Map of Reality that determines our perception, emotional responses and orientation to people and things in the External World. While the External World is only knowable through our senses and therefore can never be fully ascertained in purely objective terms (whatever this is), our challenge as evidence-based teaching practitioners is to build increasingly more useful Internal Maps of how best to facilitate learning and attainment for our

Fig. 3.2   Beliefs as a filter on reality

students (part of our External World) and improve the quality of their Inner Personal Maps of Reality through the ways we teach and interact with them.

Motivation is fundamental in activating metacognition and both are essential for learning. Most importantly, their impact is likely to be synergistic. For example, once motivation is enacted, learners are more likely to want to be successful in their learning tasks, and this is where metacognitive strategies are particularly useful. As Martinez (2006) points out, metacognitive strategies can improve persistence and motivation in the face of challenging tasks, which means the learner can maintain a strong growth mindset, achieve successful task completion that, in turn, reinforces self-efficacy. As an old saying goes, "nothing breeds success like success".

Marzano's (2007) research is of real pedagogic interest in terms of explaining how different aspects of human psychological functioning interact in terms of influencing an individual's motivation to learn. His new taxonomy focuses on three internal systems, all of which are important for learning. These are summarized below:

- **The Self-system**—This relates to the set of beliefs (and related feelings) the student holds about his or her capabilities, the meaning attributed to the task in hand, along with the perceived likelihood of success
- **The Meta-cognitive system**—This relates to the higher-level self-regulation of the student in terms being able to monitor and evaluate his or her own thinking process (e.g., setting goals, monitoring progress towards these goals and adapting to difficulties)
- **The Cognitive system**—This is the system that reasons, and thinks in specific ways (e.g., analyses, compares and contrasts, makes inferences and interpretations, and evaluates) with the information at its disposal, to achieve the desired goals.

When faced with the option of participating in a new learning project or activity, it is the Self-system which initially decides (whether consciously or subconsciously) to give attention and then activates the Meta-cognitive and Cognitive systems to provide structure and direction for the appropriate learning strategies and skills to acquire the necessary knowledge, build understanding and skills to move progressively to goal attainment. He found that teaching strategies that activated the Self-system had the greatest effect on student learning, the Metacognitive system the next most effect, and the Cognitive system least, though it is still substantial. What this means is that it is the Self-system that activates the Meta-cognitive system, which actives the Cognitive system, which creates learning. In the ideal situation for effective learning, we would like to get all systems fully 'up and running' towards meeting the demands of the desired learning goal. What we now can be reasonably sure of is that without a desire to meet a task's outcomes, belief in one's capabilities to attain the necessary knowledge and skill components and perception of likely success, there is probably little effort invested to commit to task requirements. Quite simply, unless the Self-system is firmly activated, the other important systems are not likely to be working at anywhere near optimal levels.

Schunk and Zimmerman (2012), in summarizing the research findings conclude that:

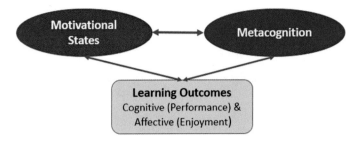

Fig. 3.3 Relationship of motivation, metacognition and learning outcomes. Adapted from the work Lens and Vansteekiste (2012)

> Clearly, motivational processes play a vital role in initiating, guiding and sustaining student efforts to self - regulate their learning. (p. 3)

The relationship between motivation and metacognition and its impact on learning outcomes is summarized in Fig. 3.3.

3.4   What Is the Difference Between Cognition and Metacognition?

Flavell's (1976) initial framing of metacognition as 'thinking about thinking' aptly captures the relationship of metacognitions to cognition. Invariably, 'thinking' and 'thinking about thinking' are essentially intertwined. However, in terms of mental activity, the main difference relates to the goal or purpose of the task in hand. Cognitive strategies 'facilitate' learning and task completion, whereas metacognitive strategies monitor the process. In most basic terms, cognitive skills are necessary to perform a task, while metacognition is necessary to understand how the task was performed.
Hence:

- *Cognitive Strategies* typically focus on acquiring, retaining and transferring knowledge for meeting specific task goals (e.g., organizing/classifying/summarizing information; transferring information from Working Memory to Long-Term Memory).
- *Metacognitive Strategies* focus on monitoring and evaluating the quality of the overall (and specific aspects) of the self-regulatory process, including the choice and application of the cognitive learning strategies.

Metacognition is necessary to understand how a task is to be performed, whereas cognition is required to fulfil the task, but both must work in unison for this to happen effectively and efficiently. For operational purposes, we can view cognition as the thinking (and strategies) employed as we mentally deal with real-world tasks (e.g., acquiring, retaining and transferring knowledge for meeting specific goals).

These cognitive strategies are conscious mental activities/operations we do to solve a problem. In the context of learning, we often refer to them as thinking or learning tools, which often involve physical tools such as Mind-mapping or concept mapping, but, as Ormrod (2011) points out, they are "…ultimately the things we do inside our heads – *thinking*" (p. 135). Hence, as with the use of tools in the physical world, the effectiveness of any cognitive/learning tool depends on how well it is used in the current learning context.

The analogy of a Toolbox is useful in terms of framing cognitive strategies in that as with the use of tools in the physical world, the effectiveness of any cognitive/learning tool depends on its appropriateness to the learning task in-hand and how well it is used. *Poor thinking, limited knowledge and skills in the application will typically produce poor results.* The cartoon below captures the point in a poignant visible form and is close to home for my competence level in the domain of household DIY.

In the practical pedagogic context, when selecting and using cognitive strategies, the following heuristic is helpful:

- What is my learning purpose (e.g., extracting information from a text; organizing information from a variety of learning resources; building understanding; cementing mental models in long term memory)
- What are useful strategies for this area of learning
- How to use the strategies effectively and efficiently
- When to use them for best results.

For example, I find mind-mapping very useful for writing research papers, as I can identify the key structure and content areas on one sheet of paper and do ongoing good thinking—critical, creative, metacognitive—to develop the paper content areas from this central advance organizer. However, there is a skill in doing good mind-mapping and one does need to know the subject domain quite well; otherwise, the mind maps can become no more than mazes, confusing rather than aiding thinking and learning.

Another strategy that I introduce to students early in a course of instruction is Retrieval Practice (see Table 3.3). This strategy helps to build understanding and cement knowledge in LTM. It involves students using their WM in response to a question posed internally or externally (e.g., what do I know about agentic engagement?) and searching LTM for what's there (or not there). This is further enhanced

Table 3.3 Retrieval practice heuristic

Strategy	How to use	When to use	Purpose
Retrieval practice	Self-reflect and test what you know—by posing questions, 'thinking aloud', 'verbalizing' to check understanding of key concepts/principles (e.g., what is important here?; how does this work?)	Ongoing—as this is crucial for building understanding and consolidation into Long-Term Memory	Create solid mental representations in LTM—build deeper understanding

when students engage in self-talk (talking aloud) and verbalizing with others as they can assess what they know already that's useful, identify gaps and/or misconceptions, and take appropriate learning action (e.g., employ specific cognitive strategies if needed). The process of extracting what's in LTM and subjecting it to scrutiny in WM, and then putting it back (often in an improved form—richer mental schemata) is a well-validated strategy for effective learning from a cognitive neuroscience perspective. Repeated retrieval practice, especially when spaced out over time, builds the learning as a solid neural network and mental model for understanding. Hence future recall, when needed, becomes increasing quicker, as well as more effective for connecting to new learning and enhancing understanding in that topic area.

Metacognition, in contrast, can be viewed as how we mentally act in relation to our cognition (e.g., monitoring and evaluating the quality of the self-regulatory process, including the choice and use of the cognitive learning strategies being employed). Martinez (2006) put it very succinctly:

> Metacognition can be seen as evaluation turned inward, especially turned toward our ideas. (p. 698)

It typically comes into play when cognition becomes problematic (e.g., when tasks are more challenging) or when a new learning challenge has been identified. Many researchers in the field acknowledge that they are probably mutually dependent on each other—hence cannot be entirely separated (e.g., Flavell 1979; Veenman et al. 2006). Figure 3.4 summarizes the key concepts and their likely relationship.

An important summary point is that students need to understand the distinction between cognition and metacognition to become self-regulated (e.g., Schraw 1998, p. 118). Schraw argues that metacognition is a multidimensional phenomenon, domain-general in nature, and that metacognitive knowledge and regulation can be improved through instructional strategies. His view is that cognitive skills tend to be encapsulated within domains or subject areas, whereas metacognitive skills span multiple domains, even when those domains have little in common (p. 116). Without a clear understanding of these distinct mental processes, what each can do for effective learning in different learning contexts, and how to work them in unison, is like driving a manual gear car without knowledge and skill in clutch control.

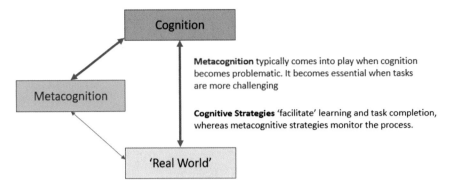

Fig. 3.4 Relationships between metacognition, cognition and the *Real World*

Teaching metacognition and teaching cognitive strategies should be mutually reinforcing as the overall aim is to develop the capability to be a self-directed lifelong learner and possess what Claxton (1998) referred to as "learning Power". Students must learn how to use metacognition to be able to select and use appropriate cognitive strategies that are most effective to successfully master learning and problem-solving challenges.

3.5 Metacognition, Cognition, and Other Types of Thinking

In Chap. 2, *Core Principle 4: Good thinking promotes the building of understanding the importance of thinking for learning* was established and explained. The difficulties of framing these internal cognitive processes were identified, as were important systemic barriers (e.g., cognitive biases, beliefs and emotions) to good thinking. However, it was also made clear that unless we can establish a sufficiently valid and practical model of what constitutes good thinking, we have little chance of establishing a sound pedagogic base for teaching and assessing it. Hence, I presented a model of thinking, evolved over some years, which has proved useful in this context. It does not claim to fully capture the range and accuracy of the cognitive (and another affect) processes involved in and impacting 'thinking', but offers useful heuristics that can be thoughtfully employed in the practices of curriculum framing and teaching.

Certainly, we are being incredibly naive if we assume that effective thinking and self-regulation will naturally occur for most students, simply by encouraging or telling them to do so. Without sound foundational knowledge and skill in good thinking, as well as an understanding on how emotions, beliefs and other vagaries of the human mind influence such capability, many will lack the necessary understanding and competence to self-regulate effectively. As Hattie and Yates (2014) summarized:

> There is a skill in knowing when to think, what to attend to, and when to stop thinking to save cognitive resources. We need to know when to think fast and when to think slowly. (xvii)

Hence, individuals who have developed high MC are likely to be more focused and successful in their approach to learning—possibly in life generally—than those lacking in this capability. They can run 'quality assurance' checks on what they know, don't know, need to find out, as well as ensure that they are utilizing, monitoring and evaluating the necessary cognitive, motivational and affective strategies needed for goal attainment. Therefore, while metacognition and cognitive (thinking) skills have a high impact on student attainment, the situation is still largely as Ambrose (2010) notes:

> Unfortunately, these metacognitive skills tend to fall outside the content area of most courses, and consequently, they are often neglected in instruction. (p. 191)

Planning Curricula to Incorporate Metacognitive Capability

In the context of teaching and learning, a major pedagogic goal is to teach students how to develop and use their MC to effectively select, use and evaluate appropriate cognitive and motivational strategies for meeting their learning goals in different situations, enhancing their well-being and developing the ability to be self-directed lifelong learners. This constitutes perhaps the *gold standard* in terms of educational aims. In practice, educational systems (and institutions) vary greatly in terms of policy, practices, and resources. Hence, Stenhouse's (1989) observation is as relevant today as it was at the time of his writing:

> …the central problem of curriculum study is the gap between our ideals and our attempt to operationalize them. (p. 3)

How this might be done thoughtfully, from an evidence-based perspective, is presented in Fig. 3.5.

The following sections explain and illustrate each of the components, and how they contribute to different aspects of the learning process. The framework is a holistic and synergistic one, in that the better each component is pedagogically addressed, integrated and contextualized to the student profile, the better the learning outcomes in terms of facilitating MC.

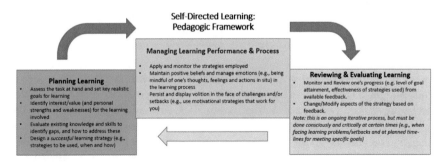

Fig. 3.5 Framework for developing metacognitive capability

3.6 The Self-regulation Cycle

Despite many approaches and models in the literature, there is general agreement that SDL (or SRL) involves the following iterative stages, irrespective of the specific terminology employed:

1. Planning Learning
2. Managing Learning Performance and Process
3. Reviewing and Evaluating Learning.

1. **Planning Learning**

As an old saying goes, 'fail to plan, plan to fail'. This applies equally to learning almost anything. Essentially, when we seek to learn something new or extend our existing learning in some way, we are going about the business of enhancing aspects of our long-term memory system (e.g., in the language of neuroscience, building on existing neural networks in terms of the richness of pathway connections). Once established, this enhances our understanding of what we are intending to learn and, hopefully, our competence in applying this new learning skilfully in real-world contexts.

Hence, in this planning stage, the learner (you, me, whoever) must do the necessary cognitive and metacognitive work, and it's necessary to have key underpinning knowledge on how we learn and how we think—as the latter involves the cognitive and metacognitive stuff.

Here's a summary of the key areas/questions to address in the planning stage:

- Assess the task at hand and set key realistic challenging goals for learning
- Identify interest/value (and personal strengths and weaknesses) for the learning involved
- Evaluate existing knowledge and skills to identify gaps, and how to address these
- Design a *successful* learning strategy (e.g., specific strategies to be used, when and how).

Set Key Goals for Learning

The importance of goals was identified and outlined in Chap. 2: *Core Principle 1: Learning goals, objectives and proficiency expectations are clearly visible to learners.*

At the initial stage, before setting clear goals, it is useful to assess the task(s) at hand and have some idea of what may be involved in terms of time and resources needed (e.g., knowledge, skills to be acquired). Goals need to be challenging but must equally be a 'viable proposition' in terms of one's life situation. If they are too challenging, one may be left with feelings of frustration and this may not be good in terms of self-efficacy in that area. This situation is well captured in an old English saying (it may be elsewhere also), "you have bitten off more than you can chew". Equally, if goals are too easy to attain, they won't stretch competence, build volition or grit, and may feed mediocrity and complacency.

Research shows that academic performance gains range from 16 to 41% in class-rooms where students are explicitly taught how and why it is important to set learning goals (e.g. Marzano 2007). Setting specific and challenging goals furthers strength-ens learning by providing students with benchmarks to measure their progress and to motivate themselves to exert the effort to accomplish their aims. As Hattie (2009) points out:

> A major reason difficult goals are more effective is that they lead to a clearer notion of success and direct the student's attention to relevant behaviours of outcomes. (p. 164)

Furthermore, when teachers help students connect to classroom goals in ways that have personal meaning for them, there is a much greater chance they will be motivated to engage in the necessary hard work often required in achieving challenging learning goals. Where possible, allowing students to set some goals for themselves adds a further intrinsic motivational element to the learning process, as student derived goals can:

- enhance one's sense of autonomy (Ryan and Deci 2017) through choice and more focused attention toward goal-relevant tasks and the most effective strategies to attain them
- increase one's effort and volition to attain them through motivation and desire for mastery
- increase one's affective reactions to targeted outcomes (i.e. the investment of emotions and desire to maintain self-efficacy).

Goals then provide the important role of framing cognitive representations of a future event and, as such, influence motivation through certain key processes. More specifically, goals, as Alderman (2008) identifies:

- Direct instruction and action toward an intended target. This helps individuals focus on the task at hand and organize their knowledge and strategies toward the accomplishment of the goal;
- Mobilize effort in proportion to the difficulty of the task to be accomplished;
- Promote persistence and effort over time for complex tasks. This provides a reason to continue to work hard even if the task is not going well;
- Promote the development of creative plans and strategies to reach them;
- Provide a reference point that provides information about one's performance (p. 107).

Identify Interest/Value and Personal Strengths and Weaknesses for the Learning Involved

As motivation is fundamental to learning, it's important to be able to be honest with oneself in terms of how important the learning goals are. We may like to achieve certain goals like getting fit, losing weight, etc., but often these don't happen, and there's a reason for this. Motivation is multifaceted, and many aspects combine to influence the level of commitment and level of effort that one may put into goal attainment. Key aspects include:

- Interest in the tasks involved (intrinsic motivation)
- Value to the individual over and above intrinsic interest (e.g., usefulness for another extrinsic personal goal that gives a feeling of satisfaction)
- Cost-Benefit Analysis and Expectations that the effort required will lead to the perceived desired outcomes
- Beliefs about the nature of intelligence and what makes people successful learners
- Self Efficacy beliefs about one's capability to be able to do what is necessary to achieve success.

Individuals are more motivated to engage in activities they find interesting. Whitehead (1967) puts a nice spin on this:

> There can be no mental development without interest. Interest is the sine qua non for attention and apprehension. You may endeavour to excite interest through birch rods, or you may coax it by the incitement of pleasurable activity. But without interest, there will be no progress. (p. 37)

Furthermore, Interest and self-efficacy were reported as significantly correlated in several studies (e.g., Zimmerman and Kitsantas 1999). Interest and self-regulation also share several characteristics. They both facilitate learning and optimal levels of performance and thus are necessary components of academic success and as the authors suggest:

The combination of interest and self-regulation has the potential to facilitate the learning of the broad range of skills and competencies students need for productive and creative futures (p. 101).

Finally, Hidi and Renninger (2006) argue that interest has important distinguishing features from other motivational variables:

1. Interest includes both affective and cognitive components as separate but interactive systems.
2. Both cognitive and affective systems involved in experiencing interest have biological roots (e.g., Hidi 2006). Neuroscientific evidence of the existence of approach circuits in the brain (e.g., Davidson 2000) and seeking behaviour in humans and animals (e.g., Panksepp 2004) indicate that interesting activities have a biological foundation in all mammals.

While interest may be situational or individual, it is the outcome of an interaction between a person and a particular content, and therefore, while the potential for interest resides in the person, the content and the environment may determine the direction of interest and contribute to its development—interest can be enhanced through pedagogical interventions.

Value, apart from interest, also plays a part in the motivational stakes. While we may seek to develop students' passion for learning (i.e., intrinsic motivation) in the subjects we teach (and this is a noble cause), we must recognize that there are equally powerful aims and considerations that are more extrinsic. Table 3.4 captures this overall framing.

Interest is, of course, impacted by value, and in many ways is part of it. Students who value the learning for its own sake are likely to have interest value, as framed

Table 3.4 Eccles and Wigfield expectancy—value model

Value component	Definition	Example
Attainment value	Importance of the activity to the individual	I value AP courses because I see myself as a capable student
Interest value	Enjoyment obtained from doing an activity	I love reading novels
Utility value	Usefulness of the activity to the individual	I am taking this math class because I want to be a doctor
Cost	Perception of amount of effort needed for the activity and how that has an impact on other valued activities	I don't have time to do my homework because I'd rather go out with my friends

by Eccles and Wigfield. While many writers focus on the importance of intrinsic motivation (e.g., Ryan and Deci 2017; Reeve and Tseng 2001), and this should be a main pedagogic focus, we must recognize that other extrinsic factors are equally important. Measuring precisely how much a persons' motivation is determined by intrinsic or extrinsic factors may be of academic interest though less impactful in terms of pedagogic practice. In many cases for people, their motivation for different activities will have extrinsic and intrinsic aspects, whose levels and focus vary over time, even over situations. In writing this book, is my motivation intrinsic or extrinsic? Both, but I would not be able to put a percentage on this. I like it when all the value components are strong, and the costs are low. I'll not argue the compositions too critically.

Cost-Benefit Analysis is a core concept in decision making, which appears to apply in some form in most contexts and cultures I have experienced. Of course, how rational is this process enacted by people is a completely different issue. Often the costs and benefits cannot be framed (or as clear as we may like) before the decision is taken, as certain information and outcomes are not available and/or cannot be predicted with high accuracy. Also, this process requires *Good Thinking*, and as we have explored in some detail earlier, which is far from a given pre-requisite in human psychological functioning.

Identifying what you know and don't know is fundamental for effective learning. Specifically, this applies to evaluate one's strengths and weaknesses, both in terms of knowledge and skill levels for a specific learning goal, as well as appraising one's dispositions for learning (e.g., belief systems; ability to maintain volition in the face of challenges). A major area of motivational theory has focused on how attributions (what we believe to be true) impact motivation and learning. For example, research (e.g., Weiner 1992) shows that students who attribute their successes and failures to internal and controllable sources (such as one's effort) are more likely to persist on difficult tasks and experience success than those who attribute success and failure to external or uncontrollable sources (such as innate ability, luck, task difficulty).

Perhaps the most notable research in the literature in recent years is that of Dweck (2006) on the impact of a Growth Mind-set (as compared to a fixed Mind-set) on student learning and attainment. This was outlined and explained earlier. Equally important are what Bandura (2004) refers to as Self-Efficacy Beliefs, which "are rooted in the core belief that one has the power to effect change by one's actions" (p. 622). As Pajares (2012) summarizes:

> People with a strong sense of self-efficacy approach difficult tasks as challenges to be mastered rather than as threats to be avoided. They have a greater intrinsic interest and deep engrossment in activities, and they set themselves challenging goals and maintain a strong commitment to them. (p. 113)

In summary, in this context, while a sizeable body of research demonstrates that students' use of learning strategies promotes academic achievement (e.g., Zimmerman and Martinez-Pons 2012), research also indicates that student self-beliefs of efficacy, causality of outcomes, and on the nature of intelligence to strategically regulate learning play a similarly crucial role in academic self-motivation (Zimmerman et al. 1992). Therefore, from a pedagogic point of view, developing students' beliefs in their learning capability, and competence in using metacognitive and cognitive self-regulation strategies may result in a synergistic impact on academic achievement and specific intrinsic motivational engagement dimensions (e.g., cognitive and emotional engagement). Hence, there is a need to infuse both metacognitive and motivational strategies into curriculum planning and facilitation, especially in the context of today's teaching challenges.

1. **Designing a Successful Learning Strategy (e.g., strategies to be used, when and how)**

Having the motivation to succeed is important, as are the beliefs that with effort and perseverance, success in many situations is possible. However, as Dweck and Master (2012) point out:

> …it is not the sheer effort that produces effective learning. Students must also learn how to select strategies that will bring success and alter their strategies when they are not working. (p. 39)

The ability to use metacognition, selecting, using and evaluating cognitive strategies is of key importance in terms of goal attainment and becoming a self-directed learner. Therefore, students must develop a deep understanding of how to use metacognitive and cognitive strategies for learning and well-being. (e.g., what these are, how they benefit specific aspects of the learning process and how to use them effectively). In generic terms, if we don't know how to do something, we are unlikely to attain competence even with much practice. Students need to learn how metacognition works in the context of good thinking and how this is part of the process of effective learning.

In terms of metacognition, most theorists would agree in distinguishing two main aspects: Metacognitive Knowledge, and Metacognitive Regulation:

- Metacognitive Knowledge refers to the information that individuals hold about their cognition and about strategies which impact on it. It also includes one's self-efficacy beliefs, motivation or interest in relation to learning demands and goals. This knowledge provides a plan or guides for processing, the rules of which may be more (explicit) or less (implicit) in terms of being amenable to conscious awareness and verbal expression.

- Metacognitive Regulation involves the execution of control strategies to regulate our cognitive activities to ensure successful planning, monitoring and evaluation of our actions towards desired goals. As we engage in metacognitive activities, based on our metacognitive knowledge, this creates 'metacognitive experiences' (e.g., Flavell 1981; Schwarz 2010). This is the conscious learning we derive from applying our metacognitive knowledge in new metacognitive regulation tasks. It is essentially a process of assimilation and accommodation of new metacognitive knowledge with prior metacognitive knowledge. For example, in writing this chapter I have probably been metacognitive throughout (not always consciously) and not just cognitively but also in terms of affect. Some days I feel more productive, even excited; others, less so. I have certainly learned that metacognition is, in fact, 'a many-headed monster of obscure parentage'. Has it enhanced my MC?—well, I finished this book. Just writing that statement enhanced my volition a bit.

Through both retrieval practice and deliberate practice in self-regulation, students will acquire both the understanding and competence relating to metacognitive knowledge and regulation, to the extent that it becomes a habit of mind (e.g., Costa) and an adaptive competence across a range of learning contexts.

Similarly, in using cognitive strategies, students will need to have gone through the same learning process, so that they have acquired:

- Strategic knowledge on how cognition works in facilitating the learning process (e.g., information processing through the working of memory systems). This enables knowing what a cognitive strategy can do in terms of assisting learning in tackling a specific learning task. For example, in using a mind map one needs to understand what the tool can do, how to use it effectively, and how this helps the learning purpose (e.g., summarize information, show relationships/connections, aid memory processing)
- Knowledge of task, which enables the ability to understand, analyse and make inferences and interpretations of what a lesson objective, activity or procedure (as explained by a teacher) specifically involves in terms of what learning (and performance) is needed
- Knowledge of self, which is essential for identifying and evaluating personal (existing) strengths and weaknesses as a learner and can choose appropriate learning strategies (knowledge of strategy) that are aligned with the task(s) at hand (knowledge of task). This will involve both an understanding of systemic barriers to cognition and learning, as well as openness of personal recognition of specific traits that may mitigate effective learning (e.g., impulsivity, procrastination).

Cognitive strategies, by definition, will focus on specific aspects of the learning process and types of learning. These will essentially fall into one or more of the following activity categorizations:

- Identifying, clarifying and verifying what is already known about a topic or activity (e.g., activating prior knowledge)
- Framing clear challenging desired goals
- Planning ways to obtain relevant knowledge, resources and develop the necessary understandings and skills to achieve desired goals—create/produce a learning plan/strategy
- Finding and accessing relevant information sources and resources
- Extracting, organizing and summarizing information
- Processing information between WM and LTM (connecting new information to prior knowledge)—thinking—to build understanding in LTM. The use of retrieval practice (pulling information back and forth from LTM to WM to check understanding and knowledge completion and cement the learning in LTM at the neural level
- Using new knowledge and skills in different situations to elaborate understanding and skill transfer
- Managing one's motivation and volition in the often frustrating and boring process of learning, especially the thinking part which is often tiring.

Strategy Evaluation Matrixes (e.g., Schraw 2006) are useful, especially for identifying, analysing and evaluating cognitive, metacognitive and motivational strategies (see Table 3.5).

As students develop both metacognitive and cognitive skills, both their mastery levels and self-efficacy in using such skills will likely increase. This, in turn, helps to maintain/enhance motivation (whether intrinsic or extrinsic) for learning.

2. **Managing Learning Performance and Process**

Good planning is crucial, and the key components have been outlined and illustrated above. However, the best plan in the world does not mean success without the appropriate action, as we all know. A great lesson plan is not a great lesson; enacted by a poor teacher it is likely to have little impact in terms of desired learning.

In this stage of the self-regulatory process, metacognition focuses on the use of cognitive and motivational strategies to ensure the successful maintenance of the overall learning strategy towards goal attainment. Hence, it is important to be mindful of:

- Ensuring the necessary skilful use and effort is put into the implementation of the cognitive learning strategies employed. Unless they are implemented effectively and conscientiously, there may be little benefit in terms of the desired learning outcomes.
- Maintaining a Growth Mindset. We are fragile creatures and can easily slip into self-doubt and loss of confidence. This happens to even the top sportspeople, so the rest of us had better be mindful.

Students need to be explicitly taught to expect challenges and setbacks that may seem to push them backwards in what they are trying to learn. This is where self-control and willpower are important to avoid or at least mitigate procrastination and maintain persistence. As Borkowski et al. (2000) made clear:

Table 3.5 Examples of strategy evaluation matrices

Learning strategy evaluation matrix (SEM—adapted for the work of Schraw)

Strategy	How to use	When to use	Purpose
Create learning plan	Identify specifically what needs to be learned (goals) and identify strategies to meet them (e.g., 5W's and H)	In planning learning, but needs to be monitored and modified throughout the learning experience	Have a structure for organizing, monitoring and evaluating the learning strategies employed
Activate prior knowledge	Pause and think about the topic/skill area—what do you already know—ask what you don't know	Prior to learning a new or unfamiliar topic (e.g., reading new content; learning a new skill)	Makes learning more effective/efficient—connecting new knowledge to old
Organize/summarize information	Creating diagrams of key concepts and their relationships (e.g., concept/mind maps)	Ongoing—to build the necessary knowledge base for what needs to be learned	Build knowledge bases related to the learning goal
Comprehension monitoring	Self-reflect by posing questions to check understanding of key concepts/principles (e.g., what is important here?; how does this work?)	Ongoing—especially when seeking integration—building understanding into Long-Term Memory	Create solid mental representations in LTM—build deeper understanding
Motivational strategies (meta cognition)	Use of self talk and keeping on task—need to maintain a positive psychological state	When needed, especially when the task becomes tougher, and during mood downturns	Keeping focused on required action, persevering, and managing anxiety

Students need to learn that classroom activities require them to work hard to achieve understanding. (p. 33)

Many terms have been used to capture the underlying traits and characteristics that facilitate self-control and willpower, including perseverance, conscientiousness, resilience, to name but a few. The new vogue seems to be 'Grit'. I remember it well in my boxing classes at school, as the instructor would say words akin to "show some grit, Dennis". I was not good enough to box professionally, but the training did develop some grit. Going to university was a relative breeze after years of getting up at 5 a.m. on cold English winter days and doing a 10 km run, followed by several hundred push-ups and sits ups. There was never any sympathy from the boxing instructor—he was a second world war veteran, and had little time for excuses not to push oneself to the limit (as he saw it). From an educational perspective, Duckworth et al. (2007) defined grit as "perseverance and passion for long-term goals" (p. 1087). Tough (2013) postulates that in the real world, learning to react to failure is as critical to success as academic achievement. Noncognitive character traits such as resilience, drive and delayed gratification are as important as cognitive skills (Farrington et al. 2012).

Of wider concern, Baumeister and Tierney (2012) conclude that:

> …most major problems, personal and social, centre on the failure of self-control: compulsive spending and borrowing, impulsive violence, underachievement in school, procrastination at work, alcohol and drug abuse, unhealthy diet, lack of exercise, chronic anxiety, explosive anger. (p. 2)

Collins (2017), based on such evidence sources, suggests that:

> Self-regulation failure is the major pathology of our time. (pp. 48–49)

However, research suggests (e.g., Baumeister and Tierney 2012) that we have limited amounts of willpower and distractions, especially in this digital age, are all around us. After studying thousands of people inside and outside the lab, researchers found that experiments consistently demonstrated two lessons:

1. You have a finite amount of willpower that becomes depleted as you use it.
2. You use the same stock of willpower for all manner of tasks. (p. 35)

As they note:

> Emotional control is uniquely difficult because you generally can't change your mood by an act of will. You can change what you think about or how you behave, but you can't force yourself to be happy. (p. 37)

Similarly, as Collins (2017) argues:

> One of the most critical self-regulatory skills for students to cultivate is impulse control. (p. 48)

From a common-sense perspective (though we know such terms are problematic), it would be nice to think that making such knowledge explicit would result in people being more mindful and responsible. However, cognitive science, as documented earlier, is making us more and more aware that the human brain is far from rational— I would go so far as saying it is not intelligent design; complicated design yes, intelligent design unlikely. As outlined earlier, there are many limits to the accuracy of our perceptions and thinking (Kahneman 2012; Willingham 2009).

While willpower or grit are challenging pedagogic goals for the above reasons, research shows that the ability to employ willpower, maintain self-control and be persistent in this capability, has positive learning outcomes. Baumeister and Tierney (2012) summarizes the implications of this, and raise the pedagogic challenges:

When researchers compared students' grades with nearly three dozen personality traits, self-control turned out to be the only trait that predicted a college student's grade-point average better than chance. Self-control also proved to be a better predictor of college grades than the student's IQ or SAT score. Although raw intelligence was an advantage, the study showed that self-control was more important because it helped the students show up more reliably for classes, start their homework earlier, and spend more time working than watching television. (pp. 11–12)

The results couldn't be clearer: self-control is a vital strength and key to success in life. (p. 13)

Rohn's reflections are also poignant in this context:

Average people look for ways of getting away with it; successful people look for ways of getting on with it.

Success lies in the opposite direction of the normal pull.

3. Reviewing and Evaluating Learning

This process is in operation throughout all stages of the self-regulation process. It is a key component/activity of MC, as we are evaluating both our thinking processes as well as those other aspects of being (e.g., beliefs; emotions; and moods) that impact self-regulation. Key questions to answer truthfully (as far as is possible) are:

- Have I met the goals I set out to achieve (yes, no, more, less)—what is the evidence? You need to be as honest and open as possible on this—good feedback is crucial.
- How effective and efficient have I been in this learning process—what strategies have worked (not worked), how well?
- Have I employed and maintained the necessary willpower, and how do I know this?
- What do I now need to modify/change and how might I get this moving as effectively and efficiently as possible?
- Do I have the motivation and volition to do what is necessary, or do I have to modify my goals and expectations, and how do I feel about that?

As in all skill development, this will require understanding, practice and perseverance over time, but with the right strategies, competence will develop. Sounds easy, and everything is easy when you know it thoroughly and can do it with expertise. However, expertise does not come easy as Gladwell (2008) pointed out. According to him, it requires considerable effort and perseverance over time, with estimates of 10,000 h being 'the magic number' with guided practice, regardless of a person's natural aptitude. With enough practice, he claimed that anyone could achieve a level of proficiency that would rival that of a professional. It was just a matter of putting in the time, around 10 years. However, the view of 10,000 h for developing high-level expertise has been challenged (e.g., Goleman 2013; Epstein 2014).

This is not surprising as expertise in different areas may not be equated so precisely in such algorithmic terms. Furthermore, there is practice and *deliberate* practice, so it is probably the case that some individuals are using qualitatively different practice activities, some favouring (or hindering) the route to expertise. Also, there are likely to be constitutional factors (e.g., psychological, physical) that come into play along the journey to expertise, as we are not born with a 'Blake Slate' as Pinker (2003) so comprehensively documented. In summary, on the question of how much time is necessary to develop expertise, there are certainly no 'quick fixes' and it will involve much time and commitment, but 10,000 h may be more of a metaphor than an evidence-based heuristic.

To develop an effective approach to developing SDL, all components of this framework need to be addressed thoughtfully in the context of student profiles. As an analogy, think of the mind like a car, which is a system, and, if we don't know the parts of the system, what they do and how they work, and their relationship to each other—we are like the poor chap in this cartoon:

This resonates with me, as I have no competence in motor vehicle maintenance. However, when I do experience problems with cars (e.g., they simply don't work), as I now know that I know nothing about motor vehicle maintenance and have no interest in this area, I now (and it took a while) behave intelligently (i.e. metacognitively) and ring a motor car mechanic—who somehow knows how to fix it!

If our aim is to develop students' capability to become self-directed learners, we must teach them what it is, how to do it, and what makes it useful for meeting their goals. This entails providing them with the essential knowledge, strategies and application opportunities through good practice over time. As Treadwell (2017) makes clear:

> Unless learners can reflect on their learning, actions, attitudes, values, motivations and think-ing, they cannot develop into independent lifelong learners who can have agency over their learning journey. Metacognitive and cognitive thinking provides the stimulus for under-standing self and others, and drives the capacity to moderate and improve our thinking and subsequently our learning. (p. 97)

On the motivational stakes, we can only provide the rationale, learning strategies and the opportunities to develop MC. It's ultimately their call.

3.7 A Curriculum Model for Developing Metacognitive Capability: The 'Thinking Curriculum'

It has been a tough call getting to grips with the many heads of the metacognition monster. Yes, it certainly is a monster, but as Arnold Schwarzenegger said in *Predator*, when he noticed the yellow 'blood' of the alien on nearby vegetation, "If it bleeds, we can kill it". Predator was a film in which a group of expert combat veterans were recruited to check out some mysterious killings in a jungle and discovered a strange presence that turned out to be an alien creature who seemed to enjoy hunting in the intergalactic context (the latter is my interpretation). Hence, once we have a clearer evidence-based frame on the nature and components of these heads, and how they work (and often don't work), we are in a better position to develop and facilitate strategies to help students manage their own 'Many-Headed Monster'. Given the above analogy, we can subject these 'many heads' to rigorous pedagogic analysis, finding ways of managing them better, and develop MC.

Veenman et al. (2006), in reviewing the literature, identified three fundamental principles that are necessary for successful metacognitive instruction:

(a) embedding metacognitive instruction in the content matter to ensure connectivity,
(b) informing learners about the usefulness of metacognitive activities to make them exert the initial extra effort, and
(c) prolonged training to guarantee the smooth and maintained application of metacognitive activity.

He suggests following the *WWW & H* rule (What, When, Why, and How to do) in implementing these principles (p. 9). This rule is essentially a cognitive strategy that creates a structured thinking heuristic in the planning and development process. To be successful, one would need to know what constitutes the What (always a big question), and Why, possess the necessary knowledge (i.e., conceptual understanding) of knowing When and Where to apply it, and the other big question—How to do this effectively and efficiently.

I have created a curriculum design model, incorporating the heuristics of Good Thinking outlined earlier—(what I euphemistically called the 'Thinking Curriculum'), (Sale 2013) that guides the whole curriculum development cycle and related processes. Figure 3.6 summarizes the key process.

In basic terms, this means that the types of thinking incorporated in the Learning Outcomes must be effectively taught through the Instructional Methods/Strategies used and accurately measured in the Assessment System.

In developing a curriculum that incorporates MC, it is first necessary to establish a sound theoretical framework that establishes what *Good Thinking* entails—as detailed in Chap. 2: *Core Principle 4: Good thinking promotes the building of understanding*. Without a sufficiently valid and practical model of good thinking, notions of effectively teaching it, let alone assessing it, are tenuous at best. Not an easy task, as we have seen, and it's not surprising that many teaching professionals are confused

Fig. 3.6 Summary frame of thinking curriculum model

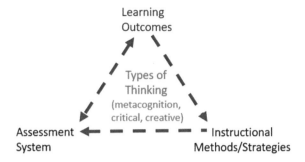

in this area. The next big question concerns the extent to which thinking is amenable to development through pedagogic means, and how best we can do this from an EBT approach.

Thinking is, essentially, a human performance activity. How we have learned to think will determine in large part how we think, much the same as for any kind of learned activity.

Furthermore, as Perkins (1995) points out, "People can learn to think and act intelligently." (p. 18). Similarly, Coles and Robinson (1989) argue that:

> The problem is not whether we can teach thinking. The evidence suggests we can. The problem continues to be whether we are willing to make the pedagogic changes necessary to do so and if we are, which changes might be the most effective. (p. 20)

Paul (1993) provides an interesting analogy between the development of mind and physical fitness. He points out that the mind, like the body, "has its form of fitness or excellence" which is "caused by and reflected in activities done in accordance with standards (critically)." (p. 103). He goes on to argue that:

> A fit mind can successfully engage in the designing, fashioning, formulating, originating, or producing of intellectual products worthy of its challenging ends…Minds indifferent to standards and disciplined judgment tend to judge inexactly, inaccurately, inappropriately, prejudicially. (pp. 103–44)

Swartz and Perkins (1990) identify six areas of 'improvement' that become apparent when students' thinking gets better:

1. Awareness of one's thinking
2. Investment of effort in one's thinking
3. Attitude towards thinking processes
4. Organization of thinking processes
5. Development of subskills
6. Smoothness in the thinking process. (p. 24)

Framing Learning Outcomes for a Thinking Curriculum

The appropriate framing of learning outcomes is central to any curriculum product. Learning outcomes can be seen analogously as the foundation construction work for building a house. When conducted by skilful and conscientious professionals using appropriate quality materials, applying correct building procedures and aligned to safety regulations, this should provide a solid base for the more aesthetic components to follow.

The learning outcomes for a thinking curriculum should significantly emphasize real-world applications of the subject matter. For this reason, I suggest writing learning outcomes, wherever possible, in terms of direct performance or competence. The curriculum focus then cues and pushes students into doing the cognitive heuristics for the types of thinking involved.

Swartz's (1987) infusion framework and Fogarty's (2009) *nested* and *threaded* approaches went a long way towards proving an effective and efficient EBT methodology for integrating thinking skills into the content curriculum. Swartz (1987) argues that both critical and creative thinking are best developed through "conceptual infusion" with the subject content. This involves identifying the ingredients of good thinking—"the skills, competencies, attitudes, dispositions, and activities of the good thinker"—and designing these into the structure of the lesson content (p. 125). In nested integration, there is a focusing on connecting specific content knowledge with a thinking skill and/or process skill (e.g., communication, teamwork); whereas threaded integration focuses on infusing a key skill area such as critical thinking, and developing it over the duration of the programme (e.g., Fogarty 2009). The essential pedagogic benefit is that good application of the thinking process and skills with the subject content mutually develop the meaningful acquisition and connectedness of knowledge to form understanding. Furthermore, specific types of thinking can be systematically developed in terms of the level of proficiency and range of context application. For example, if decision making is to be developed throughout the curriculum, then it will be important to have structured development of such subskills as 'generating and exploring options', 'gathering evidence', 'assessing evidence from different perspectives'. As Marzano (1988) pointed out:

> …we can improve students' ability to perform the various processes by increasing their awareness of the component skills and by increasing their skill proficiency through conscious practice. (p. 65)

The learning outcomes can be specially written to cue specific type(s) of thinking or in terms of direct performance, as illustrated below:

Type of Thinking

- Analyse the impact of pollution on water quality
- Compare and contrast a range of retaining structures
- Generate new design options for marketing a health food product
- Predict the outcomes of specified legal scenarios
- Evaluate a policy for animal protection.

Real Work Performance

- Conduct product packaging tests for a specified product
- Prepare a voyage passage plan

- Write a programme in Java script to animate a range of figures
- Prepare a tender report
- Produce a thinking curriculum for a humanities module.

Note: Objectives can be written at different levels of specificity and contextualized accordingly—but the general concept of focusing on the *desired performance* applies. Also, of importance, when outcomes are written in terms of a performance that does not cue the type of thinking, though clearly embedded from a pedagogic point of view, such as in "predict the outcomes of specified legal scenarios", they can be derived from asking and visualizing the answer to the following question;

How would a highly competent person think in the effective execution of this activity?

In predicting the outcomes of legal scenarios, the following types of thinking would need to be effectively employed:

- analyse the various components of a legal scenario
- make valid inferences and interpretations from legal data
- compare and contrast a range of legal scenarios.

Figure 3.7 illustrates the essential components and process.

Today we are seeing a movement from content-based curriculum to more outcomes-based curriculum approaches, especially competency-based frameworks. These descriptions of performance would mirror real-life competencies in the industrial/commercial world, at the appropriate level for designated curriculum levels. In vocational education, learning objectives should be those that are most relevant

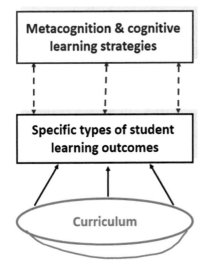

The Framing Questions are:

1. What types of student learning outcomes are most important in this curriculum?

2. What cognitive strategies will help students meet these outcomes?

NOTE: **Metacognition** runs throughout the learning process

3. When to teach these strategies in the curriculum so that they best support the subject content learning?

4. How to provide sufficient Spaced & Deliberate Practice for students to develop the skills involved?

Fig. 3.7 Summary heuristics for identifying and structuring metacognitive and cognitive strategies into a curriculum plan

to the world of work, both now and in the conceivable future. The essential thing about performance-based learning outcomes is that the subject content can be organized in ways that are most likely to be experienced as meaningful for learners. In this context, the types of thinking, as well as other process skills (e.g., communication and teamwork), can be naturally embedded in the learning tasks that students are required to master. For example, The National Vocational Qualifications (NVQ) framework, which a few decades ago revolutionized vocational education in the UK, are essentially competency-based standards derived from functional analysis of industrial roles. Their attainment is not governed by curriculum duration or modes of delivery, but solely on the learners' ability to meet the full range of competencies in a vocational area.

However, while competency models are more tailored to the present need for effective and efficient learning, there has been a criticism of these approaches. A major concern has been the extent to which learners possess sufficient depth of both content and process knowledge (i.e., accurate neural structures/mental models) of what is necessary for conceptual understanding and adaptive learning—transfer of the competences in a range of real work situations, even when achieving competence in the assessment framework. Much of the problem lies in the reduction of occupational roles into units and elements of competence, based on a methodology of functional analysis. This typically provides an atomized view of real work activity and falls into the trap of the decomposability assumption, which as Resnick and Resnick (1989) note:

> ...has been seriously challenged by recent cognitive research, which recognizes that complicated skills and competencies owe their complexity not just to the number and components they engage but also to interactions among the components and heuristics for calling upon them. Complex competencies, therefore, cannot be defined just by listing all their components. (p. 5)

The learning outcomes for a thinking curriculum, then, must avoid the decomposability trap by ensuring that performances represent holistic and complete activities, not just bits of facts and skills. The essence of this important point is captured by Fennimore and Tinzmann (1990) when they argue that a thinking curriculum must:

> ...always treat tasks as indivisible wholes; variations that acknowledge the novice status of the learner are changes the teacher can make in the environment. (p. 4)

However, the movement towards competency-based models is inevitable and necessary in higher vocational education. The challenge is to resist a reductionist position to the more easily measurable features of a performance. Highly competent performance in any vocational area is more than subject-specific collections of skill operations but includes other equally important cognitive, social and emotional skill aspects. Therefore, the performances identified as learning outcomes must embrace the integration of the whole range of skills that underpin effective performance in those areas.

In conducting a curriculum review, to construct a Thinking Curriculum to support the development of MC, firstly identify what are important learning outcomes that require metacognitive and cognitive activity (e.g., critical and creative thinking skills), and when and where they are to be taught in the curriculum. Use the framework and method outlined here as an evaluative guide—and some Good Thinking! You may need to rewrite certain learning outcomes to make specific types of thinking explicit where relevant in the curriculum structuring. Using the building analogy prior, this sets a solid foundation for the rest of the curriculum design and development process.

3.8   Developing Metacognitive Capability: Evidence-Based Strategies

The term Signature*Pedagogies* (Schulman 2005) has become popular recently, which emphasizes that different professional fields have their distinct pedagogy in terms of the appropriateness of methods used for enhancing learning. In most basic terms, there is a need to contextualize pedagogy to the subject content—what Shulman refers to as *Pedagogic Content Knowledge*. This emphasizes on domain contextualized pedagogy rather than generic pedagogy.

However, while it is necessary to select methods and facilitate learning in the context of subject domains, I see this as no more than a feature of EBT (e.g., combining high effective methods to the subject content and student profile). For example, in teaching literature there may not be many learning areas where problem-based learning would be an effective or efficient method. In contrast, in engineering, which is very much concerned with solving real-world problems, it could be seen as a 'signature method' in curriculum planning. However, having worked extensively in engineering education, I could equally make the case for case-based learning (no pun intended) or simulated practice—as well as many other method combinations. Hattie's (2009) extensive meta-analysis found little support for the use of teaching specifically focusing on subject content knowledge, as the overall effect size was relatively small (0.11). Hence, while the term signature pedagogies is useful in terms of identifying core methods for a field of study, and planning instruction accordingly, it is better seen in a wider EBT context re method blending.

There cannot be a single 'silver-bullet' (simple direct solution to a problem; derived from the folklore of killing werewolves with a silver bullet) teaching method that will work with all students in all contexts. Apart from effectiveness, students (and the rest of us) will get bored with over repetition of one method. I remember 3 decades ago, when teachers were being introduced to Project-Based Learning in a college in the UK, there great enthusiasm for its use. And, it did seem to work well initially, as students showed greater engagement and made positive responses to having more choice and variety in their learning; rather than the traditional lecture format every lesson. However, after 2–3 weeks and several projects on the go, student interest waned. I remember hearing a student saying something akin to, 'can we just have some teaching rather than another project'. The point is that any method, no matter its logical effectiveness, will lose impact when overused. Remember my Chilli Crab story from Chap. 2: *Core Principle 5: Instructional methods and presentation mediums engage the range of human of senses.*

However, students through appropriate instruction can learn to:

1. Understand essential knowledge on how the mind works in terms of human learning, our unique species-specific capability for being metacognitive and its potential for learning, well-being and life enhancement through better self-regulation
2. Thoughtfully use metacognitive, cognitive and motivational learning strategies and skills
3. Maintain volition and willpower to achieve personal goals.

3.9 The Need for Explicit Teaching of Metacognition

While there is still a current vogue for constructivism underpinning much of class-room practice, research shows that the most significant gains in student achievement result when students are taught the use of metacognitive strategies in explicit ways (e.g., Haidar and Al Naqabi 2008; Kistner et al. 2010). Kistner et al. noted that while explicit strategy instruction of metacognitive skills is positively correlated with achievement gains, as compared to implicit instruction, teachers use implicit meth-ods more often than explicit ones. In an analysis of 60 lessons, involving 20 teachers, they found that on average, teachers taught strategies through implicit instruction in comparison to explicit instruction at a ratio of 5 to 1. Implicit instruction is where an explanation is made of how the strategy is effective and students need to work out how to do it; in contrast, explicit instruction involves the simultaneous verbalizing of the teachers' thought processes and/or asking specific questions to the students during the demonstration and modelling of the strategy.

From my experience the characteristics of explicit strategy instruction can be framed to include the following integrated instructional activities:

- Direct instruction on the strategies—what they are, when and how to use them effectively. This requires a clear explanation and demonstration on how metacog-nition and cognition work, and the benefits of using the strategies for learning. This can be illustrated with a range of examples and stories (both yours, others and the students)
- Modelling the strategies—Making 'Thinking Visible' (e.g., eliciting the students' thinking, modelling expert thinking, thinking aloud and verbalizing, and mind-fulness). From this *Good Thinking* can be made explicit, modelled, practised and learned
- Retrieval, Deliberate and Spaced Practice opportunities for using the strategies in real-life/work contexts (e.g., using questions and activities to cue the different types of thinking, develop and extend their application and cement deep understanding

in long term memory; real life performance tasks that incorporate the range of MC strategies in the context of the subject content).

Direct Instruction

Direct instruction (DI) often conjures up images of teachers talking and students listening, and one in which there is little opportunity for student engagement, questions, and an inevitable loss of attention after some 10–15 min. In this situation, it may seem that students are not active and therefore are unlikely to be doing the necessary cognitive work—thinking—to make connections with prior knowledge and develop a deep understanding of the concepts to be learned. The notion that students need to be actively involved in the learning process to be more engaged and to learn better is well documented in the literature. Active learning has been identified as one of the seven principles of good practice in undergraduate education (e.g., Chickering and Gamson 1987). For learning to be active, students must do more than listen, they must read, write, discuss, or be engaged in solving problems. Most importantly, to be actively involved, students must engage in such higher-order thinking tasks as analysis, synthesis, and evaluation. Students must be doing things, and then thinking about why they are doing them. These kinds of activities can include case study, cooperative learning, debates, drama, role-playing and simulation, and peer teaching.

However, getting students to do activities while engaging—at least at the behavioural level—does not mean that they are effectively learning what is intended. Active learning methods must be aligned to desired learning outcomes and get students thinking about the key concepts to be learned. It's the thinking that does the work in building understanding—forming neurologically-based mental models in long-term memory. Equally, it cannot be assumed that students listening to a well-constituted lecture, with clear concise explanation of key concepts, illustrated with interesting real-life examples (and non-examples), striking stories and good visuals, does not get students engaged at both cognitive and emotional levels—which are more important for learning than simply behavioural engagement.

Indeed, while a long, disorganized lecture with a monotonous voice tone can be labelled a sin of teaching, so in the same vein of thinking, this can be applied to activity for activities sake.

Notions of the teacher no longer being the 'sage on the stage' but the 'guide on the side' so aptly reflects this dichotomy. Similarly, Brown et al. (2014) refer to 'student-directed learning' (a theory now current among some parents and educators), which holds that students know best what they need to study to master a subject, and what pace and methods work best for them, as "laudatory" (p. 123). It's nonsense at best, dangerous nonsense at worse. Just more excursions back into 'Educational Jurassic Park'. Expert teaching is neither of these—it is much, much more. Learning to be self-directed requires student engagement and effort, but the evidence suggests—and the very premises of teaching are based on—that they need much by way of expert help in getting there.

DI as a method, assumes that all students can learn new material when (a) they have mastered prerequisite knowledge and skills and (b) the instruction is unambiguous (e.g., Stockard et al. 2018, p. 480). The importance of student's possessing the necessary prior knowledge is a core principle of learning, essential for connecting and assimilating new knowledge to prior knowledge in long-term memory. Clear (unambiguous) explanation supports the thinking process to build further understanding and, 'mastery learning' ensures the necessary skill acquisitions. In this scenario, they will have developed the necessary neural networks and clear mental schemata (model) of what is to be learned. Furthermore, DI emphasizes the importance of testing, which uses retrieval practice, to ensure that the learning is well established in long-term memory. The same conditions apply to skill development. Recent research by Stockard et al. (2018), supported earlier reviews of the DI effectiveness literature. The estimated effects were consistently positive. The authors note that:

> ...despite the very large body of research supporting its effectiveness, DI has not been widely embraced or implemented. In part, this avoidance of DI may be fuelled by the current popularity of constructivism. (p. 502)

> Certainly, our nation's children deserve both effective and efficient instruction. As one of the anonymous reviewers of our article put it, "Researchers and practitioners cannot afford to ignore the effectiveness research on DI". (p. 503)

Hattie's (2009) framing of Whole-Class Interactive Teaching, which has an effect size of 0.81, is of interest in this context, as the key aspects of direct instruction (e.g., mastery learning and testing) have positive effect sizes. Whole class interactive teaching is a specific approach to active learning in class, which is highly teacher-led, but very active for students. This involves summaries, reviews and a range of active learning methods, including questioning. In this strategy, there is a blend of high effect size methods, which Hattie refers to, in analogy, like 'Russian dolls'. In the strategy, there is direct instruction, which acts as both providing key subject content (e.g., key concepts fundamental to understanding; examples, analogies and stories for illustration) as well as facilitating the context for interleaving with appropriate active learning methods to encourage types of thinking and practice modes for skill development. It is the evidence-based and creative blending of the various methods, tailored to the learning outcomes, student profile, and situated context (e.g., mood of the group, time of day) that underpins ***Evidence-Based Creative Teaching***.

Figure 3.8 identifies the basis of an instructional strategy I often use in my workshops. I may add other specific methods (e.g., Signature Pedagogies), typically in the form of performance tasks that provide an authentic and experiential learning experience for the student group. For example, in business studies or life science lesson, I may use a case study or a problem-based learning task. I always use retrieval practice to facilitate ongoing two-way feedback.

I typically employ the following broad strategies:

- Evoke or reinforce a specific thinking process/type of thinking (e.g., compare and contrast, make inferences and interpretations)
- Feel, from observation and dialogue with students, that the group need some motivating or change in lesson structure and method
- In teachable moments—where an event/the discourse in a lesson is ripe for making a big learning point/key concept, either for introduction or reinforcement.

Underpinned by the planned and situated application of Core Principles of Learning

Fig. 3.8   Example of a whole-class interactive 'Russian Doll' method blend

In this process, I can check their understanding, deal with knowledge gaps and misconceptions, and provide the retrieval practice for them to cement this learning in long-term memory. The important point here is that the explicit teaching of MC is not a passive lecturing scenario, but a highly effective and active teacher-led instructional approach.

Modelling MC Strategies

Schraw (1998) specifically emphasizes:

> Modelling of regulatory skills such as planning, monitoring, and self-evaluating, is especially important. Every Teacher should make a concerted effort to model explicitly these behaviours. (p. 122)

Modelling MC strategies extend the knowledge from the explicit instruction to the how of using the strategies, showing what is to be done at the specific strategy level. For example, if mind mapping as a cognitive strategy is explained and illustrated, modelling goes through the essential stages, procedures of how to do it, from the central framing of the learning area to the selection of the superordinate branches, and then to extensions. Other features that enhance the mind map, such as colour and images are them modelled and demonstrated.

Using a Whole-Part-Whole Strategy (see Fig. 3.9) is useful here, as it provides a key structure to the process, utilizes retrieval practice and reduces cognitive load.

The WPW strategy is analogous to completing a jig-saw puzzle, in that we can continually refer back to the whole picture while grouping various components based on similarity (e.g., in a jig-saw puzzle, there are always specific features, such as sky, objects, people, animals). As we use the strategy, we are doing ongoing retrieval practice, and as more of the segments are put correctly into place, the task becomes easier and the cognitive load less. One could say that life is one big jigsaw puzzle, albeit a more difficult one than a country scene, and there always seem to be pieces missing but we don't know what they are.

As students work through this learning strategy, metacognitive, cognitive and motivational strategies can be systematically infused and facilitated as students tackle

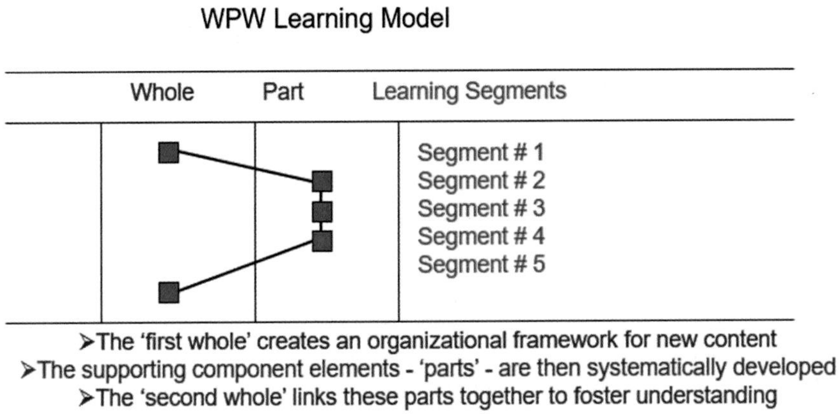

WPW Learning Model

Whole	Part	Learning Segments
		Segment # 1
		Segment # 2
		Segment # 3
		Segment # 4
		Segment # 5

➤The 'first whole' creates an organizational framework for new content
➤The supporting component elements - 'parts' - are then systematically developed
➤The 'second whole' links these parts together to foster understanding

Fig. 3.9 The whole-part-whole strategy

different learning challenges. Educational research (e.g., Schraw 1998) and practice support the generic teaching of metacognitive and cognitive skills, though customizing cognitive skills to the subject domain context. This makes perfect sense as we want our students to be mindful not just in the biology lesson, but across all subjects; in-fact much more than that, in their lives holistically. There are serious limitations in displaying good thinking in one life situation, but not in others.

Making Thinking Visible

In the process of helping students to build an understanding of what constitutes good thinking and how to develop MC, making thinking visible (both student thinking and teacher thinking) is essential. Questioning is a key strategy in developing all types of thinking. As the famous success coach Anthony Robbins (2001) wrote:

> Questions are the primary way we learn virtually everything. Thinking itself is nothing but the process of asking and answering questions. (pp. 179–180)

The use of questions creates a stimulus to get students searching (both consciously and sub/unconsciously) through their LTM system to find appropriate responses to the stimulus question. The process of questioning, when done skillfully, makes the invisible internal neural representations occurring when thinking visible to some extent. As Ritchhart et al. (2011) point out:

> We need to make thinking visible because it provides us with the information we as teachers need to plan opportunities that will take students' learning to the next level and enable continued engagement with the ideas being explored. (p. 27)

In this process, it's necessary to ensure that good thinking is clearly and explicitly explained, modelled and then practised through application in a range of relevant activities that incorporate key subject content learning outcomes. Students must develop their accurate representation of the heuristics of good thinking, which involves the forming of rich integrated neural networks in the brains to have useful models for doing good thinking in real-life contexts. As Sheppard et al. (2009) recognized in advocating that:

> …teachers have to make their intellectual processes (their performances) visible. This means that the teacher-expert has to make visible to learners the otherwise invisible processes of thinking that underlie complex cognitive operations …
>
> Teachers have to articulate and demonstrate rather than assume the thought processes they want students to learn. (p. 188)

Hattie (2009) provides the final summary in this context:

Visible Teaching and learning occurs when learning is the explicit goal, when it is appropriately challenging, when the teacher and the student both (in their various ways) seek to ascertain whether and to what degree the challenging goal is attained, when there is deliberate practice aimed at attaining mastery of the goal, when there is feedback given and sought, and when there are active, passionate, and engaging people (teacher, student, peers, and so on) participating in the act of learning (p. 23).

> …the model of visible teaching and learning combines, rather than contrasts, teacher-centred teaching and student-centred learning and knowing. Too often these methods are expressed as direct teaching versus constructivist teaching (and then direct teaching is portrayed as bad, while constructivist teaching is considered to be good. (p. 26)

As outlined earlier metacognition is often framed in terms of two main components (e.g., Brown 1987; Flavell 1979): (1) the knowledge component, which refers to knowing one's cognitive processes, such as knowledge about oneself as a learner, characteristics of the tasks to be undertaken to meet learning goals, and the appropriate effective strategies that are required to result in effective performance; (2) the regulation component, which refers to the actual strategies one uses to manage the cognitive processes, such as planning a strategy to meet the demands of a task, monitoring one's understanding of how well one is tackling the task, and evaluating progress and performance. In practice, metacognitive knowledge, as for knowledge about anything, will guide action. Through action, this knowledge can be evaluated in terms of usefulness and improved, hence a better understanding can emerge over time. In the language of cognitive neuroscience, an enhanced neural network is formed in long-term memory and this facilitates a better mental model for future learning. In lay terms a better level competence/expertise is coming about. In this way, the knowledge and regulation components supplement each other and are both essential for optimal performance. Ku and Ho (2010) point out:

> While participants should be encouraged to reflect on their cognitive activities, through checking and questioning to build up the habit of metacognitive regulation, it is important to explicitly teach the related metacognitive knowledge when necessary. Without the support of metacognitive knowledge, metacognitive regulation will not be effective in enhancing thinking performance. (p. 264)

Many studies (e.g., Pintrich and De Groot 1990; Schraw and Dennison 2004) have found that having metacognitive knowledge, such as knowing what factors affect one's thinking (person variables), how to make sense of a problem or how different problems demand different cognitions (task variables) and knowing when and why to use a skill (strategy variables) facilitate metacognitive regulation. However, having only an awareness for the need to apply metacognitive strategies is not enough for good performance; one must also know when, how and which strategy to use in different learning contexts. The same reasoning applies across most learning fields and domains.

To facilitate this effectively, the key terminology relating to the various heuristics of the types of thinking needs to become part of the language of learning (to consolidate a language of thinking). For example, I often hear teachers, who I assume are seeking to encourage student thinking, use terms like "What are your comments on this" or "Let's discuss this"—even "I want good thinking on this". This latter example assumes that students have a mental model (schemata) of what constitutes thinking. However, in practice, this may vary widely (e.g., Fig. 3.10). If they have no prior useful mental model, then they are either blank or in the process of just commenting, which typically results in statements like, "It's ok", or "I don't like it much" that has little underpinning thoughtful analytical or evaluative base to it.

In contrast, when students understand the different types of thinking and the cognitive heuristics involved, they can respond thoughtfully (no pun intended) to their teacher's systematic use of language to specifically cue these types of thinking. This provides essential modelling and practice to develop competence over time. This is illustrated in Table 3.6.

The use of appropriate curing questions, in which the types of thinking are naturally infused into the content of the topic, helps students to become familiar with the 'language of thinking'. For example, when asked to evaluate options, whatever the subject context, they will have already internalized that this requires the deriving of relevant criteria to be used in evaluation, the likely prioritizing of these criteria in terms of relative importance in making the decision, and finally to apply the criteria, based on the available information, to the option or range of options.

In teaching this strategy, I am often posed the scenario that this takes time and eats into the teaching of content. However, a medical friend of mine has frequently

Fig. 3.10 Mental models of thinking

Table 3.6 Examples of language to cue types of thinking

Often used teacher language	Cueing types of thinking
Comment on these two proposed solutions	Let's compare and contrast these two solutions…to identify what is similar and different for each…then we can apply our understanding of what may be most useful in some specific situations
What's your view on this?	What do these data sources suggest about the likely causes of this particular accident…what other information would your need to help make an even more accurate assessment?

Note Use of language needs to be contextualized to the learner group

said to folk something akin to, 'eating better and taking exercise is, in the longer term, less work and cheaper, than treating type 2 diabetes or a stroke'. Poignant, I know, but the analogy is there. Teaching the heuristics of good thinking, does not, in practice take that long, and it can be done incrementally, using spaced, retrieval and deliberate practice. Once students are familiar with the terms, it's then just a bit of good EBT pedagogic design to infuse this effectively into a lesson structure. For example, for a couple of years, I was given (though I did not ask for it) the role of AIDS educator/counsellor, back in the days when the condition, its causes and treatment were ambiguous. In this role, I had to do the necessary research to be as current as possible. In my teaching role in this area, I would typically infuse some specific thinking questions, tailored to the learning outcomes, as illustrated below:

- What are the similarities and differences between Hepatitis A and HIV?
- In what ways are these differences significant?
- What is the relationship between HIV infection and poverty?
- What inferences and interpretations can be drawn from the data on HIV infection in Asia?
- How might we evaluate the effectiveness of the present HIV prevention programme?
- What other ways might we make people more aware of HIV infection?

Self Talk and Verbalization

I recall revising for my final exams at university, having been told by one of the psychology professors that talking to yourself while going through notes helps the learning process. Initially, I thought this was somewhat strange. However, having used self-talk and verbalization for a few decades, as well as teaching my students to use it, I am assured to know that if there is a valid cognitive-neuroscience base to what I was doing. For example, Treadwell (2017) points out:

> Throughout our lives, we are conscious, non-consciously and subconsciously creating, refining and modifying our identity through our self-talk. The influence of self-talk is foundational in establishing who we are, how we learn, and what we expect of our-self. (p. 40)

It is an easy to use metacognitive strategy as it runs a quality check on what you know and   dont know, pulling information out of long-term memory into working memory and then doing retrieval practice with it. My many hours of walking around the kitchen talking to myself about different personality theories, their underlying premises, the similarities and differences, and making inferences and interpretations as to their applicability in different psychological cases, was time well spent—even if anyone looking in—listening in—thought I was just another 'strange', psychology student. Doing verbalization with a friend (a group of friends, can be even better) has similar and potentially greater benefits. Apart from the sharing of perspectives, and wider knowledge bases that become available in a collaborative learning context, it can relieve the boredom of study—even add some fun, and we know this helps. In doing my Post-Graduate Certificate in Education, I introduced such practices to my soccer team-mates. We employed these methods, with others (what I would now refer to as EBT methods—but had no idea of such terms then). The format was:

- Obtain and analyse collectively the past papers for the exam and look for the most salient question areas. Note this was done at a local eatery so more palatable (again no pun intended) than alone at home
- Delegate, through consensus (more or less), who was going to research each area and produce useful revision notes (to be done in pairs or triads)
- Agree on a timeline for doing this and the next meeting date
- In the meetings, pairs/triads would share their notes and explain any key points that they think will help others learn (e.g., key points of definition, similarity and differences between various perspectives on the topic areas)—there are always these in education
- Have a 'good old chat'—sorry, critical dialogue on the topics involved. This would be done over several days.

In doing such activities, you are doing what is referred to as Elaboration, which has benefits in terms of consolidating, extending and facilitating the transfer of learning. As Brown et al. (2014) explain:

> Elaboration is the process of giving new material meaning by expressing it in your own words and connecting it with what you already know. The more you can explain about the way your new learning relates to your prior knowledge, the stronger your grasp of the new learning will be, and the more connections you create that will help you remember it later. (p. 5)

It certainly made the revision process less tedious, the actual learning of content seemed quite easy, and we all did well in the exams. Of interest, many components of this broad process have been refined and formalized in 'Flipped Classroom Learning' and the now-famous method of 'Peer Instruction' developed by Mazur (1996). The typical format is as follows:

1. Students are posed a challenging (key concept question) based on prior learning
2. Ask students to provide their answers (e.g., capture through an e-tool, response cards, even show of hands) and display the results:

- If most answer correctly, the discussion can be brief (1–2 min)
- If most answer incorrectly, it may be necessary to backtrack and reteach the key ideas being addressed in the question or problem
- If students are split, ask students to find a peer with a different answer and discuss the basis/reason for their answers. Make explicit that they can change their mind about the answer and there will be a re-vote on the correct answer
- Provide students with an opportunity to re-answer the question and show results
- Follow the revote with a whole class discussion, explaining what makes the correct answer correct and what makes the others incorrect.

Testing, Testing and More Testing

It is now recognized that **Retrieval Practice**, which involves the retrieval of information from Long-Term Memory to Working-Memory and then back to Long-Term Memory, is a key cognitive and metacognitive strategy (e.g., Lang 2016). The process of reciprocal retrieval, apart from gap analysis, and clearing up misconceptions, strengthens the connections between neurons; hence building stronger neural networks for this area of learning. In doing so this creates increasingly accurate and useful mental models for the application of learning. As Brown et al. (2014) summarizing from the research concluded:

> In virtually all areas of learning, you build better mastery when you use testing as a tool to identify and bring up your areas of weakness. (p. 5)

Retrieval practice can be done anytime, in most places, and as one becomes more familiar and fluent with content (e.g., understanding is becoming richer) the process can be super-fast. As Willingham (2009) demonstrated, the more factual knowledge you have in long-term memory, the easier it is to acquire more factual knowledge. To quote:

> It means that the amount of information you retain depends on what you already have. So, if you have more than I do, you regain more than I do, which means you gain more than me. (p. 34)

The importance and workings of Deliberate and Spaced practice were outlined and illustrated in Chap. 2: *Core Principle 7: The development of expertise requires deliberate practice.*

Also, according to Lang (2016), interleaving helps to learn in the overall practice process. Interleaving refers to the practice of spending some time learning one thing and then pausing to concentrate on learning a second thing before having quite mastered that first thing, and then returning to the first thing, and then moving on to a third thing, and then returning to the second thing, and so forth. This utilizes the brains sub/unconscious processing in terms of the dynamic of memory systems. It also reduces 'habituation'—attention disengagement/loss of interest in an activity—factors that typically creeps in when staying with a single task for too long. Moving to a new task introduces some novelty and may energise conscious thinking. At the same time our brains are still indwelling (e.g., Moustakas 1990) sub/unconsciously on the previous task, but with a different neural focus in a more relaxed manner, as documented by Claxton (1998).

In writing, I use interleaving quite extensively. I rarely finish a chapter before moving on to the next one. In space/time between the first and making some inroad into the second, I typically come up with new ideas and areas of content for enhancing sections of the previous chapter—and so on.

Practice doesn't always make perfect. However, the thoughtful use of the types of practice outlined above certainly enhances learning capability and makes future learning easier with a 'bit of practice'—again, no pun intended. They can be infused into any instructional strategy, as well as used situationally when needed. These practice methods are certainly a well-constituted Russian Doll in themselves—in terms of Hattie's analogy.

Real World Performance Tasks

Performance-based learning tasks are perhaps the most authentic means for developing MC as key selected skills (e.g., metacognitive, cognitive, volitional) can be embedded in the task to be accompanied. These skills would have already been decided and infused into the content curriculum and should be familiar to both faculty and students. In this way what is to be learned, as well as what will be assessed is made explicit and transparent. As Tombari and Borich (1999) explain:

> A performance assessment is a test that tries to determine if a learner "really knows" about something, or has a deep understanding. It does this by challenging learners with tasks that ask them not simply to recall knowledge but to construct or organize it, not just to solve problems but to demonstrate a disciplined approach requiring strategic thinking and metacognition. (p. 148)

In specific terms, as Collins (2017) argues, this:

> …embodies authentic tasks and assessments, a dual focus on the teaching of particular competencies in the context of accomplishing meaningful tasks, peer teaching and mentoring, and a learning cycle of planning, doing and reflecting. (p. 16)

The goal is to develop schooling that will have a major impact on student motivation and learning, and will better prepare students for the complex world they are entering (p. 9). The extent that performance tasks reflect real work activity, the greater the opportunities to develop the selected twenty-first century competencies and to what level in what vocational context. It is for this reason that more and more vocational and higher education institutions are using internships as an essential part of the learning experience for their programmes. Internships can take many forms, in terms of format and duration, but the goal is to provide learners with direct authentic work experiences in a specific vocational sector where they can acquire workplace experience, key functional competencies and greater knowledge of that industry. Invariably, this relies heavily on the arrangements between the educational institution and the employing industry firm. It is to be noted that an internship is not a novel learning model. The notion of students having work experience has long been part of the curriculum landscape. However, this can be largely a 'hit-or-miss' experience, for both students and employers. Often students don't feel they learn much of value, apart from doing photocopying, making tea and generally running errands. Equally, some employers feel that these novices offer little value to the company effort, especially when they don't have a 'good attitude', and are more trouble than what they are worth. In contrast, of course, many students have felt that this was the most valuable and enjoyable part of the curriculum programme. Similarly, many employers have been delighted with some interns and have offered them employment post-graduation. Having been involved in this work for many years, this is my experience and, from numerous conversations with teaching faculty and students, it's pretty much the norm.

However, increasingly, the internship is being used in more pedagogically-structured ways. This involves certain important planning and conduct features:

- Close liaison between the educational institution and employing industrial personnel. It is important to identify, clarify and agree on what types of learning experiences the company can provide, and the relationship between these work functions and curriculum learning outcomes
- Close liaison between the educational institution supervisor and the interns. Interns must be aware of their role and responsibilities in this learning event. It is useful to clarify with interns what they are likely to have to do in the company and work role, and how this relates to other aspects of their course, as well as the specific learning outcomes that can be met during the internship period
- Ongoing communication and collaboration between all stakeholders to monitor and evaluate the learning experience and ensure the smooth running of the internship. This involves many aspects, including how interns are progressing in terms of key learning outcomes, work performance and attitude, and agreed assessment arrangements. It is also important to keep regular contact with workplace supervisors to ensure that interns are meeting expectations and taking any necessary corrective action if needed.

Apart from internships, a wide range of methods can be used to provide students with meaningful real-work learning activities, these include:

- Projects
- Problem-Based Learning
- Case Studies
- Simulated Practice
- Role-Play
- Presentations and Critique
- Experiments.

Essentially, any activity that reflects real work/life skills can be used. What is essential is that it relates to curriculum outcomes and employs some skill(s) and/or attribute(s) of desired twenty first-century competencies. For example, in the infusion model introduced earlier, metacognitive strategies should run throughout the activities.

In my experience, projects and problem-based learning activities are popular and viable methods for providing interesting, real-world/work-related tasks that can be used as vehicles for the development of skills subsumed in most twenty-first century competencies. If well-constructed, they provide:

- authentic learning tasks that mirror real world activities and make learning more meaningful
- an integrated learning activity that naturally combines subject knowledge, types of thinking and other process skills
- performance-based assessment opportunities that enable a more authentic assessment of actual competence than traditional pencil and paper tests
- a means of developing SDL by offering students more autonomy and choice in their learning
- a framework for a collaborative and experiential learning approach, whereby students are required to solve problems through information resourcing and types of thinking.

Note that the terms project-based learning and problem-based learning are often used to refer to the same method, and are similar in most features, though it is possible to have projects that are not focused on solving an ill-defined problem, which is the hallmark for problem-based learning 'purists'. There are other similar methods, such as 'scenario-based learning' and 'challenge-based learning' within this broad genre of *Inquiry-Based Learning*, which can be employed to add variety and focus on specific attributes. All can be effectively employed to develop metacognitive and cognitive strategies, communication and collaboration skills, and related attributes. Once the twenty-first century competencies are infused in the curriculum plan, it's then a question of good task design and facilitation, whatever the delivery context (e.g., school, workplace, online). Sounds easy, and it is—when you can do it. For projects, and similar method variants, I have evolved the following guiding heuristics:

Step 1: Identify clearly the knowledge, skills and attributes to be incorporated into the project task

For this step it is important to:

- Choose topic areas in your curriculum that encompasses key underpinning knowledge (e.g., central concepts, principles, procedures) and skills essential for understanding and performance in real-world applications.
- Identify the types of thinking (i.e., specific components of metacognition, critical and creative thinking) that are important for promoting understanding and competence in these topic areas. For example, in promoting critical thinking, analysis, comparison and contrast, inference and interpretation, and evaluation are typically infused.
- Identify other skills and attributes that are part of the wider competency area (e.g., team-working, self-regulation components) that are important for competent performance in the identified areas.

Table 3.7 Project components

Subject knowledge	Types of thinking	Other process skills
– Circuit design and integration principles – Circuit building – Use of sensors	– Generating possibilities relating to circuit design – Analysis—part-whole relationships of sensors in an integrated circuit – Compare and contrast—previous options and new options generated – Making inferences and interpretations from data relating to the behavior of sensors in an integrated circuit – Evaluation of interesting options in relation to derived criteria	– Oral and written communication – Teamwork

An example of framing the main knowledge and skill components for an electrical and electronic engineering project is presented in Table 3.7.

Step 2: Produce the learning task

It is important that the task:

- Involves the application of the knowledge, skills and attributes identified from Step 1
- Is sufficiently challenging, but realistically achievable in terms of student's prior competence, access to resources, and time frames allocated
- Successful completion involves more than one correct answer or more than one correct way of achieving the correct answer
- Clear notes of guidance are provided, which:

  - Identify the products of the task and what formats of presentation are acceptable (e.g. written report, learning materials, portfolio, oral presentation)
  - Specify the parameters of the activity (e.g. time, length, areas to incorporate, individual/collaborative, how much choice is permitted, the support provided)
  - Cue the types of thinking and other desired skills and attributes
  - Spell out all aspects of the assessment process and criteria.

Table 3.8 is an example illustrating some of the above design features. However, like a Lego set, more features for specific purposes can be added as the learning needs arise (e.g., additional skills and attributes, a higher level of expertise, team-based, level of guidance provided).

It is possible and desirable to design learning activities that incorporate the use of MC strategies and skills as an essential part of task completion. Also, motivational aspects can be enhanced through the provision of student choice (as much as possible) without compromising the learning outcome standards. The more that

Table 3.8 Draft learning task incorporating types of thinking

Design A food package
Select a food product and design the packaging that you think will give it best marketability. You must be able to identify the product attributes, protection and enhancement needed to satisfy the functional and marketing requirements, and use suitable packaging material(s) and package type. The work produced should reflect the quality of your thinking in the following areas:
• Identify the criteria for evaluating the marketability of a product
• Analyze the components of a product that constitute an effective design
• Generate new ways of viewing a product design beyond existing standard forms
• Predict potential clients response to the product given the information you have
• Monitor the development on the group's progress and revise strategy where necessary

learning (and assessment tasks) are perceived as meaningful and useful from the students' perspective, greater is the potential for intrinsic motivation to be evoked. In this situation, coupled with good teaching, there is the potential for creating synergy in the utilization of both metacognition and motivation. As an *Advance Organizer* for the next chapter, which addresses the meaty challenge of making learning more motivating for students, especially intrinsic motivation, here's a positive frame from Wlodkowski (1999) who suggests:

> …if something can be learned, it can be learned in a motivating manner…every instructional plan also needs to be a motivational plan. (p. 24)

3.10 Summary

In this chapter, I have attempted to get to grips with Metacognition; to repeat, with some linguistic rejigging, Brown's (1987) reference to it as a '*many-headed monster of obscure parentage*'. One purpose has been to provide the reader, without the pain of going through the extensive and mazy literature on this 'creature', and with as much clarity as possible on its nature, evolution and role in the world of education and learning. The major purpose though is to propose Metacognition, framed here as **Metacognitive Capability** as the *most important* twenty-first century competence. Increasingly included in definitions of so-called twenty-first century competencies, for the reasons explained and illustrated extensively in the chapter, it is still relatively under-developed in schools. There is an urgent need to address such a significant curriculum area. Hopefully, this chapter helps to better frame how to go about doing this from an evidence-based perspective.

MC is both an essential teaching competence as well as a generic competence for enhancing learning capability, well-being and better citizenship (however defined). The potential results of this are, as Treadwell (2017) summarizes:

> Once we learn how to learn, and we achieve that efficiently, we can learn independently and anything becomes possible. (p. 49)

References

Adler H (1996) NLP for managers. Judy Piatkus, London

Akturk AO, Sahin I (2011) Literature review on metacognition and its measurement. Procedia Soc Behav Sci 15:3731–3736

Alderman MK (2008) Motivation for achievement: possibilities for teaching and learning. Routledge, London

Ambrose SA et al (2010) How Learning works: 7 research-based principles for smart teaching. Jossey-Bass, San Francisco

Bandura A (1997) Self-efficacy: the exercise of control. Worth Publishers, New York

Bandura A (2004) Swimming against the mainstream: the early years from chilly tributary to transformative mainstream. Behav Res Ther 42:613–630

Baumeister RF, Tierney J (2012) Willpower. Penguin, London

Biggs J (1988) The role of metacognition in enhancing learning. Aust J Educ 32(2):127–138

Borkowski JG, Chan LK, Muthukrishna N (2000) A process-oriented model of metacognition: links between motivation and executive functioning. Issues in the measurement of metacognition. University of Nebraska Press, Lincoln, NB, pp 1–41

Bransford JD, Brown AL, Cocking RR (1999) How people learn: brain, mind, experience, and school. Committee on Developments in the Science of Learning, Commission on Behavioral and Social Sciences and Education. National research Council

Brown AL (1987) Metacognition, executive control, self-regulation, and other more mysterious mechanisms. In: Weinert FE, Kluwe RH (eds) Metacognition, motivation, and understanding. Lawrence Erlbaum Associates, Hillsdale, New Jersey, pp 65–116

Brown PC, Roediger HL IIII, McDaniel MA (2014) Make it stick: the science of successful learning. Harvard University Press, MA

Chickering AW, Gamson ZF (1987) Seven principles for good practice in undergraduate education. Am Assoc High Educ Bull 3–7

Claxton G (1998) Hare brain tortoise mind: why intelligence increases when you think less. Fourth Estate Limited, London

Coles MJ, Robinson WD (1989) Teaching thinking: a survey of programmes in education. The Bristol Press, Bristol

Collins A (2017) What's worth teaching: rethinking curriculum in the age of technology teachers. College Press, New York

Costa A (2019) https://artcostacentre.com/html/habits.htm. Last accessed 20th Oct 2019

Davidson RJ (2000) Affective style, psychopathology, and resilience: brain mechanisms and plasticity. Am Psychol 55:1196–1214

Dignath C, Buettner G, Langfeldt H (2008) How can primary school students learn self-regulated learning strategies most effectively? A meta-analysis on self-regulation training programs. Educ Res Rev 3:101–129

Duckworth AL et al (2007) Grit: perseverance and passion for long-term goals. J Pers 92(6):1087

Dweck CS (2006) Mindset: the new psychology of success. Ballantine, New York

Dweck CS, Legget EL (1989) A social-cognitive approach to motivation and personality. Am Psychol Assoc 95(2):256–273

Dweck CS, Master A (2012) Self-theories motivate self-regulated learning. In: Schunk DH, Zimmerman BJ (eds) Motivation and self-regulated learning: theory, research and applications. Routledge, New York

Eccles JS, Wigfield A (2002) Motivational beliefs, values, and goals. Annu Rev Psychol 53:109–132

Epstein D (2014) The sports gene: inside the science of extraordinary athletic performance. Penguin, London, UK

Ericsson KA, Krampe RT, Tesch-Romer C (1993) The role of deliberate practice in the acquisition of expert performance. Psychol Rev 100:363–406

Fadel C, Trilling B (2015) Meta-learning for the 21st century: what should students learn? Center for Curriculum Redesign, Boston, MA

Farrington CA et al (2012) Teaching adolescents to become learners: the role of noncognitive factors in shaping school performance: a critical literature review. The University of Chicago Consortium on Chicago School Research (CCSR), Chicago

Fazey MA, Fazey JA (2001) The potential for autonomy in learning: perceptions of competence, motivation and locus of control in first-year undergraduate students. Taylor & Francis, London, UK

Fennimore TF, Tinzmann MB (1990) What is a thinking curriculum? Web Page: https://www.ncrel.org/sdrs/areas/rpl_esys/thinking. Last accessed 22nd Nov 2019

Flavell JH (1976) Metacognitive aspects of problem solving. In: Resnick LB (ed) The nature of intelligence. Erlbaum, Hillsdale, NJ

Flavell JH (1979) Metacognition and cognitive monitoring: a new era of cognitive developmental inquiry. Am Psychol 34(10):906–911

Flavell JH (1981) Cognitive monitoring. In: Dickson WP (ed) Children's oral communication. Academic Press, New York, pp 35–60

Flavell JH (1987) Speculation about the nature and development of metacognition. In: Wernert FE, Kluwe RH (eds) Metacognition, motivation and understanding. Lawrence Erlbaum, Hillsdale, NJ

Fogarty R (2009) How to integrate the curricula. Sage, London

Fromm E (1987) The anatomy of human destructiveness [1973]. Penguin Books, Harmondsworth, UK

Gladwell M (2008) Outliers: the story of success. Little Brown and Company, New York

Goleman D (2013) Focus: the hidden driver of excellence. HarperCollins, New York

Guglielmino LM (1978) Development of the self-directed learning readiness scale. Doctoral dissertation, University of Georgia, 1977. Dissertation, Abstracts International, 38, 6467A

Haidar AH, Al Naqabi AK (2008) Emiratii high school students' understanding of stoichiometry and the influence of metacognition on their understanding. Res Sci Technol Educ 26(2):215–237

Hattie J (2009) Visible learning. Routledge, New York

Hattie J, Yates GCR (2014) Visible learning and the science of how we learn. Routledge, New York

Hidi S (2006) Interest: a unique motivational variable. Educ Res Rev 1(2)

Hidi S, Renninger KA (2006) The four-phase model of interest development. Educ Psychol 41:111–127. https://doi.org/10.1207/s15326985ep4102_4

Jones MG, Farq JD, Surry D (1995) Using metacognitive theories to design user interfaces for computer-based learning. Educ Technol 35:12–22

Kahneman D (2012) Thinking fast and slow. Penguin Books, London

Kistner S et al (2010) Promotion of self-regulated learning in classrooms: investigating frequency, quality, and consequences for student performance. Metacogn Learn 5(2):157–171

Kluwe RH (1987) Executive decisions and regulation of problem solving behavior. In: Weinert FE, Kluwe RH (eds) Metacognition, motivation, and understanding. Lawrence Erlbaum Associates, Inc., New Jersey

Knowles MS (1975) Self-directed learning: a guide to learners and teachers. Association Press, New York

Ku KYL, Ho IT (2010) Metacognitive strategies that enhance critical thinking. Metacogn Learn 2010(5):251–267

Lai ER (2012) Metacognition: a literature review. https://images.pearsonassessments.com/images/tmrs/Metacognition_Literature_Review_Final.pdf. Last accessed, 15 Nov 2019

Lang JM (2016) Everyday lessons from the science of learning. Jossey-Bass, San Francisco, CA

Mango C (2010) The role of metacognitive skills in developing critical thinking. Metacogn Learn 2010(5):137–156. https://doi.org/10.1007/s11409-010-9054-4

Marcus G (2008) Kluge: the haphazard evolution of the human mind. Mariner Books, New York

Martinez ME (2006) What is metacognition? Phi Delta Kappan 87(9): 696–699. https://doi.org/10.1177/003172170608700916

Marzano RJ et al (1988) Dimensions of thinking: a framework for curriculum and instruction. ASCD, Alexandria, VA

Marzano RJ et al (2007) Designing and teaching learning goals and objectives: classroom strategies that work. Marzano Research Laboratory, Colorado

Maudsley DB (1979) A theory of meta-learning and principles of facilitation: an organismic perspective. Thesis (E.D.), University of Toronto

Mazur E (1996) Peer instruction: a user's manual. Prentice Hall, NJ, USA

Mlodinow L (2012) Subliminal: how your unconscious mind rules your behaviour. Vintage Books, New York

Moustakas C (1990) Heuristic research. Sage, London

Neubauer S et al (2018) The evolution of modern human brain shape. Sci Adv 4(1)

Noushad PP (2008) Cognitions about cognitions: the theory of metacognition

Ormrod JE (2011) Our minds, our memories: enhancing thinking and learning at all ages. Pearson, London

Pajares F (2012) Motivational role of self-efficacy beliefs in self-regulated learning. In: Schunk DH, Zimmerman BJ (eds) Motivation and self-regulated learning: theories, research, and application. Lawrence Erlbaum Associates, NJ

Panksepp J (2004) Affective neuroscience: the foundations of human and animal emotions. Oxford University Press, Oxford

Paris SG, Winograd P (1990) Promoting metacognition and motivation of exceptional children. Remedial Spec Educ 11(6):7–15

Paul RW (1993) Critical thinking: foundation for critical thinking. Santa Rosa, CA

Perkins DN (1995) Outsmarting IQ: the emerging science of learnable intelligence. The Free Press, London, UK

Pinker S (2003) The blank slate: the modern denial of human nature. Penguin, London

Pintrich PR, De Groot EV (1990) Motivational and self-regulated components of classroom academic performance. J Educ Psychol 82:33–40

Reeve J, Tseng C (2001) Agency as a fourth aspect of students' engagement during learning activities. Contemp Educ Psychol 36(2011):257–267

Resnick LB, Resnick DP (1989) Assessing the thinking curriculum: new tools for educational reform. In: Gifford BR, O'Connor MC (eds) Future assessments: changing views of aptitude, achievement, and instruction. Kluwer Academic Publishers, Boston

Ritchhart R, Church M, Morrison K (2011) Making thinking visible. Jossey-Bass, San Francisco

Robbins A (2001) Unlimited power. Pocket Books, London

Rohn J (2019) Quote available at: https://www.quoteswise.com/jim-rohn-quotes-5.html. Last accessed on 3rd Dec 2019

Ryan RM, Deci EL (2017) Self determination theory: basic needs in motivation, development, and wellness. The Guilford Press, New York

Saks K, Leijen A (2013) Distinguishing self-directed and self-regulated learning and measuring them in the e-learning context. In: International conference on education & educational psychology

Sale D (2013) The challenge of reframing engineering education. Springer, New York

Schank R (2011) Teaching minds: how cognitive science can save our schools. Teacher College Press, New York

Schraw G (1998) Promoting general metacognitive awareness. Instr Sci 26:113–125 (Kluwer Academic Publishers, The Netherlands)

Schraw G, Dennison RP (2004) Assessing metacognitive awareness. Contemp Educ Psychol 19:460–475

Schraw G et al (2006) Promoting self-regulation in science education: metacognition as part of a broader perspective on learning. Res Sci Educ 36:111–139

Schulman LS (2005) Signature pedagogies in the professions. Daedalus 134(3):52–59

Schunk D, Zimmerman BJ (2012) Motivation and self-regulated learning: theory, research, and applications. Routledge, New York

Schwarz N (2010) Meaning in context: metacognitive experiences. In: Mesquita B, Barrett LF, Smith ER (eds) The mind in context. Guilford, New York, pp 105–125

Sheppard SD et al (2009) Educating engineers: designing for the future of the field. Jossey-Bass, San Francisco

Stenhouse L (1989) An introduction to curriculum research and development. Heinemann Educational Books, Oxford

Stockard J et al (2018) The effectiveness of direct instruction curricula: a meta-analysis of a half century of research. Rev Educ Res 88(4):479–507

Swartz RJ (1987) Teaching for thinking: a developmental model for the infusion of thinking skills into mainstream instruction. In: Baron JB, Sternberg RJ (eds) Teaching thinking skills: theory and practice. Freeman, New York

Swartz RJ, Perkins DN (1990) Teaching thinking: issues approaches. Critical Thinking Press & Software, CA

Tombari M, Borich G (1999) Authentic assessment in the classroom. Prentice-Hall, New Jersey

Tomlinson CA (2005) The differentiated classroom: responding to the needs of all learners, 2nd edn. ASCD, Alexandria, Virginia

Tough P (2013) How children succeed: grit, curiosity, and the hidden power of character. Random House Books, London, UK

Treadwell M (2017) The future of learning. The Global Curriculum Project, Mount Maunganui, NZ

Veenman VJ et al (2006) Metacognition and learning: conceptual and methodological considerations. Metacogn Learn 2006(1):3–14

Weiner B (1992) Human motivation: metaphors, theories and research. Sage, Newbury Park, CA

Whitehead AN (1967) The aims of education. Free Press, New York

Willingham DT (2009) Why don't students like school: a cognitive scientist answers questions about how the mind works and what it means for the classroom. Jossey-Bass, San Francisco

Winne PH (1995) Inherent details in self-regulated learning. Educ Psychol 30(4):173–187

Wlodkowski RJ (1999) Enhancing adult motivation to learn: a comprehensive guide for teaching all adults. Jossey-Bass, San Francisco

Zimmerman BJ, Kitsantas A (1999) Acquiring writing revision skill: shifting from process to outcome self-regulatory goals. J Educ Psychol 91(2):241–250. psycnet.apa.org

Zimmerman BJ, Martinez-Pons M (2012) Student differences in self-regulated learning: relating grade, sex, and giftedness to self-efficacy and strategy use. J Educ Psychol 82(1):51–59

Zimmerman BJ et al (1992) Self-motivation for academic attainment: the role of self-efficacy beliefs and personal goal setting. Am Educ Res J 29(3):663–676

Chapter 4
Motivation and Well-Being: An Evidence-Based Frame

Abstract This chapter addresses the challenge of enhancing student's intrinsic motivation in school-based learning environments and, in essence, motivation generically. Motivation underpins learning in that without motivation, people don't bother to learn (or think) too much. It connects extensively to the pedagogic framework developed in Chap. 2, as well as the superordinate twenty-first century competence of Metacognitive Capability explored and illustrated in Chap. 3. Metacognition and Motivation are extensively and dynamically interlinked. Without motivation people are unlikely to do much by way of metacognition. However, without metacognition and the related thinking and self-regulatory skills, people may not fare well in the learning stakes as tasks become more complicated—and that's an existential problem we face today. Intrinsic motivation is explored in detail, especially the Self-Determination Theory (SDT) of Ryan and Deci, as it offers much potential for enhancing learning and well-being in educational contexts.

4.1 Introduction

From a social science perspective, motivation is a concept used to explain why humans behave as they do, what factors determine or shape this activity, both from internal sources (e.g., neural processing in the brain, unconscious wants) and external socialization factors (e.g., life experiences, beliefs). It is a complex and dynamic process, as captured by the following definitions:

> *Motivation* refers to the process whereby goal-directed activities are energized, directed, and sustained. Contemporary cognitive theories of motivation postulate that individuals' thoughts, beliefs, and emotions are central processes that underlie motivation.
>
> (Schunk and Usher 2012, p. 13)
>
> Motivation is the force or energy that results in engagement. In a classroom, the complex interaction of teacher, student, and curriculum helps to create motivation that yields high engagement.
>
> (Gregory and Kaufeldt 2015, pp. 9–10)

© Springer Nature Singapore Pte Ltd. 2020
D. Sale, *Creative Teachers*, Cognitive Science and Technology,
https://doi.org/10.1007/978-981-15-3469-0_4

Certainly, motivation initiates, directs and maintains all human behaviour. It is inseparable from learning in that without some motivational base, limited attention and effort will be given to that area of human activity. However, attention, as we experience it, is a complex and variable process, as it is not something that we can simply switch on and off like a light bulb. While we can consciously choose to give attention to a specific phenomenon in our world at any given time, in reality (unless it is so striking in some way—good or bad) we typically drift away from what we were paying attention to and start attending to something else. This may be prompted by a change in the external environment, as captured in the saying, "that caught my attention". Mostly, we don't need a change in the external world, as things will happen internally which will change the focus and degree of attention in significant ways.

What this means is that while we may make conscious plans about what we want to achieve and formulate strategies to meet such goals, in everyday life we typically flit from one mind-state to another and, over the course of even a day, may go through a whole range of emotions from feeling confident to angry to tired to anxious—all depending on what's going on in both the external world, as well as our inner mind dynamics. This is what is typically referred to as our psychological state (e.g., Robbins 2001) which is how we think, feel and perceive at any given moment. The ability to manage one's psychological state is a key capability for achieving peak performance—an integral part of MC. Our psychological state is largely determined by our internal representations (e.g., our learned maps of the world, beliefs, and how we have come to organize experiences) and physiology (e.g., present biochemistry—especially neural activity in the brain).

These work together to filter the way we process things occurring in our here-and-now environment and our response/approach to such situational events. For example, when we use terms such as, "my head is in a real mess at the moment", we are referring to a feeling of a bad state of mind. In such states, rational thinking often gives way to anxiety—even fear—and typically leads to poor performance in whatever activity we are involved in doing. This is noticeable among sportspeople in critical moments in game situations. Even the very best can get into a poor state, and this is often reflected in almost novice-like mistakes. However, the best tend to do this less frequently and/or have the mental resources to quickly get out of a bad state, than those lacking such skills/attributes. Of course, you need the potential performance capability in the first place to do well in an activity, but how well you perform in a given situation (e.g., at your very best, poorly, or somewhere in between) is often determined by your state of mind. Such 'mental interference' affecting our state of mind has been famously documented by Gallwey (1987), firstly about playing tennis; then in many other performance contexts. In our everyday lives, the same chaotic processes are going on, continually impacting our motivation and performance.

In most everyday contexts (and classrooms are everyday contexts for students) the notion of consistently maintaining high levels of attention may be unrealistically optimistic—the competition for it is too great, both from outside and within.

Often, as teachers, what we get is varying amounts (and levels) of attention from students. Unfortunately, without sustained focused attention on a topic being explained, students may remember very little. Attention is, to all intent and purpose, both a quantitative and qualitative human capability and from a pedagogic point of view—a big challenge. As Miller (2016) States:

> Optimizing your learning experiences for attention is the first step towards optimizing it for long-term memory and higher thought processes. (p. 8)

Motivation and attention are very much connected in the world of the classroom, as in all areas of human activity. When students are motivated, they are much more likely to give a higher level of attention and engagement than in situations where motivation is poor. They are also more likely to put effort into the learning process, especially when difficulties are encountered. In such scenarios, especially if supported by a good teacher, students will typically experience better learning outcomes which, in turn, provides the basis for further motivation, as well as enhancing self-efficacy. Over time increased mastery is likely to be achieved, which further reinforces positive beliefs, confidence and motivation in such activities or area of learning.

However, while motivation is recognized as fundamental to learning, there is much debate about how it works and, more significantly, how we as teachers can harness such human energy in the pursuit of educationally desired learning goals. The literature is rich in terms of theories and models of human motivation (e.g., Maslow 1962; Herzberg 1966; Ryan and Deci 2017; Dweck 2006) but, as Kim (2013) argues:

> Existing theories on motivation bear three limitations. First is the vagueness of the concept of motivation. It is practically impossible to draw a clear line between motivation and other concepts such as drive, need, intention, desire, goal, value, and volition. Due to this conceptual vagueness, it is difficult to come to a consensus on whether motivation refers to a psychological state or process, let alone the definition. Various constructs in different theories of motivation are overlapping and often create confusion. For instance, the vague conceptual distinctions between intrinsic motivation and interest, self-efficacy and perceived competence, value and reward, self-regulation and volition hinder effective communication and constructive arguments on the identical phenomenon of motivation.

I would concur on the above and have some empathy with the frame of the management guru Peter Drucker, who made the challenging assertion that:

> We know nothing about motivation. All we can do is write books about it.

Indeed, this may seem to have a fair measure of face-validity at least in terms of widespread practice in educational institutions, as Levin (2008) concluded:

> ...boredom and lack of engagement remain endemic in schools around the world, and seemingly unmotivated students are a main complaint of teachers. (p. 99)

Certainly, whatever the underpinning bases of human motivation entail, especially in the context of the school environment, there seems to be a real problem which has not been sufficiently addressed to date. For example, Wagner (2010) made the point that:

In countless focus groups, I've conducted with high school students, "boring classes"- which include so-called advanced classes – are among the main complaints about the school. (p. 114)

It is not surprising, therefore, as Gregory and Kaufeldt (2015) noted:

Motivation, enthusiasm, perseverance, drive, grit, and tenacity are currently hot topics in education. Understanding how to get students to pay attention and engage in rigorous tasks is something every teacher desires. (p. 1)

4.2   The Components of Motivation

On many occasions, I have heard teachers being told by various sources that they should 'ignite the passion for learning in every child'—or something akin to this. The assumption is that by activating passion (e.g., a strong emotional desire for something—in this case learning) it will provide the sustained attention and motivation to do the necessary hard work that learning often entails. A nice ideal and it should be a goal we seek to attain. However, it's a bit like saying doctors should be able to cure all diseases and sickness, and this may be a goal that many seek. I would particularly like that, especially if they can reverse the ageing process also. However, it's not the world as I know it. The evidence would also support this, as people are still getting sick and dying, and I am not getting any younger. Referring to motivating students— is it possible to ignite a passion for learning in all students? Well, I'm going to play my 'get out of jail' card (this is a card used in the game of Monopoly to enable your moving counter icon to immediately get out of jail when it, unfortunately, through the throw of the die, lands on a space that denotes 'Go to jail'). I don't know the answer to the question, and I'm not sure there is one. It is like asking the question of whether people are born *good* or *neutral* in terms of dispositions, or are some simply badly wired to be difficult or even dangerous?

However, while we might like algorithmic answers to our big questions, whether it's how best to motivate our students or other areas of life that are meaningful to us, in reality we may have to settle for a well-framed evidence-based set of useful heuristics; otherwise we may simply go with personal preference or fad, albeit more philosophical than empirical. In most basic terms, from my experience, I would not dispute the English philosopher Jeremy Bentham's framing of human motivation in terms of:

Nature has placed mankind under the governance of two sovereign masters, pain and pleasure.

Invariably, what is a pleasure to one person may be a pain for another, but little in my life has seriously questioned these underlying premises. Indeed, such a perspective, with the additional component of 'novelty' has been supported by the field of cognitive neuroscience (e.g., Cloninger 1997). The avoidance of pain is inherent in the notion of punishment, whether for the infraction of laws, regulations or mores. Punishment is based on the fact that folk won't do certain acts of deviance for fear of the painful consequences of such acts. For example, long prison sentences are framed both in terms of being retributive and justice providing, as well as a deterrent to others who might otherwise transgress. When I was at school, the cane was used for specific misdemeanours such as fighting, bullying, being disrespectful—whatever the school establishment framed as meriting such treatment. There was also some notion of severity; the worse the infraction, the more strokes of the cane administered. Did it work?—probably for most of us. However, I remember one of my classmates who seemed intent on being caned by all teachers in the school, probably as a mark of masculinity. The point is that most students, whether they would admit it or not, don't enjoy failing in their learning. Those who want to learn the schoolwork are likely to experience pain in failure, and this can come in many forms (e.g., loss of self-esteem and self-efficacy, punishment by parents).

Let's move on to the pleasure aspect of motivation. People like doing things that lead to pleasurable feelings—albeit variation in the things deemed pleasurable. We can, therefore, assume that people are more likely to be interested in things that provide pleasurable feelings for them; hence interest drives, in no small part, motivation. We actively seek to do stuff we like. It's as simple as that. As explored in the previous chapter interest is fundamental to motivation and has biological roots.

Futhermore, interest is very much bound up with the experience of a reward. For example, I have a passion for soccer—which means I am interested in playing it, watching it—even talking about it. I also often experience a range of emotions when playing and watching soccer, which can change from negative to positive emotions—especially the latter when the game goes well (e.g., the team wins, especially when the quality of the soccer is good); it's psychologically rewarding. Rewards are, according to Kim (2013), one of the most powerful variables influencing motivation, irrespective of reward type (physical or social reward; extrinsic or intrinsic). The main function of a reward is to induce positive emotions, which then releases the neurotransmitter dopamine in the brain—this creates a great emotional feeling of pleasure. Emotions, whether positive or negative, are highly impactful on the learning process, as Wlodkowski and Ginsberg (2017) summarized:

> Research in the neurosciences and the field of intrinsic motivation indicates that emotions are critical to learning (Reeve and Lee 2012). Not only do emotions largely determine what we pay attention to and help us to be aware of our mind-body states also affect what we remember. We are much more likely to remember things that engage us emotionally. (p. 18)

Kim (2013), from a neuroscience perspective, argues that a more evidence-based approach to motivation should focus on:

> ...pleasure, value, and goal as principal units of analysis on the psychological level because their underlying neural mechanisms have been heavily investigated and relatively well-identified.

Value and goal are concerned with the importance of the activity. For example, if a highly valued goal is being sought, there is more likelihood of sustained motivation being activated and maintained. Invariably, of course, other factors come into play, but that's what makes predicting human behaviour heuristic at best; never fully algorithmic.

A more neuroscientific model of motivational processes suggests certain very practical educational implications to enhance motivation to learn. As reward is an essential driving force in the learning environment, there is a need to ascertain what aspects of the learning arrangements can be made more rewarding and interesting for the students we teach. Rewards can be both extrinsic and intrinsic, though there is more long-term merit in intrinsic motivation. I have noticed that chocolate is quite an effective extrinsic situational motivator—it's a reward right! (most students like chocolate and its less messy in class than ice cream). However, it's an expensive motivator from the teachers' point of view, and it's not creating much by way of a passion for learning.

Intrinsic reward in the classroom can be any stimulus which has a positive expected value, including quality feedback on how students are learning, genuine praise for effort, interesting and meaningful learning activities, humour and fun activity, social support, and relatedness. Much of this is known, but not always utilized by some teachers (and this may not be their fault). It makes sense, therefore, to find out and incorporate those things that students see as rewarding from their perspective, providing they are viably implementable in the educational context. This is likely to activate the reward circuits of the brain—creating that pleasurable feeling through the release of dopamine. Note, there is little worry of students becoming badly affected by addiction to learning.

However, even pleasure itself is not a constant entity, but a variable and changeable one at the level of subjective experience. For example, the repetition of the same stimulus tends to reduce the positive rewarding impact of that stimulus. In neuroscience terms, there will be less dopamine released and the feelings of pleasure less. (e.g., there's only so much chocolate students will eat before the pleasure largely abates). Furthermore, the psychological process of habituation comes into play, or in lay terms 'boredom has set in'. Even the chocolate loses some impact when continually presented in highly predictable circumstances. Such processes are aptly captured in the economists' concept of *Marginal Utility* (the law of marginal utility states that the first x is worth more than the second x—be it dollars, hours of free time, pieces of food, etc.) and, in the worst scenario, starkly captured in the folk saying, "familiarity breeds contempt". The message is clear, though a challenging one: It is desirable to introduce various reward contingencies unexpectedly to sustain motivation. This calibrates very much to *Core Principle 5: Instructional methods and presentation mediums* engage the range of human senses and the ability to create *Von Restorff* effects in the classroom—remember this? (i.e., the tendency of the brain to respond activately to novel stimuli).

Such considerations raise questions on the extent to which the experience of school learning is interesting for students, as experienced by them. Of course, there will be some that enjoy all of it (okay, most of it) and others that probably loathe all of it. In between, and this is perhaps where most students fall, and where the high-level pedagogic battles need to be fought (metaphorically speaking) in the learning stakes. In most basic terms, can we make the learning experiences more pleasurable (at least in some significant part) that lead to better engagement in the learning process? In Chap. 1, the purpose was to make a clear case for evidence-based practice in teaching. The evidence is clear that what teachers do and how they do it significantly impacts student's attainment. Does the same logic and heuristics apply to enhancing students' intrinsic motivation? According to Ryan and Deci (2017):

> School curricula or materials are often not packaged to be intrinsically motivating, nor in any way made to be particularly meaningful or relevant to the students' daily lives or purposes. In addition, under various top-down policy pressures, many modern schools have become extremely focused on a *very* narrow set of cognitive goals, often to the neglect of the varied interests, talents and more holistic psychological and intellectual needs of students. (pp. 352–353)

In educational contexts, few would disagree that students who perceive classroom learning as painful and boring are unlikely to contribute much, except to absenteeism rates and disruptive behaviour. The converse is also true. When students experience the learning as personally interesting, or place value on the outcomes to be obtained from successful completion of a programme (e.g., qualification/certification), they are more likely to participate meaningfully in the learning activities. For many adult learners, there are clear goals associated with their learning. These may have both extrinsic and intrinsic motivational components. Extrinsic motivation typically refers to the motivation coming from external factors to the activity (e.g., money, status or power, rather than the specific work activity). In contrast, intrinsic motivation is where motivation is derived from doing the task itself (e.g., passion for teaching). For example, having conducted more than 100 teacher education programmes, it was apparent that many participants had joined the programmes largely for purposes of accreditation (e.g., no certificate, no job). However, even for such extrinsically motivated persons many did, over the duration of the programme, find intrinsic interest which resulted in added value to their overall learning experience.

Where there are strong extrinsic motivators, it is always likely that adult learners, even with little intrinsic interest will try to maintain a level of attention and the necessary volition to achieve success on the programme (typically certification). Even for non-adult learners, grades and passing the examinations are strong extrinsic motivating anchors. However, for many school students, there may be limited extrinsic motivators (e.g., passing the exams does not get them a desired job) as well as little or no intrinsic interest in the content of the school subjects. This makes teaching such students highly challenging and potentially frustrating. I need say no more on this.

4.3 Evidence-Based Heuristics on Human Motivation and Well-Being

The seeking of pleasure, the avoidance of pain, and a desire or need for variation and novelty are existential aspects of motivation, in that they drive people to do things—often referred to as needs. Indeed, we need to eat, and when feeling hungry, we have the drive to find food. It may also be pleasurable and novel—even painful. I have many favourite foods, but will usually try something new when in different countries—I won't bore you with examples—but I have found some pleasurable, others not pleasurable (painful would be a bit of an exaggeration) but certainly novel. I retain the motivation to explore eating new food items in different countries and contexts.

As noted, prior, there are many models and theories of motivation. One that has been probably the most prevalent for the past three decades has been that of Abraham Maslow's Hierarchy of Needs, summarized in Fig. 4.1.

However, despite its appeal in terms of providing a clear and appealing structure for framing human motivation in terms of progressive need enhancement towards self-actualization, it has been criticized on a number of counts; notably its absence to include social cohesion and collaboration, which are fundamental to human survival, as well as the notion of hierarchy (e.g., Rutledge 2012). Most significantly, as Ryan and Deci (2017) summarize:

> ...despite the appearance of Maslow's (1943) pyramid-shaped hierarchy in almost every introductory psychology textbook and its intuitive appeal, empirical evidence for his hierarchy of psychological needs is quite thin. (p. 93)

In contrast, the Self-Determination Theory (SDT) of Ryan and Deci (2017) is perhaps becoming the most validated current motivational theory that offers evidence-based heuristics for framing motivational strategies that enhance learning in educational contexts. As the authors state:

> SDT is not a relativistic framework; it hits bedrock in its conception of certain universals in the social and cultural nutrients required to support healthy psychological and behavioural functioning. (p. 4)

The theory posits only three basic and universal psychological needs: competence, relatedness and autonomy. As Ryan and Deci (2017) explain:

> ...a basic need is essential for growth, wellness and integrity. Accordingly, optimal development, supported by basic need satisfaction, will be manifested in the motivational process of (1) intrinsic motivation, a fundamental psychological growth process; (2) the internalization and integration of behavioral regulations and social prescriptions and values, which results in psychological coherence and integrity and (3) an experience of vitality and wellness. (p. 98)

Competence refers to feeling effective and impactful in one's actions. As Ryan and Deci (2017) state:

Fig. 4.1 Maslow's hierarchy of needs

As a psychological need, competence is not only functionally important but is also experi-entially significant to the self. Phenomenally, feelings of effectance nourish people's selves, whereas feelings of effectance threaten their feelings of agency and undermine their ability to mobilize and organize action. (p. 95)

Relatedness refers to the feeling of belonging and being significant in the eyes of others. Ryan and Deci (2017) define it in terms of a need to:

...feel responded to, respected, and important to others, and, conversely, to avoid rejection, insignificance and disconnectedness. (p. 96)

Autonomy refers to the need to experience personal control and self-endorsement of one's behaviour. In the context of educational settings, Reeve and Lee (2012) suggests that:

Students experience autonomy need satisfaction to the extent to which their classroom activity affords them opportunities to engage in learning activities with an internal locus of causality, sense of psychological freedom, and perceived choice over their actions. (pp. 153–154)

SDT is a psychological theory that focuses on people's subjective experience as the key factor that determines their behaviour. It is how people interpret external and internal stimulus inputs, and the meaning they attribute to them in terms of need fulfilment, that is key to predicting behaviour and its consequences. The relevance of SDT to learning and well-being is captured by Ryan and Deci (2017):

...aspects of a social context that are likely to support satisfaction of the fundamental psy-chological needs are predicted to promote effective functioning and integrated development, whereas features of a social context that are likely to thwart need satisfaction are predicted to diminish effective functioning and to support non-optimal developmental trajectories.

...classroom climates supporting autonomy, providing high structure, and conveying relat-edness and inclusion foster personal well-being and feelings of connection to one's school and community. (p. 12)

While what constitutes well-being is open to contestation, the framing of the likely preferences of people by Pinker (2003) makes perfect sense to me:

Most people agree that life is better than death. Health is better than sickness. Sustenance is better than hunger. Abundance is better than poverty. Peace is better than war. Safety is better than danger. Freedom is better than tyranny. Equal rights are better than bigotry and discrimination. Literacy is better than illiteracy. Knowledge is better than ignorance. Intelligence is better than dull wittedness. Happiness is better than misery. Opportunities to enjoy family and friends are better than drudgery and monotony. (p. 51)

Such a viewpoint can be supported in many similar veins and contexts. For exam-ple, Harari (2016) argues that "the most real thing in the world is suffering" (p. 307). This may sound somewhat extreme—even shocking—but I don't think many of the inmates of Auschwitz would take issue with such a claim. And, we don't need to look for archetypal examples, suffering is embedded in most peoples' lives at some time, in various ways. The point is not to be morbid here, but if suffering has a realism beyond the arguments of social relativists, then there is value in considering *Well-Being* as a necessary consideration in framing twenty-first century competencies.

In the moral domain, there is much that reason and science can contribute to our understanding of well-being. Pinker (2019) argues, with a few unexceptional convictions, that:

> ...all of us value our welfare, and that we are social beings who impinge on each other and can negotiate codes of conduct – the scientific facts militate toward a defensible morality, namely principles that maximize the flourishing of humans and other sentient beings. (p. 395)

Similarly, Harris (2010) argues the case for looking at science for ways to better understand human values and what aspects of the moral domain may enhance human action, relationships and well-being. He takes as a premise that:

> Human well-being entirely depends on events in the world and states of the human brain. Consequently, there must be scientific truths to be known about it. A more detailed understanding of these truths will force us to draw clear distinctions between different ways of living in a society with one another, judging some to be better or worse, more or less true to the facts, and more or less ethical. Such insights could help us to improve the quality of human life – and this is where the academic debate ends and choices affecting millions of lives begin. (p. 3)

Conceptions of well-being can vary, and there is subjectivity here, but extensive research on the impact of a whole host of physical, social and emotional experiences have massive implications for brain development, physical and mental well-being. For example, Swaab (2015) summarizing the evidence, highlights:

> Children who are seriously neglected during their early development ...have smaller brains; their intelligence and linguistic and fine motor control are permanently impaired, and they are impulsive and hyperactive. (p. 27)

Certainly, we would not argue the case that such experiences and outcomes contribute to the well-being of these children. As Harris (2010) points out:

> (1) Some people have better lives than others, and (2) these differences relate, in some lawful and not arbitrary way to states of the human brain and states of the world. (p. 15)

He also goes on to argue that:

> Kindness, compassion, fairness and other classically "good" traits will be vindicated neurologically – which is to say that we will only discover further reasons to believe that they are good for us, in that they generally enhance our lives. (p. 180)

Hence, the pedagogic framing of twenty-first century competencies must go beyond considerations of cognitive attainment. Of course, attainment and employability are central concerns, but so is well-being—especially in a society of increasing volatility and change. In an overview of SDT, Vallerand et al. (2008) suggest that "SDT represents a theory with great heuristic power" (p. 257). In summarizing the research findings in this area, the authors point out:

findings in all articles underscore the fact that environments that provide autonomy support lead to qualitatively superior forms of motivation characterized by high levels of self-determination (i.e., intrinsic motivation and identified regulation) that, in turn, are conducive to more adaptive cognitive, affective, and behavioural outcomes. (p. 257)

In contrast, as Ryan and Deci (2017) note:

when there is little intrinsic motivation for learning and no inherent interest and excitement in what is going on in the classroom, then both learning outcomes and student wellness are in jeopardy, as longitudinal data confirm. (p. 354)

From an evidence-based perspective, SDT focuses on psychological processes, as these relate directly to the subjective experiences of persons, which deal with the dynamics of external events in their environment and the internal forces (both conscious and non-conscious). As Ryan and Deci (2017) state:

...the level of analysis that is most needed for scientific understanding of motivation and behaviour change is the level encompassing the psychological processes operating within the individual and the variables and influences within the social context that activate or diminish those processes. (p. 7)

Most importantly, as the authors summarize, it is:

...the most practical level at which we can intervene in human behavioral affairs. (p. 7)

In terms of practical interventions for enhancing teaching and learning, this is where we can directly impact such processes as perception, thinking and meaning-making, as it is at the level of subjective experience where people make sense of their world, their selves, others in the world, and respond accordingly.

As Ryan and Deci (2008) point out:

Well-being, mental health, and a life well-lived are all about experiencing love, freedom, efficacy, and meaningful goals and values, all of which are psychological phenomena.

Thus, within, their "objective" circumstances (*the external world - my inset*), the most important feature in people's lives is their experience of living, so enhancing that experience, with its various consequences, is an important focus for psychological interventions. (pp. 654–655)

4.4 Perception, Thinking, and Meaning Making

As the world-famous psychologist Edward De Bono stated:

Most of the mistakes in thinking are inadequacies of perception rather than mistakes of logic.

Indeed, from my life experiences, I would say that everything is a matter of perception, though it is probably more often than not a lack of good perception. Unfortunately, what we see is shaped by what we know, and this will determine how we think about it, and the meaning we make of it. Chapter 2 outlined many of the limitations, systemic biases—if you like 'bugs' of thinking—and their impact on behaviour.

In the previous chapter, I argued that Metacognition—more specifically Metacognitive Capability is the superordinate twenty-first century competence, as its significance for developing self-directed lifelong learners is paramount. However, there is little point in having a Rolls Royce car in the garage if you loathe driving and travelling in cars. Equally, we may have the best curriculum program in a particular topic area, and the best teachers to deliver it, but if there are no students who are interested in the offering, it is essentially redundant. Hence, motivation remains the most fundamental concern for learning in formal educational settings. It's not a concern when motivation is intrinsic. As an adolescent I was fascinated with astronomy, and I learned much more about this on my own volition, than I did for most of my school subjects, where I was largely motivated by the desire to avoid punishment (e.g., redoing the boring work, not being able to play soccer) for unacceptable performance.

Furthermore, as explained in Chap. 3, beliefs are fundamental in shaping perception, as they are no more than prior perceptions that have become strong neurologically wired mental schemata in long term memory. The work of Dweck (2006) on the importance of developing a Growth Mindset in students, documented previously, is especially important for developing positive beliefs that are conducive to students' motivation to learn. For example, Saphier (2017) concludes that the evidence suggests that the variables that appear to have the most significant impact on a person's development and achievement extend well beyond any measurement of innate intelligence. He identifies that:

> These variables appear to include the quality and quantity of schooling one experiences, the amount and kind of effort one invests, and the beliefs one holds in the individual's capability to grow ability itself. (pp. 30–31)

Most significantly, and specifically, as Schunk and Zimmerman (2012) point out:

> …unless people believe that their actions can produce the outcomes they desire, they have little incentive to act or to persevere in the face of difficulties. (p. 113)

It is not difficult to understand how beliefs profoundly affect the way people approach their learning and the subsequent impact on motivational (and eventually attainment) levels. Beliefs act as major neurological filters that determine how we perceive, and respond to, external reality (the world around us and the people in it), and as Henry Ford famously wrote:

> If you think you can or think you can't, you're right.

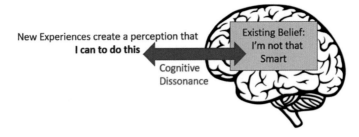

Fig. 4.2 Simple model of cognitive dissonance

How to Change Beliefs—especially those that limit motivation, learning and well-being

Just as beliefs originated as perceptions, they can be changed by new perceptions that significantly challenge the existing beliefs. As a kid, I believed in Tooth Fairies, Father Christmas and the Bogeyman under the bed. Since I sawed the legs off my bed, I've stopped believing in the latter. Just joking—and I think you will have made the relevant inferences and interpretations by now.

Invariably, people can be highly resistant to belief change, especially to deeply held ones that define life-meaning and personal identity. However, in many cases, it can be a quick process when a new experience significantly challenges the belief and creates *Cognitive Dissonance* (i.e., a real challenge to the reality/validity of the existing belief). Figure 4.2, illustrates how this can happen in terms of changing one's perception of smartness.

For many years, I worked in educational institutions where many students had little belief in their intellectual capabilities, and perhaps even less in the usefulness of teachers to do much to help them. In this situation, my priority is to bring about changes in their perception of themselves as learners, and this can only be achieved by achieving some degree of mastery in learning tasks meaningful to them. However, this firstly will typically involve getting some positive reframing by them on me as a person, not their wider perception of teachers per se. Unless I can get their attention, build some rapport, and 'prove' that there is some meaning and value to the learning experience, it's going to be a tough time as Michelle Pfieffer learned in 'Dangerous Minds', 1995, an American drama film in which she faced a very challenging class, but eventually got their attention and this made the difference. You can watch the film to find out how she did it, in the context of the strategy above. I was not able to get students to reframe their beliefs about teacher's in general, but I influenced them to see me; firstly as ok and not like other teachers they had experienced; secondly as someone they got on with and trusted; and finally, someone they liked and who could help them meet desired personal goals. In that last scenario, the new perception had replaced the prior belief—at least on me anyway, and that's what was important (for me) at that time (see Fig. 4.3). This story is extended in Chap. 5, so there is more to come, and there's a happy ending.

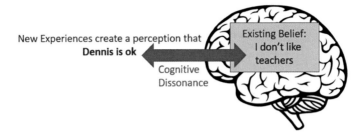

Fig. 4.3 Dennis is ok

Notions of purpose and meaning are of course central to learning and existential to well-being. For example, Pink (2009), while using Autonomy and Mastery as key psychological needs, also includes *Purpose* in his motivational framework. One could argue, that to some extent at least, purpose gives meaning to one's activities—even life itself. Frankl's (2014) account of how he managed to survive the experience of Auschwitz through his belief that one day he would share his psychological theories (e.g., logotherapy) to the world and his ability to maintain a sufficiently enduring positive psychological state (in the face of the terrible external realities around him), defines "Man's Search for Meaning" (the title of his book).

However, purpose and meaning, in the everyday context, can be interpreted in different ways with different levels of importance, which may also be highly situational. Students are likely to experience and frame them in terms of what they feel are important at that time in their life-worlds. For example, being in the school soccer team was highly meaningful to me at around 14 years of age, and a major purpose at that time was to have my hair straightened—how I hated that wave in my hair. Was I indeed a younger version of 'Shallow Hal', a 2001 American romantic comedy film starring Gwyneth Paltrow and Jack Black about a shallow man who, after hypnosis, begins to see people's inner beauty reflected in their outward appearance? Certainly, helping students to find purpose and meaning in their schoolwork and facilitating their well-being at this important developmental period are important aspects of a good holistic education (e.g. supporting their purpose and meaning in school life and future planning). Good career mediation is facilitative in this context. Indeed, purpose and meaning are very much bound up with need satisfaction as students who, in their everyday lives, are experiencing mastery, relatedness and autonomy are likely to have a positive sense of well-being. This may give them sufficient purpose and meaning to be both settled and proactive in their learning and life generally. They may even be happy. Of course, these are generalizations, and personality and other factors inevitably come into play in such existential matters. Ryan and Deci (2017) illustrate the likely linkages in terms of these human states:

> Empirical studies also show a linkage, with basic need satisfaction reliably predicting meaning. Recent research suggests that one's sense of meaning in life is largely accounted for by SDT's basic needs, along with feelings of benevolence, which itself is need-satisfying. (p. 253)

There are evidence-based practical strategies that foster these psychological aspects of intrinsic motivation, growth mindset, well-being, as well as the wider process of learning and how this happens at the level of subjective experience and brain functioning. These strategies may have very quick positive impacts, or it may take a while, and teachers may need some deliberate practice and persistence in making them work in unison. They are not separate, as they have an underpinning syntax in terms of psychological functioning and processes—which are evidence-based. In practice, they are highly synergistic in terms of an overall instructional strategy and impact on motivation, learning and well-being:

- Use an *Autonomy Supporting Style of Teaching*
- Teach students about *How Humans Learn*
- Communicate *Positive Growth Mindset Messages* in everyday teaching
- Use instructional strategies that support *Mastery Learning*
- Use *Teachable Moments* to reinforce aspects of students' learning experiences that support a Growth Mindset.

Use an Autonomy Supporting Style of Teaching

Reeve (2015) defined Autonomy Support (AS) as:

> …a coherent cluster of teacher-provided instructional behaviours that collectively communicate to students an interpersonal tone of support and understanding, such as "I am your ally; I am here to support you and your strivings." (p. 407)

Behaviours conducive to establishing this 'interpersonal tone of support and understanding' include:

- Using informational, non-controlling language
- Communicating the purpose/value of the learning (e.g., explanatory rationales)
- Acknowledging and accepting students' expressions of negative effect
- Listening to students and encouraging them to ask questions
- Allowing students choices/ preferences wherever possible on how they learn and the context of learning.

In that learning experiences can meet a learner's basic psychological needs, there is potential for enhancing intrinsic motivation. From an SDT perspective, the communication styles adopted by teachers play a fundamental role in shaping how students experience their learning and the extent to which such needs are likely to be met. Reeve et al. (2004), for example, who compared the communication styles of teachers in terms of a continuum ranging from highly controlling to autonomy-supportive, noted that:

> In general, autonomy-supportive teachers facilitate, whereas controlling teachers interfere with, the congruence between students' self-determined inner motives and their classroom activity. (p. 148)

In more basic terms, Deci et al. (1991) reported that:

Students in classrooms with autonomy-supportive teachers displayed more intrinsic motivation, perceived competence, and self-esteem than did students in classrooms with controlling teachers. (p. 337)

Furthermore, and of particular interest, Ryan and Deci (2017) noted:

…when teachers are autonomy supporting, they are typically also supportive of the students' need for competence and relatedness. (p. 369)

An AS style of teaching or facilitation is congruent with *Core Principle 9: A psychological climate is created which is both success-orientated and fun.* Teachers who systematically and consistently utilize such behaviours in their everyday interactions with students are likely to develop good relationships and rapport with students. Hence, in meeting the needs of mastery, relatedness and autonomy, which creates such a psychological climate, they are supporting key aspects of the learning process (e.g., attainment opportunities; engagement) and student well-being. Note that while I use the term *intrinsic motivation*, the focus is essentially on student *engagement* in the learning process and how this impacts their feelings of being competent and confident in their learning (i.e., self-efficacy). The distinction between these two constructs is that motivation is a private, unobservable, psychological and neurological (unconscious) process that serves as an antecedent cause to the more publicly observable behaviour that is engagement. Furthermore, engagement is an essential educational outcome in itself, as a marker of students positive functioning, as it predicts highly values outcomes such as students' academic progress and achievement (e.g., Ladd and Dinella 2009). As Zepke and Leach (2010) summarize, from an extensive review of the literature, "…teaching and teachers are central to engagement." (p. 170). Without meaningful and sustained engagement, there would be underlying poor attention, and learning would be limited at best.

Engagement, in the context of classroom learning, is student motivated action toward meeting desired educational goals. There is a consensus that engagement has the following 3 component constructs/dimensions:

- Behavioural (on-task attention, effort, persistence)
- Emotional (presence of interest, enthusiasm)
- Cognitive (use of thinking, learning strategies) While teachers are tasked with designing and facilitating instructional strategies that engage students in the learning process and optimize their attainment opportunities, learning is also enhanced when students are actively involved in shaping the instructional process themselves, making it more of a collaborative rather than a one-way process.

Reeve and Tseng (2011) referred to such student involvement as *Agentic Engagement*, suggesting it as a fourth aspect to student engagement, which they define as:

…a students' constructive contributions into the flow of the instruction they receive. What this new concept captures the process in which students intentionally and somewhat proactively try to personalize and otherwise enrich both what it is to be learned and the conditions and the circumstances under which it is to be learned. (p. 258)

This is where students take a consciously active role in shaping the learning context and instructional activities through collaborative learning relationships with teachers. Agentic Learners are confident to:

- Offer input into the lesson content (e.g., based on a diagnosis of what they are finding difficult and what might help their understanding for a topic area)
- Express preferences in terms of how they learn best (e.g., what methods/learning resources are most useful)
- Communicate their thinking or learning need (e.g., ask specific questions about what to learn and how)
- Communicate their level of interest (e.g., provide negative effect constructively when bored or frustrated)
- Solicit resources to help their learning (e.g., feel comfortable in asking for help).

Agentic Learners can be framed as students who have learned how to learn and have developed appropriate habits of mind (e.g., Growth mindset) that support active engagement and autonomy in their learning relations and strategies. Framing the characteristics of agentic students is not a difficult task, and these fully calibrate with the components and features of MC. For example, we would expect agentic learners to have the following skills and attributes:

- Aware of being metacognitive and what this entails in terms of personal self-regulation and the ability to be self-directed
- Understand the key aspects of how people learn (e.g., cognitive scientific principles from a user's perspective)
- Competence in using a range of cognitive strategies; knows when to use them and for what specific learning purpose and goals
- Frame clear, challenging realistic goals
- Plan, implement, monitor and evaluate a learning plan/strategy for meeting learning goals
- Aware of the range of existential and personal barriers to maintaining motivation, effort, and belief in meeting challenging goals
- Strategize and use techniques and tools for maintaining self-regulation in-the-face of inevitable challenges
- Understand that failure is part of the learning process, and sees this as important feedback for future learning
- Active seeking of feedback from a range of sources (teachers, peers, anyone who might know), and ability to ask thoughtful focused questions
- Aware that showing empathy, respect and interest in the goals, thoughts and feelings of others builds rapport, and that this is a key life skill
- Can use motivational strategies to keep a balanced perspective on managing self in situations of poor mood, boredom, or challenges to a growth mindset.

Agentic engagement adds much of value to the learning process, as agentic students elicit (demand) greater mediation between the teacher and themselves, facilitating collaborative two-way feedback on what instructional strategies are best supporting their learning. Treadwell (2017) argues that empowering learners with the competency to take agency over their learning is something every school and educational system needs to address (p. 9). He goes on to argue:

> If we give people agency over their world they generally rise to the occasion and increasingly manage their world more successfully, but only if we provide them with the underlying competencies to achieve this. (p. 9)

> One of the key aspects of increasing agency in school is having the learners take responsibility for their learning, where they can be aware of, and learn about, their mind-set and capacity to show 'grit' (determination). Learners in schools get the opportunity to 'play' with taking agency over their world and this is one of the most powerful learning lessons we can gift them. (p. 146)

The importance of developing agentic learners as outlined above is fundamental to the wider aims of developing self-directed lifelong learners. Agentic students are both a product and the 'co-facilitators' (e.g., with their teachers and peers) in an ongoing creative evidence-based learning approach, which also includes good collaboration and communication skills as central to such learning and well-being. As Reeve and Tseng (2011) suggest:

> The reason why agentic engagement contributes uniquely to achievement is presumably that it is through intentional, proactive, and constructive acts that students find ways to improve their opportunity to learn by enriching the learning experience and by enhancing the conditions in which they learn. (p. 263)

Finally, as Ryan and Deci (2008) point out, AS is not only beneficial to students learning and engagement but has more generic benefits:

> In short, both giving and receiving autonomy support in a close relationship are related to relationship quality and well-being. (p. 671)

Teach students about How Humans Learn

Throughout this book, there has been, and this will continue in subsequent chapters, an extensive focus on the nature, underpinning components, features—even nuances—of the learning process and that's because this is the *core business* of education and training.

In becoming competent and confident self-directed learners, students need to have key understandings of how humans learn and are competent in using metacognition, cognitive and motivational strategies. In this way, both the teacher's and the students *thinking becomes visible* and they have a shared *language of learning*, which makes the mediation of learning between stakeholders (e.g., teacher, student, peers) much more effective.

Teaching students the principles of learning (Saphier 2017; Sale 2015) is the essential primary strategy for helping students to understand how their minds work (in the context of school learning and well-being), enhancing their learning capability and, over time, empowering them to become self-directed lifelong learners. Saphier (2017) recommends that educators:

> …Share with them teaching and learning strategies we use ourselves. By including them in the secret knowledge of teaching and learning strategies we give students choices, power, and license to control their learning.
>
> Principles of learning should be explicitly taught to students as they can use them to be more powerful learners. (pp. 114–15)

Such knowledge and its application to everyday learning and life goals can be effectively taught. This is 'The Future of Learning', as Treadwell (2017) described it in a book of that title.

Communicate Positive Growth Mindset Messages in everyday interactions with students

To develop good thinking, I emphasized the importance of making the cognitive heuristics of the specific types of thinking explicit for students—visible—and through retrieval practice in a range of application contexts, this becomes part of *the language of learning*. The same pedagogic process is necessary to make growth mind-set concepts visible and part of everyday classroom interactions. As Saphier (2017) points out:

> It is through daily interactions in everyday classroom life that we can:
>
> - Convince students that "smart" is something they can get,
> - Show them how, and
> - Get them to want to. (p. 40)

It is important not to confuse such a style with giving excessive or gratuitous praise. It is most appropriate to give praise when students, especially those who have worked hard and struggled more than others to achieve mastery, finally making those neural connections and suddenly have that 'aha' feeling of 'I get it'. This is a direct experience of the underpinning logic of a growth mindset, which is inherently simple: more effort, with expert teaching and appropriate practice leads to mastery learning, which enhances self-efficacy—and so the synergistic cycles moves-on—of course with blips and challenges—but that's life also. The essence here, though, is that if students understand the nature of this scenario; how it works; and that it is manageable; and they know they have the strategies to do this effectively—then they also know they are becoming Self-Directed Learners. This is also the way to attack (and eventually destroy) limiting beliefs.

Teaching Expertise involves the capability of *Mindfulness* (Langer 2016) Essentially, in this context, mindfulness refers to being aware—observing, listening and reflecting—of the interactions occurring in the here and now classroom situation, and what this may mean in terms of student learning and well-being. For sure, students pick up on our language codes, vocal tones, and body language and form both impressions of us as teachers (as people), and of course how they think we perceive them. Tomlinson (2005) captures this poignantly:

> My students hear every message I send – whether overt or implied – about their capability to learn and succeed. (p. 76)

Use Instructional Strategies that promote Mastery Learning

Teachers are often pressurized into completing the curriculum in specified time frames. However, students who lack prior knowledge and/or skills often fall behind, have difficulties with subsequent topics, and risk becoming labelled (and may self-identify as) 'weak learner's'. *Mastery Learning* focuses on ensuring that all students master a topic area at a specified level of conceptual understanding or skill, before moving on to a more advanced one (allowing students to move at their own pace in this process). Hence, when students master the content, they experience a sense of being competent, which is fundamental to supporting a growth mind-set as it provides a direct experience that perseverance and effort lead to success. Simultaneously mastery promotes and reinforces self-efficacy in learning the topics, and the more this is experienced across topics, enhances the possibilities of transfer and intrinsic motivation.

Facilitating mastery learning inevitably involves some differentiation in the instructional strategy but, with good pedagogic design and the thoughtful application of technology, it is an achievable and desirable educational goal in the present educational context.

Use Teachable Moments to reinforce aspects of students' learning experiences that support a Growth Mindset

A *teachable moment* is the time at which learning a specific topic or idea becomes possible or easier. Great teachable moments for supporting a growth mind-set are when students directly experience a learning breakthrough (e.g., achieve a learning goal—a new understanding or skill level) and have that 'aha feeling'—which essentially means that they have 'got it'—and will have developed a useful addition to their neural networks in long-term memory. This is especially the case when this new learning can be directly linked to their positive learning actions, such as extra effort, persistence, using a learning strategy well. Nothing convinces like direct experience. Once students understand how learning works, and how their actions impact the results, they are more likely to adopt the beliefs and behave in ways consistent with a growth mindset. The cartoon below is just for amusement.

Hence, from an EBT approach, first and foremost, the design and facilitation of learning experiences must involve, as an essential design heuristic, ways to generate and sustain learner motivation. The Core Principles of Learning, outlined and illustrated in Chap. 2, are a significant part of the instructional design heuristics for enhancing motivation to learn as they constitute the key underpinning knowledge—*Pedagogic Literacy*—for making teaching more effective, efficient and engaging. However, it is how the core principles are employed in the design and facilitation of learning—*the situated context of teaching*—that is most important for maximizing student motivation. This entails the creative blending of strategies and methods, utilizing technologies, and teachers being active agents in the motivational process. As Brophy (1987) makes explicit:

> …teachers are not merely reactors to whatever motivation patterns their students had developed before entering their classrooms but rather are *active socialization agents* capable of stimulating the general development of student motivation to learn and its activation in particular situations. (p. 40)

Therefore, as Alderman (2008) argues:

> …teachers have a primary responsibility in education to help students to cultivate personal qualities of motivation that can give them resources for developing aspiration, independent learning, achieving goals, and fostering resilience in the face of setbacks. (p. 3)

This may just be the gold standard (e.g., Adaptive Expertise, Hatano and Inagaki 1986) of expert teaching, or what I refer to as **Creative Teaching Competence**.

For the purposes of this chapter, we can conclude that while motivation is a complex and dynamic entity, influenced by a range of external environmental factors and internal features of the human brain and mind (and not necessarily working in unison) as well as much of its processing operating unconsciously (the extent of which is contested)—we can derive useful heuristics for enhancing motivation and well-being As Miller (2016) summarizes:

> …the combination of an atmosphere of self-determination and a connection to student's personal goals and values is a potent formula for motivating college students. (p. 73)

4.5   Summary

Considerations of motivation are fundamental to all aspects of planning and facilitating the practices of teaching. Creating and facilitating motivationally enriching experiences, as well as providing high attainment opportunities, requires a high level of creative teaching competence. The ability to do this consistently may no longer be a 'nice to have' capability for the few who can do it, but rather a necessary competence and proficiency level for mainstream teaching and training personnel This is especially challenging—even for creative teachers—in that, as Zig Zigler famously stated:

> People often say that motivation doesn't last. Well neither does bathing – that's why we recommend it daily.

Finally, in this framing of student motivation, don't forget yourself in this endeavour. As professional educators, while we are paid to do this challenging work, there is no harm, in fact, massive benefit, in enjoying the experience. There is little pleasure or novelty, and certainly considerable pain in teaching groups of unmotivated learners. However, when we have learners who show interest in what we are teaching (not necessary all the time), positively interact with us as human beings, and are successful in the attainment stakes, it is a highly rewarding experience, and it's why many of us do this job. As Levin (2008) summarized:

> Greater engagement is a vehicle that improves students' work and makes teachers' lives easier as well.

> …increased student motivation is very positive for teachers' experience of their work. (p. 99)

Similarly, as Ryan and Deci (2017) conclude:

> Having teachers experience need satisfaction and be autonomously motivated to teach and having students experience need satisfaction and be autonomously motivated to learn is the optimal situation in classrooms. (p. 374)

References

Alderman MK (2008) Motivation for achievement: possibilities for teaching and learning. Taylor & Francis Inc., Milton Park, Oxfordshire, UK

Bentham J (1789) Quote. Available at www.goodreads.com/quotes/1013033-nature-has-placed-mankind-under. Last accessed, 3rd Nov 2019.

Brophy J (1987) Synthesis of research on strategies for motivating students to learn. Association for Supervision and Curriculum Development, Alexandria, Virginia

Cloninger CR (1997) A psychobiological model of personality and psychopathology. Jpn Psychosom Med 37:91–102

De Bono E (2019) Quote available at https://www.brainyquote.com/quotes/edward_de_bono_124615. Last accessed on 2nd Dec 2019

Deci EL et al (1991) Motivation and education: the self-determination perspective. Educ Psychol 26(3 & 4):325–346

Dweck CS (2006) Mindset: the new psychology of success. Ballantine, New York

Frankl VE (2014) Man's search for meaning. Beacon Press, Boston, US

Gallwey TW (1987) The inner game of tennis. Jonathan Cape, London

Gregory G, Kaufeldt M (2015) The motivated brain: improving student attention, engagement, and perseverance. ASCD, Alexandria, Virginia

Harari YN (2016) Homo Deus: a brief history of tomorrow. Vintage, Penguin, London

Harris S (2010) The moral landscape: how science can determine human values. Free Press, New York

Hatano G, Inagaki K (1986) Two courses of expertise. Child development and education in Japan, pp 262–272

Herzberg F (1966) Work and the nature of man. World Pub. Co., Cleveland

Kim SI (2013) Neuroscientific model of motivational process. Front Psychol 4:98. https://doi.org/10.3389/fpsyg.2013.00098

Ladd G, Dinella LM (2009) Continuity and change in early school engagement: predictive of children's achievement trajectories from first to eighth grade? J Educ Psychol 101(1):190–206

Langer HJ (2016) The power of mindful learning. Da Capo Lifelong Books, Boston, US

Levin B (2008) How to change 5000 schools. Harvard Education Press, Cambridge

Maslow A H (1943) A theory of human motivation. Psychol Rev 50(4):370–396

Maslow A (1962) Toward a psychology of being. Van Nostrand, New York

Miller MD (2016) Minds online: teaching effectively with technology. Harvard University Press, Cambridge, MA

Pink DH (2009) Drive: the surprising truth about what motivates us. Riverhead Books, New York

Pinker S (2003) The blank slate: the modern denial of human nature. Penguin, London

Pinker S (2019) Enlightenment now. Penguin, UK

Reeve J (2015) Giving and summoning autonomy support in hierarchical relationships. Soc Pers Psychol Compass 9(8):406–418

Reeve J et al (2004) Enhancing students' engagement by increasing teachers' autonomy support. Motivation Emotion 28(2)

Reeve J, Lee W (2012) Neuroscience and human motivation. In: Ryan RM (ed) Oxford library of psychology. The oxford handbook of human motivation (pp 365–380). Oxford University Press

Reeve J, Tseng C-M (2011) Agency as a fourth aspect of students' engagement during learning activities. Contemp Educ Psychol 36(4):257–267. https://doi.org/10.1016/j.cedpsych.2011.05.002

Robbins A (2001) Unlimited power. Pocket Books, London

Rutledge P (2012) Social networks: what Maslow Misses. https://www.psychologytoday.com/gb/blog/positively-media/201111/social-networks-what-maslow-misses-0

Ryan RM, Deci EL (2008) Self-determination theory and the role of basic psychological needs in personality and the organization of behaviour. In: The handbook of personality: theory and research. Guilford Press, New York

Ryan RM, Deci EL (2017) Self determination theory: basic needs in motivation, development, and wellness. The Guilford Press, New York

Sale D (2015) Creative teaching: an evidence-based approach. Springer, New York

Saphier J (2017) High expectations teaching. Sage, New York

Schunk DH, Usher EL (2012) Social cognitive theory and motivation. In: Ryan RM (ed) The Oxford handbook of human motivation. Oxford University Press, Oxford, UK

Schunk DH, Zimmerman BJ (2012) Motivation and self-regulated learning: theory, research, and applications. Routledge, New York

Swaab D (2015) We are our brains: from the womb to Alzheimer's. Penguin, London

Tomlinson CA (2005) The differentiated classroom: responding to the needs of all learners, 2nd edn. ASCD, Alexandria, Virginia

Treadwell M (2017) The future of learning. The Global Curriculum Project, Mount Maunganui, NZ

Vallerand RJ, Pelletier LG, Koestner R (2008) Reflections on self-determination theory. Can Psychol/Psychol Canadienne 49(3):257–262

Wagner T (2010) The global achievement gap. Basic Books, New York

Wlodkowski RJ, Ginsberg MB (2017) Enhancing adult motivation to learn: a comprehensive guide for teaching all adults. Jossey-Bass, San Francisco

Zepke N, Leach L (2010) Improving Student Engagement: ten proposals for action. Act Learn High Educ 11:167

Zig Zigler https://www.brainyquote.com/quotes/zig_ziglar_387369. Last Accessed 30th Nov 2019

Chapter 5
Creative Teaching Competence: The SHAPE of Creative Teachers

Abstract This chapter analyses the psychological capability for creative thinking and how this is most usefully contextualized to everyday teaching and learning contexts—what I refer to as *Creative Teaching Competence*. As creativity or creative thinking appears in most frameworks of twenty-first century competencies, it is necessary to reduce misconceptions and frame this area in more evidence-based specific practical terms. While creative teaching can take many forms, and be both a planned experience and/or a situated invention in the teaching context, there is an underpinning syntax to how it works and what are typical outcomes of such behaviour in terms of specific practices. In summary, creative teaching can be demystified, explained and illustrated. Hence, it is a learnable competence/expertise for motivated teaching professionals.

5.1 Introduction

Chapter 1 argued for teaching to become evidence-based, as is the case for other professions such as engineering and medicine—you may remember the Jurassic Park analogy! Chapter 2 established cognitive scientific principles—the core principles of learning—as a foundational base for an emerging and increasing *Pedagogic Literacy*. Chapter 3 argued that Metacognitive Capability is the superordinate twenty-first century competence, and what makes it an essential component of student learning and expert teaching. Chapter 4 unpacked motivation, especially intrinsic motivation, to better understand how it works at the level of subjective experience and human psychological needs.

However, while the core principles of learning provide evidence-based heuristics for the design and facilitation of learning experiences, they do not dictate the form or structure of any specific learning event. This is always mediated by a whole host of situated factors, as portrayed in the 'fly fishing analogy' (e.g. learning outcomes, subject content, student competence, resource access). Hence, teaching is not a highly prescriptive activity, governed by a finite range of algorithmic teacher behaviours; rather it is an act of situated design—but not determined by whim or fashion. Design in teaching is essentially similar to design in other fields in that it must involve

D. Sale, *Creative Teachers*, Cognitive Science and Technology,
https://doi.org/10.1007/978-981-15-3469-0_5

aspects of a structured process based on a range of known heuristics, but invariably open to new forms of reality within the context of the field. For example, an engineering product such as a hand-phone must, by definition, have certain features to enable the process of communication between people, but the shape, colour, additional features and aspects of functionality are variables that can be 'played with' to generate novelty and aesthetic appeal. Maintaining the status quo function with perceived added features and functionality is what enables competitive advantage and typically encompasses some creativity within these parameters. Design, in any field of application, is by nature both a systematic and creative process, aptly captured by Beetham and Sharpes' (2007) definition:

> …a systematic approach with rules based on evidence, and a set of contextualized practices that are constantly adapting to circumstances. It is a skilful, creative activity that can be improved on with reflection and scholarship. (p. 6)

In this chapter, building on the key heuristics outlined in previous chapters, a frame on creative teaching is developed and illustrated using Beetham and Sharpe's heuristics on design. It seeks to establish what I refer to as **Creative Teaching Competence (CTC)**, which can move us in some useful practical way along the funnel of knowledge, from mystery to heuristics, in terms of what creative teachers do and how they do it. It will constitute a key competence, along with **Metacognitive Capability (MC)**, in the framing of expert teaching in the coming decade or so. It also addresses some questions that I have been asked many times over the years, which essentially can be summarized as:

- Is creative teaching different from effective teaching?
- Is creative teaching better than effective teaching?

You may already have an opinion on this or maybe framing one as you reflect on these questions. You may also be thinking of the well-worn discussion of whether or not, or to what extent, teaching is an art or science.

There is little doubt that our increased understanding of human psychological functioning is providing a strong evidence base for the practices of teaching. In that sense, we can certainly talk about the science of teaching. However, we should also recognize the importance of artistry (however defined) in teaching, as Eisner (1995) so boldly asserted:

> …artistry in teaching represents the apotheosis of educational performance and rather than try to diminish or replace it with rule-governed prescriptions, we ought to offer it a seat of honour. (p. 96)

Art and creativity have much in common and are sometimes seen as synonymous. Indeed, art by its very nature is always seeking new forms and genres, and the most valued in financial terms are often those exhibits that are deemed 'creative'. However, in the context of teaching, while recognition has long been there, creativity has proved elusive in terms of clear framing and how its works pedagogically. As Schon (1987) notes:

> …outstanding practitioners are not said to have more professional knowledge than others but more "wisdom", "talent", "intuition", or "artistry". (p. 13)

In the following sections, I will explore the notion that the art of teaching can also be better understood from an EBT approach in that aspects of human psychological functioning, which appears elusive—almost ephemeral—but impactful in terms of learning and well-being, can be effectively modelled, learned and applied like other more overt aspects of behaviour. It just may be the case that this applies to all forms of art, and *works* as a result of its subtle impact on human sensation and perception, creating positive aesthetic and affective responses (both conscious and sub/unconscious). The dichotomy of science and art of teaching may cease to be a useful one; we may simply talk about Creative Teaching or Creative Teaching Competence.

5.2 Teaching as a Systematic Approach with Rules Based on Evidence

In the context of teaching, key heuristics such as the core principles of learning provide the systematic approach with rules, based on evidence. When planning the design of any learning experience (e.g., module, workshop) these heuristics should be at the forefront of planning decisions. For example, in planning a professional development workshop, I will typically generate and address the following kinds of questions:

- What are the learning goals and key outcomes for this learning group?
- What prior knowledge do these learners already have, and what activities might best capture their present understanding of these areas of learning?
- What are the key concepts that need to be negotiated to facilitate understanding, and how is this best organized and managed with this group of learners?
- What essential questions connected with this topic might get them thinking in critical or creative ways about areas of interest or problem-solving?
- What presentation mediums and resources do I have that will generate and maintain interest?
- How do I organize the learning sequences to avoid cognitive overload, facilitate the 'digesting' and understanding of the content to enable effective transfer into long-term memory?
- How best can I make the learning experience active and experiential for supporting application-based learning outcomes?
- What specific skills need to be developed and how might effective spaced and deliberate practice be organized?
- Are there areas where feedback is likely to be most critical for effective learning; and what strategies will I employ to maximize efficiency?
- How best do I present myself to this group of learners to encourage the building of rapport, positive interpersonal interactions and a 'can do' feeling for the tasks in hand?

- What activities, stories, examples, artefacts do I have that will generate interest and engagement around the key areas of learning?

Ok, I'll confess, the process may not always be as explicit, meticulous or pristine as the above description depicts. Furthermore, as with all learning over time, especially with expertise, it becomes a largely unconscious competence and can be done very quickly. The key point is that there is a systematic process in design; it is far from ad hoc.

5.3 Teaching as a Set of Contextualized Practices Constantly Adapting to Circumstances

In teaching, the term practices can relate to many activities, resources and arrangements designed to enhance desired learning outcomes. At a technical level, and typically for competency-based teacher education programs, practices are often seen in terms of specific functional competencies such as:

- Write clear and appropriate learning outcomes
- Prepare a teaching/training plan
- Produce learning resources
- Conduct teaching/training
- Design differentiated learning activities
- Produce a scheme of assessment
- Conduct assessment
- Produce an online module
- Facilitate an online module.

Such functional competencies are foundational to teaching/training as they comprise—metaphorically speaking—the 'tool-box' and underpinning knowledge relating to key technical practices. However, I have conducted numerous teaching observations in which the designated competencies were technically met, but I did not perceive the actual learning experience as being effective in terms of student learning—especially concerning cognitive and emotional engagement. I might sum up the experience as 'just ok'. The reason is that practices cannot just be seen solely in technical terms, but also in the way teachers conduct themselves, interact with learners and mediate the situated learning experience to engage, motivate and, on the best occasions, inspire them. As Andrews et al (1996) pointed out:

…the hallmark of excellent teaching is more than adequate content expertise and effective technical performance. (p. 82)

This is similarly echoed by Bain (2004), who emphasized that:

…the best teaching can be found not in particular practices or rules but the attitudes of the teachers. (p. 78)

Andrews et al. (1996) develop this further in their description of excellent teachers:

…excellent teachers seem to want to facilitate a meaning approach (deep) to learning rather than a reproducing (surface) approach. Moreover, they tend to engage in instructional processes that are congruent with their preferred approach and have values and beliefs, and characteristics (for example, honesty, integrity, genuineness and respect for self, students, materials and the process of teaching) that are considered foundational to a meaning approach to teaching. (p. 101)

While considerations of human conduct may be contested in certain situations, there is strong evidence of core universal principles that transcend cultural norms and rituals. Nucci (2001) from extensive research of the literature noted:

…the domain of morality is structured around issues that are universal and non-arbitrary. The core of human morality is a concern for fairness and human welfare. Thus, there is a basic core of morality around which educators can construct their educational practices without imposing arbitrary standards or retreating into value relativism. (p. 19)

The practices of teaching, therefore, involves both a range of technical functional competencies as well as 'social and emotional' competencies and underpinned by core principles of human conduct. The social and emotional aspects of life are embedded in all human encounters, and teaching is no exception. Furthermore, while this provides a systematic approach to the overall design of learning experiences, it will always require thoughtful contextualization and adaptation to the particular learning group and context, as outlined previously. This also applies to the use of specific practices. For example, teachers often debate the merits or otherwise of different teaching methods. It is as though some are looking for a pedagogic 'silver bullet'; an approach or strategy that will engage and motivate all students and meet the desired learning outcomes. Such a wish is akin to alchemy. There are many reasons for the necessity to contextualize and adapt methods, including their appropriateness to outcomes and learners, the basic human desire for novelty and variation, and even the situated mood of the class at a particular time. Teaching methods are *structures* that deal with the delivery of content to help students acquire knowledge, build understanding and develop skills and competence. As is now well documented, some methods are more effective than others in terms of their effect sizes on student attainment, and this should be a key consideration in method selection. However, most methods can have benefits in terms of student learning when used skillfully and appropriately in context. The relative merit of different instructional methods has been well captured in an analogy by Bransford (1999):

Asking which teaching method/technique is best is analogous to asking what tool is best – a hammer, a screwdriver, a knife, or pliers. In teaching, as in carpentry, the selection of tools depends on the task at hand and the materials one is working with. (p. 22)

Equally the most powerful methods, in terms of their potential for enhancing student attainment, may be ineffective when employed by less competent practitioners, just as the best tools are often wasted in the hands of DIY (do-it-yourself) novices—as I have learned from personal experience.

Similarly, while universal human conduct principles such as equity, fairness, respect, concern for the person are fundamental to the practices of teaching, the actual style and the content of human interaction often requires much contextualization and adaptation. Even simple greetings, which are pretty much universal, need careful contextualization across cultural and ethnic groups. A kiss on the cheek when being introduced to a lady may be expected in France, but it would be highly risky in many other countries. Fortunately, we are not confronted with such decision-making in class on an everyday basis.

5.4   Teaching as Skillful Creative Activity

When people are skillful in an area, the inference is that they can perform a range of specific activities in a highly effective and efficient manner. For example, to say that the soccer player, Cristiano Ronaldo, is very skillful means that he is considered highly proficient or expert in employing such skills as controlling, passing, heading, shooting and dribbling with a soccer ball when playing soccer. However, while such skills are essential for a high level of performance in an activity, they are not the only components in determining an individual's actual performance in real-life situations. Other attributes, such as aptitude, personality traits, and attitudinal components also play an important part in determining performance. To capture the range of attributes involved in a complex performance the term competency is often used (Fig. 5.1), as it attempts to capture the wider configuration of attributes that determines actual performance in real work and life contexts.

Aptitude

This term is generally used to refer to a person's innate tendency for competence in an area. For example, a female basketball coach once asked me, perhaps with some humour in mind, if I knew any 6′6″ women who have an athletic physique. Interesting question. My answer was "none", but I did ask her, "why did you ask?" The reply was something like 'if I had one in my team, she would probably become a millionaire'. Interesting, I thought, so I asked, "How does that work?" The rationale provided was simple; very few women reach that height, so given some motivation, good training, and playing time for a year or two, she is likely to be recruited by a top

Fig. 5.1 Generic competency pyramid

team, and top players earn large salaries. How many 5'2" professionals basketball players have you seen?

A similar framing can be applied in terms of psychological/neurological aspects related to learning. For example, in the contested field of defining human intelligence, Spearman (1950) proposed a two-factor theory of intelligence that posited that all cognitive performance can be explained by two variables: one general ability (g) and the many specific abilities (s). The General ability or G factor intelligence refers to inborn aptitude, which varies from individual to individual and facilitates mental operations across all situations. People vary in terms of this G factor intelligence, and it underpins the basis of much of present-day intelligence testing. However, individuals typically possess specific abilities (S factors), also innate, and these facilitate their learning in these areas (e.g., singing, painting, drawing). People can also have more than one special ability and in varying degrees of innate potentiality. However, of interest, one form of specific ability may not help in the development of others. For example, If you are very good at painting and you can also sing well, this does not mean that your ability in painting helps you sing—or vice-versa. The extent to which this intelligence can be impacted by instructional interventions has been argued extensively over the years—often depicted in terms of the nature-nurture debate. The stance taken here is not to engage in discussions of extent, as clearly both are important determiners of behaviour, and we can only impact what is impactable—so to speak. Most significantly, in this context, is the notion that intelligence can be developed through a positive mindset (e.g., Growth Mind-set, Dweck 2006), student effort and perseverance and, of course, good instruction. There is, little doubt that an innate disposition for a specific S factor, other things being equal, has a natural advantage over someone not possessing that S factor. For example, in my case, I have tried to learn to sing, and while I have improved, it's been minimal at best. I would not claim even the most basic of competence (however defined). In contrast, I have worked extensively in the Philippines and am astonished at how well so many of the locals can sing. How much is innate, and how much is cultural, I don't know. I quake in fear (as novices do) when they ask me to sing a song after a workshop session in a social event. In summary, from Spearman's perspective, most tasks you perform are likely to be influenced by G factor, and all individuals have some specific abilities, which may help in task success, though varied in nature and intensity. Most importantly, however, both G and S intelligence are highly subject to positive environmental influence. As Bloom (1985) concluded:

> After forty years of intensive research on school learning in the United States as well as abroad, my major conclusion is: What any person in the world can learn, almost all persons can learn if provided with appropriate prior and current conditions of learning. This generalization does not appear to apply to the 2% or 3% of individuals who have severe emotional and physical difficulties that impair their learning. At the other extreme, there are about 1% or 2% of individuals who appear to learn in such unusually capable ways that they may be exceptions to the theory.

He goes on to illustrate:

> This middle 95% of school students become very similar in terms of their measured achievement, learning ability, rate of learning, and motivation for further learning when provided with *favourable learning conditions*.
>
> One example of such favourable learning conditions is mastery learning where the students are helped to master each learning unit before proceeding to a more advanced learning task. In general, the average student taught under mastery-learning procedures achieves at a level above 85% of students taught under conventional instructional conditions. (p. 4)

Another, more recent framing of intelligence is that of *Multiple Intelligences* (Gardner 1983), which posits a number-of different kinds of intelligences. He initially proposed that there are eight main types of intelligence:

- Visual-Spatial
- Linguistic-Verbal
- Interpersonal
- Intrapersonal
- Logical-Mathematic
- Musical
- Bodily-Kinaesthetic
- Naturalistic.

In 2009, Gardner suggested that existential and moral intelligence may also be worthy of inclusion. However, Gardner's theory, much like that of the learning styles proponents, has come under criticism from psychologists who argue that his definition of intelligence is too broad and that his eight different "intelligences" simply represent talents, personality traits, and abilities. Gardner's theory also suffers from a lack of supporting empirical research and as Hattie (2009) argues:

> …the pursuit of multiple intelligences has a limited return. Realizing that students have different abilities, talents, and interests is critical, but there is no need for a rhetoric of multiple intelligences that goes beyond this well-argued, well known, and almost simplistic (but powerful) message. (p. 90)
>
> …it is desirable to have multiple ways of teaching and there is no need to classify students into different 'intelligences'. (p. 91)

Traits/Dispositions

The area of traits and dispositions often creates some conceptual confusion, not dissimilar to that of comparing & contrasting metacognition, self-regulated learning, self-directed learning and meta-learning. The term trait is used to describe a typical characteristic that a person has or is perceived to possess. For example, we may refer to a person in such terms of 'friendly', 'selfish', 'unreliable', etc. Sometimes the term personality is attached to the description, e.g., she's an extrovert personality or he is introverted. Essentially, personality is generally framed, though with variation and focus, as the more generalized set of key traits that a person displays. In mainstream psychology, these are often referred to as the Big 5:

- introverted or extroverted
- neurotic or stable
- incurious or open to experience
- agreeable or antagonistic
- conscientious or undirected.

Personality traits are hereditary and are relatively permanent and stable over time. The innate basis of personality traits has been extensively validated, especially through the study of monozygotic (identical) twins separated at birth (e.g., Mittler 1971; Bouchard 1990). Such twins show remarkable similarities, even though reared in very different environments. Hence, people often postulate on how much of personality, our trait configurations, are hereditary and how much is influenced by environmental experience. This nature-nurture debate has raged (note: I do not use this term loosely) over the decades, swinging back and forth depending on the dominant ideology of the time. Over the past two decades the dominant 'social science model' (i.e., focus on culture and socialization as the main determinants of behaviour, and perhaps most prevalently voiced in terms of gender and race issues) is now being challenged by a wide range of research findings (e.g., Pinker 2003; Swaab 2015). The important issue from a pedagogic point of view is that while we have finally 'put the learning styles issue to bed', we cannot avoid the reality that our learners have different personalities that can vary greatly—and some configurations may be easier (much easier) to teach than others. O'Connor and Paunonen (2007) have noted that:

> …there are behavioural tendencies in personality traits that can affect certain habits that influence academic achievement (e.g., perseverance, conscientiousness, talkativeness)…While cognitive ability reflects what an individual can do, personality traits reflect will do. (p. 40)

In an unfavourable scenario, you simply may have a disproportionate number of neurotic, incurious, antagonistic and undirected learners. Even worse, have you had managers who have such personality configurations? Sadly, whether we like it or not, and it may not be the person's fault, as this is how genetics and early experiences play out (even from inside the womb according to Swaab 2015) in creating micro-wiring neurologically in the brain that determiners such behaviour (largely sub/unconsciously). Such framing invariably raises questions relating to free will, or our lack of it, aptly elucidated by Harris (2010). These are deep philosophical issues, under increasing scrutiny by cognitive scientists and philosophers alike. However, from an educational point of view, this only heightens the importance of MC, as it is perhaps our best—maybe only means for enhancing conscious (more rational) self-direction.

It can be argued that such hereditary factors influence what is often referred to as 'dispositions', which are generally framed in terms of a person's tendency to exhibit a consistent response to a particular stimulus situation, which may not be within their control (e.g., Katz 1993). As Hattie (2009) summarizes:

> …students not only bring to school their prior achievement (from preschool, home, and genetics), but also a set of personal dispositions that can have a marked effect on the outcomes of schooling. (p. 40)

For example, we may notice that some students tend to be impulsive when given a challenging task—which can be framed as a maturational issue or a disposition. It's probably some of each, and there is little point in trying to ascertain how much to allocate to either aspect. What is important is that dispositions can easily lead to the forming of habits, which are particular-behavioural sets, often involuntary and automatic. We tend to think of habits in negative terms, such as 'that person has some shocking-habits'. In practice habits can be productive for learning, as in the case of students who have the habit of doing key reading and critical thinking when doing their assignments. MC can help individuals to analyse their dispositions, identify what one's constitutes strengths and weaknesses for learning and well-being, and then take the necessary self-regulatory action. Hence, there is little need to change positive dispositions. These include what Costa (1991) referred to as *Habits of Mind* or *Intelligent Behaviours*, which include such dispositions as 'persisting', 'managing impulsivity', 'listening with understanding and empathy', 'creating', 'innovating', 'thinking flexibly, 'thinking about thinking', 'taking responsible risks', 'striving for accuracy', 'finding humor', 'questioning and posing problems', 'applying past knowledge to new situations', and 'remaining open to continuous learning'.

Whatever the terminology employed, Positive Dispositions, Habits of Mind or Intelligent Behaviours, they are essentially referring to the same attributes (traits/dispositions) that facilitate better learning and well-being. However, equally, many negative dispositions can mitigate one's learning capability and in doing so, impact self-efficacy and well-being. These can be seen in terms of the absence of the above positive dispositions/ habits of mind. For example, impulsivity, poor thinking, sloppiness in doing work, etc. are not good on the learning front, and are likely to have negative impacts across all of life's domains. However, many poor dispositions can be addressed through specific cognitive strategies and tools, supported by metacognitive strategies. Over time, better habits will form, and these will develop neurological correlates that will further support the behaviour—even reducing the innate valency of the disposition over time.

Skills

Skill in the most basic sense is the ability to do something well. For example, one could say that a person has a good skill in heading a soccer ball, presenting a speech, creative thinking, cake-baking, etc. Skill is often confused with competence, as we could equally apply very similar terminology to the above examples (i.e. competent at heading a football, etc.). However, as we know, a person may be very good at doing something (i.e., skilful), but not choose to (or be able to) display his/her skill in real work or life situations, for whatever reasons (e.g., impulsivity, introverted). Hence, the term competence is generally used to describe the work-related performance or complex performances in any field that involve more than skilful activity, but also other important attributes for successful performance—as identified and explored above.

In the case of teaching, and this would apply to any complex activity, there are many skills involved as part of the overall performance (i.e., in the observable behaviour, from which we can infer and interpret a person's competence), and these can vary significantly in terms of complexity. For example, if we talk about presentation skills, which are important in teaching, one aspect of the skill is voice tone. Mlodinow (2012), for example, quoting research by Apple et al. (1979), highlighted the power of voice on person perception. This involved an experiment in which volunteers were asked to judge the attributes of speakers' voices (without seeing them), who were using the same content, but with a specific variation in the vocal qualities. In this way, the listeners' assessments would be based on the influence of those vocal qualities and not the content of the speech. Mlodinow's summary captures the key results:

> ...speakers with high-pitched voices were judged to be less truthful, less emphatic, less potent, and more nervous than speakers with lower-pitched voices. Also, slower-talking speakers were judged to be less truthful, less persuasive, and more passive than people who spoke more quickly. ...And if two speakers utter the same words but one speaks a little faster and louder and with fewer pauses and greater variation in volume, that speaker will be judged to be more energetic, knowledgeable, and intelligent. Expressive speech, with modulation in pitch and volume and with a minimum of noticeable pauses, boosts credibility and enhances the impression of intelligence. (p. 113)

However, it is not just the use of a voice that is impactful in terms of influencing how people experience a speaker, but also a whole host of non-verbal aspects of interpersonal communication, many operating unconsciously, that significantly affect perception and attention. For example, a key behavioural aspect of human interaction which I have long believed to be fundamental in creating a good psychological climate, whether in the classroom or the local coffee shop, is Smiling. Not surprisingly, cognitive neuroscience is providing a strong evidence base for its impact, which has been summarized by Hattie and Yates (2014):

> ...the smile is one of the most powerful tools to use in interpersonal teaching situations. (p. 259)

They go on to note that research:

> ...has documented that a split second's exposure to a smiling face can gently alter people's minds with attitudes to neutral objects becoming more positive, as well as other people being rated more favourably. (p. 259)

The above analysis has important implications for identifying, defining, and subsequent professional development in building teacher expertise. Firstly, it is valid and useful to conclude that voice tone and smiling can be framed as skills (e.g., she is skillful in voice modulation; he has a skill in the way he smiles). While not typically found in competency-based teacher education frameworks, I see them as key human communication (even human conduct) skills, and given that teaching is essentially an act of communication, they are definitely in my toolbox of skills for teacher expertise.

5.5 Competency-Based Teacher Education

An influential approach to framing competencies in teaching is what is known as Competency-Based Teacher Education (CBTE), which works by identifying the range of demonstrable competencies in the teaching role through a process of functional analysis and thereby define standards of quality in teaching. A crucial assumption of C.B.T.E., as Tuxworth (1982) pointed out, is that:

> …the role of the teacher can be described in terms of specific acts, functions or competencies which are observable and which can be learned by the student-teacher or in-service teacher. (p. 5)

In today's context, there is an increasing focus on Outcomes-based Education, of which Competency-Based Training is a key framework. According to Blank (1982):

> The competency-based approach will work equally well in any occupational training area… as well as at any level…vocational, technical, or professional. (p. 19)

> It is simply not true that the competencies that make up competency-based training programs have to be low level or basic skills. If the trainee needs to be able to 'solve quadratic equations,' or 'diagnose the patient's condition,' or 'land a 747 without power'…then it's simply a matter of saying so. (p. 20)

Quality in teaching, within this approach, is defined and assessed in terms of meeting 'agreed' competencies. A teacher's quality can then be judged by the number of competencies met and to what extent. The appeal of this approach is that it provides a clear framework from which to develop teaching competencies, as well as for purposes of appraisal. It identifies performance criteria across specified important areas of teaching (e.g., lesson planning, method use, presentation and explanation, producing and using learning resources, and assessment). Indeed, we would expect teachers to be able to demonstrate these competencies consistently. Such competency areas are systemic to the act of teaching and learning, though with some variation in the framing of content areas and the specific skills emphasized.

Competency frameworks are typically framed in terms of broad competency standards, comprising specific units of competence, which contain more specific subsumed elements of competence. For each element of competence, key underpinning knowledge and performance criteria are stated as requirements for meeting the competence. There is also guidance on the range and context in which the competency is to be demonstrated and what constitutes valid and sufficient evidence sources for deciding that a person has met the competency. Figure 5.2 illustrates the generic structure.

However, the key question is whether or not CTC can be, (1) sufficiently subjected to functional analysis and framed in clear, valid and useful competency-based terms and; (2) learnable by other motivated teaching professionals? For many writers in the field, the CBTE approach is severely limited in terms of addressing quality in teaching because of its narrow and simplistic conception of what teaching is and what teachers do. As Elliot (1991) pointed out:

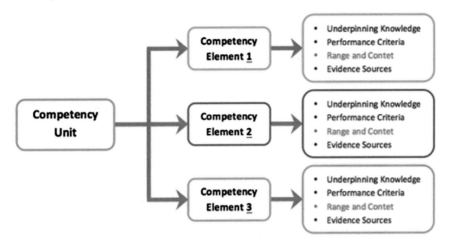

Fig. 5.2 Singapore workforce skills qualification competency-based system

> They are told it will deliver quality. They believe it will deliver managerial control over their performance, leave less room for professional judgement and reduce their status to that of a technical operative. (pp. 118–119)

I have empathy with such critiques of CBTE, as having used this for over a decade in the UK context, I often experience cognitive dissonance in making judgements of teacher competence based on the criteria provided, which often seemed to focus on easy to assess aspects like the font size in PPT slides, positioning in the class-room, rigid sequencing of content, etc. Now, these things impact learning and are important. However, I found from much experience, that while I was accrediting competence based on the performance criteria provided on the assessment proforma, my personal feelings were that this did not constitute good teaching in many cases, especially in terms of emotional, cognitive and agentic engagement. Yes, they met the stated criteria in minimalistic terms, and they were not incompetent, but I would not have employed them. CTC must extend beyond functional competence, as there are essential qualitative aspects to expert teaching, though more difficult to frame and, therefore high inference in terms of assessment.

The Reflective Practitioner Model

Carr (1989) called for a fundamentally different approach to addressing quality in teaching from that of a competency-based model, focusing on teachers' reflecting on their practice. He argues that:

> When teaching is interpreted in this way, 'quality' has little to do with the measuring up to a list of performance criteria but instead is something that can only be judged by reference to those ethical criteria which teachers tacitly invoke to explain the educational purpose of their teaching. This means that teaching quality cannot be improved other then by improving teachers' capacity to realize educational values through their practice. It also means that this process of improvement can be nothing other than a research process in which teachers

reflect on their practice and use the products of their reflections to reconstruct their practice as an educational practice in a systematic and rational way. (p. 11)

The Reflective Practitioner approach to improving teaching is popular and consonant with a constructivist perspective as it focuses on individuals making personal meaning of their learning through their experiences and reflections on those experiences. As Dewey wrote: "We do not learn from experience… We learn from reflecting on experience".

It is argued, from this approach, that conceptions of clearly defined and organized knowledge, which can be applied in universal and systematic ways to the problems of practice, do not capture in the words of Schon (1991):

…the complexity, uncertainty, instability, uniqueness, and value conflicts which are increasingly perceived as central to the world of professional practice. (p. 14)

Also, and most importantly, it does not recognize what Schon (1987) refers to as artistry:

…the kinds of competence practitioners sometimes display in unique, uncertain and conflicted situations of practice. (p. 22)

Teaching from this point of view cannot be seen solely in terms of the systematic application of universal or licensed competencies; rather it involves a process of social construction whereby acquired professional knowledge is dynamically negotiated concerning the unique contexts of practical situations. Furthermore, implicit in such a model is the notion of professional practice driven by professional judgement and ethics. Brophy and Evertson (1976) richly captured the wide range of attributes involved, and it's our challenge to unpack this and provide teacher education that facilitates the learning of such capability:

Effective teaching requires the ability to implement a very large number of diagnostic, instructional, managerial, and therapeutic skills, tailoring behaviour in specific contexts and situations to the specific needs of the moment. Effective teachers not only must be able to do a large number of things: they must also be able to recognize which of the many they know how to do applies at a given moment and be able to follow through by performing the behaviour effectively. (p. 139)

Having worked with both frameworks, there are merits and limitations in each. The competency-based approach provides a systematic structure for teachers to learn pedagogic competencies such as lesson planning & preparation, the delivery of learning, design and using teaching/resources, classroom management, assessing learning, etc. The terminology may change but it's essentially planning, teaching and assessment, and there's nothing wrong with that in broad terms. However, given the focus on clear measurable aspects of these activities, it often does not capture the more qualitative aspects of human interaction that influence student's engagement, beliefs and feelings, and ultimately the quality of their learning. However, I am not arguing that a competency-based approach cannot do this, but in practice, it often results is a more reductionist focus on the easier to define—more quantitative technical aspects of classroom management and methods—missing the more difficult to define qualitative interactional aspects that shape perception and behaviour. Similarly, working with reflective, rather than highly prescriptive criteria, has the benefit of a more holistic and interpretive framing of what is occurring in the classroom, focusing more on how students are engaged emotionally and cognitively, rather than just behaviourally (though not minimizing this aspect of engagement). The major limitation of this approach is that without a strong underpinning pedagogic literacy, the reflection may simply result in what Hattie (2009) described as 'post hoc justification' (e.g., superficial coverage and generalities, lacking sound pedagogic analysis and evaluation against evidence-based criteria—*my interpretation*).

5.6 Expertise and Creativity in Teaching

The work of Hatano and Inagaki (1986) has generated interest in terms of differentiating performance at the highest level of competence, which is expertise. They distinguish between two broad categories of expertise, "routine expertise" and "adaptive expertise". Routine expertise is characterized by a high level of technical proficiency across the typical range of real-world problem-solving contexts. However, as problems become less familiar or novel, the performance of routine experts can dip significantly. In contrast, adaptive experts can reframe problems in different ways, modify or invent strategies and combine skills to deal much more effectively in solving such problems. They suggest a range of factors that encourage adaptive expertise in the context of education and may support creative teaching competence. These include:

- the extent to which the situation has flexibility of options, rather than rigid procedures, to enable exploration of new approaches
- the degree to which people can tackle problems with a degree of playfulness and acceptance of some risk in terms of it not working out in practice
- an organizational culture that encourages better practice from professionals as a key goal. For example, Hatano and Inagaki note:

> …they are invited to try new versions of the procedural skill, even at the cost of efficiency. (p. 270)

Creativity is one of today's global buzzwords and figures high on lists of so-called twenty-first-century competencies and skills. Furthermore, if it is such a necessary attribute, so essential in the worlds of engineering, business and medicine, it should also be similarly valued in education—and *teaching* is the core activity of education. It is important, therefore, to make the best sense we can on what creativity entails in the context of teaching, and the ways in which it can be utilized at the level of practice. As a basic assumption, it would seem logical to approach creativity in teaching as analogous to creativity in any domain, as it involves combining existing knowledge in some new form to get a useful result. As Amabile (1996) suggested:

A product or response will be judged creative to the extent that (a) it is both a novel and appropriate, useful, correct or valuable response to the task at hand, and (b) the task is heuristic rather than algorithmic. (p. 35)

In any attempt to define creativity there are inevitable questions about what constitutes the novel, in whose eyes, by what criteria, and to what extent? Furthermore, the notion of useful, correct or valuable also involves subjectivity. Rap music may meet certain criteria of novel, but in no way does it feel useful or valuable to me.

However, because something is difficult to define in precise and uncontested terms should not detract one from working towards useful heuristics and their practical application, as outlined earlier. Most things at some point in time were a mystery but eventually move down the knowledge funnel (Martin 2009) as a result of systematic inquiry and evidence-based practice—as illustrated in the case of HIV in Chap. 1.

If novelty, in some form and at some level, is foundational to creativity, then effective teaching may not entail creativity. For example, a teaching professional applying the core principles of learning, selecting high effect size methods and using them thoughtfully to the situated context may be teaching very effectively, even perhaps a "routine expert"—though not an "adaptive expert" in terms of the differentiation of experts by Hatano and Inagaki (1986). To frame creativity in the context of teaching, it is essential to identify, in realistic and specific terms, what this might entail in both the design and facilitation of learning experiences. For example:

- What specifically can be considered novel and useful in the context and practices of teaching?
- What are the processes and activities that can generate novel resources for incorporation into the design and facilitation of learning experiences?

Firstly, creativity, like wealth and beauty, are value-laden and relative. When I travel to some countries, in certain locations I get a sense that some people think I am very wealthy in financial terms. However, in certain social circles in Singapore (and this is not specific to Singapore) I could feel relatively impoverished, as I don't own the condominium I live in and only have one modest car. In making sense of creativity the same framing applies, in that novelty and usefulness is relative and one of extent. Fasco's (2006) *creativity continuum*, (identified prior, and refreshed here for context), in which creativity can extend between two poles: *Big C* for 'extreme forms of originality' (e.g., Nobel-prize winners in science) and *Little c* for 'everyday creativity' (e.g., adding butter to coffee to make it tastier), has usefulness in framing creativity for practical purposes. For example, if creativity is framed primarily in *Big C* terms, then notions of developing a better *creative competence* for any professional group (teachers included) becomes a very tall order indeed. In contrast, if we see creativity in terms of such a continuum, then we enter a completely different arena for conceptualizing creative teaching; one that is both challenging and realistically achievable for any motivated teaching professional.

Secondly, it is important to understand how the creative process works, especially the underpinning thinking processes, as these are fundamental to producing creative outcomes. There is certainly an extensive research literature base on all aspects of creativity and many factors have been identified as contributing to such outcomes, including biology, biography and the systematic use of creative techniques and tools. Changing biology is difficult, and past biography is exactly that, hence a focus on the creative thinking process and how the mind works may be the best avenue for enhancing creative capability in practical ways.

I like travelling to different countries and have been fortunate to work and participate in a wide range of cultural contexts. I have also seen most of the acclaimed tourist sites, especially in Asia. However, what I find most interesting is talking to people in local eating places, sharing stories and finding mutually meaningful humour. For me, stories provide the key narrative to understanding the human condition and building rapport with people, irrespective of culture and location. How the creative process works can also be framed in terms of stories, as each creative act has a story to tell. One story that comes readily to mind and it fully fits a *Big C* categorization is that of Percy Shaw and his invention in 1933 of the *cat's eye*, a road stud for lighting the way along roads in the dark. While there are a number of stories on what led to him inventing the cat's eye, a popular version (one that I like anyway) is that on a foggy night in 1933, when he was driving back to his home in the Boothtown area of Halifax from nearby Bradford, he hit a perilous stretch of road with a sheer drop down a hillside to the right of the road. It was very dark and Percy could not see where the road ended and the hillside began, until suddenly he spotted, in the darkness, the reflections of his car headlamps in the eyes of a cat sitting by the road. It is then that he is said to have hit upon the idea of replicating the reflection of a cat's eyes to guide drivers along dark and dangerous roads.

The main purpose of this story is to illustrate some important aspects of creativity. Firstly, once invented, a creative act often seems so simple and logical. How many people before Percy Shaw had, on a dark foggy night, noticed the reflective power of cat's eyes to light, but failed to make the internal neural connections to create a new perception which may have generated the idea of a reflective road-stud? As De Bono (2003) emphasized: "…every valuable creative idea must always be logical in hindsight." (p. 24).

Thirdly, creativity is not a process of 'thinking out of the box', which is impossible, I think. Rather it is more useful to think of it as a process that changes internal neural connections and creating new representations 'in the box', which is, of course, the human brain. The process moves on further when these 'new representations' become conscious and surface as a new perception of phenomena in the external world—as happened to Percy Shaw above. Typically, without disruption, the brain will process information through established learned neural networks, and that makes good sense as life would be chaotic without a high degree of perceptual consistency. As De Bono (1992) summarized:

> What it all amounts to is a system in which incoming information sets up a sequence of activity. In time this sequence of activity becomes a sort of preferred path or pattern. (p. 17)
>
> So whenever we look at the world we are only too ready to see the world in terms of our existing patterns… (p. 18)

For many people, everyday life is a fairly ordered series of activities in which existing neural networks fire in relation to well-known and predictable stimulus events, which further reinforce those connections. There is little need for creativity, or the likelihood of it occurring. An interesting question is whether or not continually enhancing one's knowledge will eventually result in creativity. Such activity will certainly increase neural density and elaboration in long term memory and the notion would seem to have face validity in that many big C people fit this description. Leonardo Da Vinci, for example, was no sloth on the knowledge stakes, frequently referred to as a *polymath* (i.e., a person whose expertise runs across several subject domains and professional fields).

However, creativity involves more than having rich knowledge bases and expertise. Many experts are not noted in the creativity ratings. Hence, while expert knowledge bases may enhance the likelihood of a new perception that results in a creative outcome, it is far from guaranteed. Other factors are invariably important, such as personality, other neurological features, belief systems, effort, and typically, some luck. For example, as with successful learning generally, what may be of particular significance is the desire and belief in one's creative capability, and the persistence to keep going with a problem scenario until a creative perception occurs. As Albert Einstein is famously quoted:

> It's not that I'm so smart, it's just that I stay with problems longer.

Furthermore, unless situations dictate or there is a process of continuous deliberation to break up or at least challenge existing perceptions, additional information will still largely be processed within the existing neural organization. For Creativity to occur, it is necessary to be able to perceive some aspect of reality in a different light, and that requires some internal neural restructuring of existing knowledge. As Mauzy and Harriman (2003) describe:

> …breaking and making connections is where the fundamental action of the creative process takes place, and what's known in the fields of psychology and brain physiology lines up with this. (p. 22)

As a result, De Bono's (2003) is correct in arguing that:

> We need creativity to break free from the temporary structures that have been set up by a particular sequence of experience. (p. 27)

This can be facilitated by deliberative interventions in which existing neural pathways ("main track") are disrupted and new ones created ("sidetrack"), especially through what he refers to as the techniques of provocation:

They are methods of helping us to escape from the main track to increase our chances of getting to the sidetrack. That is also the basis of the expression *lateral thinking*. The 'lateral' refers to moving sideways across the pattern instead of moving along them as in normal thinking. (p. 24). The purpose of this is to take us out of the normal perceptual pattern and to place our minds in an unstable position from which we can then "move" to a new idea. (p. 71).

De Bono challenges the view that the brain is naturally creative. He acknowledges, however, that:

New ideas may be produced by an unusual coming together of events. New ideas may be produced by a chance provocation provided by nature… (p. 67)

In terms of explaining Percy Shaw's creative act of generating the idea of the 'cat's eye', the notion of 'an unusual coming together' and 'chance provocation provided by nature' seems to fit nicely. Who knows, if the cat had not been there, on that night, would the new perception have emerged? How methods of provocation can relate to producing creative instructional strategies for teaching (i.e. developing creative teaching competence) will be explored later. The important point to emphasize here is that novel perceptions must inevitably have, as their basis, the elaboration and restructuring of neural configurations. Creativity is essentially just another aspect of the generic process of learning, but with a different cognitive spin (so to speak); the building of more differentiated frames on reality. As de Bono makes fully explicit:

In my view learning, creative thinking is no different from learning mathematics and any sport. (p. 57)

A similar analogy can be applied to the naturally funny people in life, whether professional comedians or otherwise. Do such people have funnier experiences than those lacking humour, or do they actively look for the funny side of the experience, provoking new ways of experiencing everyday reality, hence creating the conditions in which funny outcomes are more likely? For example, in East London humour (well when I lived there some 30 years ago), there was a type of humour which was referred to as 'selling a dummy'. This involved making a silly statement in jest and waiting to see the response of others. For example, if someone talks perhaps too positively about someone else (e.g., they are very talented, kind, generous, etc.), a listener may respond by saying, "Well, she must have some good points as well". If this results in the speaker, taking this seriously and responding with some mild annoyance, he/she has been "sold a dummy". Now, what's clever, if the person who was being sold the dummy initially does not falls for it, and sells a dummy back, it's called a 'double dummy' and that's a very witty thing to be able to do. Can you do this?

As a Cockney from East London, I think I know the answer to the question posed in the last paragraph. If you look at things in the same way and do the same things in the same situations, you will typically (unless there is a chance provocation) get the same results. To get different results, it is necessary to do something differently. Hence, it is not surprising that people who desire and persevere in deliberately connecting things that may not initially seem to be naturally connected and look for new ways of perceiving aspects of reality are likely to produce more creative outcomes, whether in engineering design, teaching, or in making people laugh.

Finally, in terms of fostering creative outcomes, there appear to be recognizable phases, mindsets and activities that can be systematically employed to stimulate and enhance creativity. For example, Petty (1997) described the creative process as consisting of six interrelated phases: inspiration, clarification, distillation, perspiration, evaluation and incubation (p. 15). He also highlighted:

One of the main difficulties for creative people is that the different phases require radically different, even opposite 'mind-sets', each of which is difficult to sustain without deliberate effort. (p. 19)

Most significant in the context of this chapter is moving the focus of one's mind through the different stages, from generating new possibilities and applying more critical thinking frames (e.g., analysis, comparison & contrast, inference and interpretation, and evaluation) until the idea reaches fruition and practical application. This is in many ways the result of good thinking, which not only involves managing the thinking process (cognition) but the whole swirl of beliefs, emotions and other vagaries of the human mind. Perspiration, which is massive effort over time by another name, is an expected necessity in most cases, especially for coming up with something exceptionally novel and useful (e.g., a big C creativity outcome) as this is far from easy as we all know. Thomas Edison made the point most bluntly:

Genius is one percent inspiration and ninety-nine percent perspiration.

Of interest is the phase Petty referred to as 'incubation'. Creativity, in terms of creative outcomes, cannot be summoned up at will over a designated period-of-time (e.g., let's be creative in the next 3 h). We may focus our minds on various phases in the process, but incubation has its patterns of behaviour and they are outside of our conscious control. As Petty (1997) pointed out:

Many brilliant ideas have occurred in the bath or traffic jams. If you can stop work on a project for a few days, perhaps to work on other things, this will give your subconscious mind time to work on any problems encountered, and will also distance you from your ideas so that you are better able to evaluate them. (p. 8)

Claxton's (1998) analysis of the interplay between our fast conscious mind (which has some similarities to Kahneman's, 2011, description of "System 1 Thinking", outlined in Chap. 2) and a slower more fluid 'undermind' of "unconscious awareness" (p. 10), which acts as an "intelligent unconscious" (p. 133) is particularly interesting in this context. It sees creative ideas as being slowly and unconsciously brewed in the neural re-configurations of long-term memory and when sufficiently structured, flashing from the unconscious mind into conscious thought. In other words, while we are not consciously seeking a creative solution our mind slows down, becomes more relaxed and uninhibited, enabling it to do such creative work in its way, and eventually switching on that elusive new perception on reality. Claxton (1998) may have captured this internal process accurately:

Interesting intuitions occur as a result of thinking that is a low focus, capable of making associations between ideas that may be structurally remote from each other in the brainscape. (p. 148)

He went on to summarize the wider process of creative thinking:

The creative mind possesses a dynamic, integrated balance between deliberation and contemplation. It can swing flexibly between its focused, analytical, articulated mode of conscious thought and its diffused, synthetic mode of intuition. (p. 96)

The above analysis on how creativity works in terms of psychological and neurological functioning, and factors that may promote its development and capability, helps to similarly frame creative teaching from a more evidence-based perspective. Firstly, using Amabile's (1996) definition earlier, I previously offered (Sale 2015) the following operational definition of creative teaching:

> Creative teaching occurs when a teacher combines existing knowledge in some novel form to get useful or valuable results in terms of facilitating student learning and attainment. This may be either planned before the act of teaching or *invented* as a response to the demands of the here and now learning situation. (p. 100)

Secondly, in terms of Fasco's (2006) *creativity continuum*, we are looking more towards *Little c*. However, while it may be *Little c* in that it may go unnoticed except by those directly influenced, the cumulative impact of such teaching over time will significantly impact the perception, beliefs and actions of students towards better learning experiences and attainment. It might only be *Little c* in the world stage, but *Big C* for those students who get inspired to learn, attain better grades, achieve goals that are meaningful to them and experience well-being—*it changes their lives*.

Thirdly, ensuring a high level of competence with the longer-term aim of developing "adaptive experts" rather than "routine experts" (Hatano and Inagaki 1986), provides a clear viable goal for framing the creative teacher. Creative teaching is, therefore, different from effective teaching, but both involve a high level of pedagogic literacy and proficiency in terms of technical competence in the practices of teaching, as outlined earlier in the chapter. However, creative teachers have the added capability of combining existing knowledge to produce novel and useful learning experiences as well as being able to reinvent their pedagogic strategies in situ, to meet changing demands in different learning contexts. They can see more flexible connections between the technical skills they possess and the range of resources that can be accessed and weaved together to create a better-situated strategy for supporting learning then. Hattie's (2009) concept of "Russian Dolls" has relevance again in this context as it captures the ability to creatively combine several high effect size methods and e-tools into 'optimal' instructional strategies. Creative teachers are more able to make connections between methods, activities and resources that may not always seem to be logically connected, but in practice make significant impacts on aspects of the learning process. This is lateral thinking (De Bono) in operation, and it will often involve teachers consciously provoking themselves to create new ways and activities for teaching difficult concepts or processes. Teaching from this perspective is outlined in Fig. 5.3. Note: this is a *conceptual model*, not hierarchical in that one stage must be achieved before the next. It is essentially iterative. However, Competent and Creative teachers employ a strong Pedagogic Literacy—whether explicitly or tacitly.

Some teachers often complain that their students are bored in certain lessons and don't show any interest in learning. This is a legitimate complaint but often, from my experience, they still employ the same instructional strategy for those topics over again, and the inevitable happens—the results are usually the same. Creative teachers in such situations do things differently and most importantly what they do

Fig. 5.3 Summary frame on teaching expertise

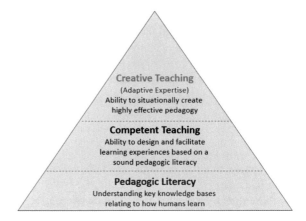

differently tends to work better. Over time, using an EBT approach with lateral thinking (e.g., deliberate practice using lateral thinking), most teachers, those motivated to do so, will get better and quicker at coming up with more interesting and effective components to infuse into their lessons. They will be developing ***creative teaching competence***.

5.7   The Syntax of Creative Teachers

For over two decades, I had responsibilities for the mentoring and coaching of under-performing teachers, and many of them were 'conscripts' (e.g., sent to me for improvement). Over a decade ago, having worked with hundreds of such folk, in many different contexts, I noticed patterns of behaviours and thinking that made understanding what ineffective teachers do quite easy to work out. It has variation in nature and form, but there is an underlying syntax of ineffective teaching behaviours, irrespective of whether they are intended or not, and in most cases, it is not intended. Essentially, in the context of EBT, they violated many of the core principles of learning very consistently and to some measure of negative effect. Eventually, having cracked the 'code' of poor teaching—so to speak—it occurred to me, that it is equally tenable to be able to unpack what effective creative teachers do and how they do it (e.g., the work of Bain 2004, was pioneering in this context). Hence, for more than a decade, I have continued in this vein as a researcher in Singapore. I must confess, that developing the creative teaching framework documented in Creative Teaching: An Evidence-Based Approach (Sale 2015) was not a question of switching on a 'creative thinking switch', but the very opposite; many thousands of hours of hard work, probably building up—incubating—over many years (much occurring unconsciously)—as documented earlier in the chapter. It was like that. As Winget (2007), in his book titled "It's Called Work for a Reason"—fits my experience well.

In observing teachers who consistently get high attainment results and positive feedback (both quantitative and qualitative) I typically notice high levels of engagement and, in particular, good rapport with students. Most significant, was that while they might have different teaching styles, there is an underpinning syntax in what they do in terms of certain behavioural elements that positively shape student experiences, which lead to such results. While people make sense of the world based on prior experience and selective perception, our common human apparatus and need orientation typically results in shared ways of experiencing the world. Indeed, without this commonality, the inter-subjectivity of everyday life would be even more problematic than it is already. As Marton (1981) argued:

> …we have repeatedly found that phenomena, aspects of reality, are experienced (or conceptualized) in a relatively limited number of qualitatively different ways. (p. 181)

You may recall from Chap. 1 that even the defining of effective teaching has proved contentious and problematic over the decades, so what chance is there of achieving an adequate definition of creative teaching? I can only offer a frame on this and let's initially recognize fully that we have little choice but to live with a fair measure (whatever that is) of subjectivity in making sense of the world. For example, we have beauty contests and there are judges, contestants, decisions made, and winners identified. Well, how does that work when supposedly, "Beauty is in the eye of the beholder?" Well, it is and it isn't—right? Yes, beauty is subjective, but there's a lot of common agreement, explicit or otherwise, about what its key features are, at least at a heuristic level; and that's why I have yet to win one.

Now, what's important is that in the mediation between teacher and students, and between students, some things facilitate learning and well-being much better than others—as we have documented, and this is the basis of EBT. However, much is subtle and working unconsciously, and appears almost ephemeral—but that is because the mind 'is as the mind is'—essentially chaotic and open to subtle influences in terms of impacting mood or psychological state. However, combinations of certain behaviours, often linked to specific activities, create positive perceptions and feelings about what is occurring, and we know what positive feelings do—create a nice little shot of dopamine.

The following sections outline and illustrate some specific notable features and aspects of CTC, which I have captured in the acronym SHAPE (Stories, Humour, Activities, Presentation Style and Examples). This acronym provides a useful and easy to remember metaphor as it is so much a term in our everyday vocabulary for qualitative descriptions of things. For example, when a person has attained a high level of physical fitness, which is visible in terms of muscle tone, etc., we might say to the person that he or she is in "good shape". The converse is also true at the level of perception, though we are highly unlikely to say this to the person. In the context of creative teaching, SHAPE was just something that came into my conscious mind as a result of indwelling about what creative teachers do (based on students' qualitative feedback on these teachers) at the behavioural level in their classrooms. There were so many references to 'stories', 'humour', 'interesting activities', 'personality of the teacher' and 'good examples' in the students' qualitative responses about teachers that were perceived as creative or interesting. These same teachers also got responses very much in tune with Willingham's (2009) description of the teacher as 'being a nice person'. It is interesting that Willingham, in reviewing researchers' analysis of feedback questionnaires to figure out which professors get good ratings and why, noted that one of the interesting findings is that most of the question items are redundant. He suggests that:

> A two-item survey would be almost as useful as a thirty-item survey because all the questions really boil down to two: Does the professor seem like a nice person, and is the class well organized…Although they don't realize they are doing so, students treat each of the thirty items as variants of these two questions. (p. 50)

While academics may break up the components of highly effective teaching into a wide range of sub-components or constructs, this is not how students perceive and apprehend the experience of their teachers; rather their perception is based on more holistic generic constructs such as *personality* and *organization*. As Willingham summarized:

> When we think of a good teacher, we tend to focus on personality and on the way the teacher presents himself or herself. But that's only half of good teaching. The jokes, the stories, and the warm manner all generate goodwill and get students to pay attention. But then how do we make sure they think about meaning? That is where the second property of being a good teacher comes in - organizing the ideas in the lesson plan in a coherent way so that students will understand and remember. (p. 51)

Furthermore, it is highly likely that this evaluation process by students will be based on unconscious as well as conscious processing. Students are likely to relate the experience of their teachers to key aspects of motivation—pleasure, pain avoidance and novelty in relation to their need orientation. From this they will derive varying degrees of meaning, which will then translate into ratings for the teachers being evaluated. The disorganized and dull, even mean, teacher, is going to fare badly in most cases. The interesting question becomes, "*What are the components of the experience that lead many students towards perceptions of a very well organized and nice teacher?*".

From an EBT approach it is not too difficult to largely answer this question, as a teacher who can effectively employ the core principles of learning thoughtfully in the situated context, is likely to be perceived in such ways—though not necessarily by all the students all of the time, as that's how life plays out.

Furthermore, once framed in the wider context of an evidence-based approach to teaching, SHAPE seemed so simple as a metaphor on creative teaching. In retrospect, for most of the least effective and dullest of teaching observations I have been involved with, the description of 'poor SHAPE' fits the overall experience very accurately. Although, unlikely to figure on a formal evaluation form for teachers, a teaching session with no stories, no humour, no engaging activity, poor presentation style and no examples, is most likely to be a drab and ineffective affair by any criteria.

5.7.1 Stories

Human history is a collection of stories of how we have attempted to make sense of the world around us, find solutions to a whole range of existential problems and even explain the nature of our very existence. As Schank (2011) argued:

> Human beings understand stories because stories resonate with them. Characters have dilemmas that readers or viewers themselves have had. Stories appeal to emotions rather than logic, and emotions are at the heart of our pre-7-year old unconscious selves. (p. 42)

When there are no more stories to tell, we may be in that perfect world where thinking is redundant and there are no problems to solve. At the personal level, we communicate our experience through the stories we tell: they reflect who we are, the sense we have made of our experiences and they become a stimulus for other people's perception and the quality of attention they are likely to give us.

Watch the very best speakers in any field and there is typically a story in their presentation, invariably a very poignant one to their main purpose and audience. From the perspective of enhancing a learning experience, stories connect powerfully with others as they immediately associate with their own experiences, especially at the emotional level. From an evidence-based perspective, as the cognitive neuroscientist Willingham (2009) suggested:

> The human mind seems exquisitely tuned to understand and remember stories – so much so that psychologists sometimes refer to stories as "psychologically privileged," meaning that they are treated differently in memory than other types of material. (p. 51)

Learners may forget the factual content of our lessons but stories that are embedded with meaning, especially when it connects to their own experiences, needs and interests, are committed easily to memory and provide a powerful anchor for recall. Stories can also be transformative in that they connect with people emotionally and are a key means of enhancing positive beliefs. As noted previously, beliefs are no more than perceptions that have been around a long time, but they are real to the believer. There is often little point in telling students that they are smart when they believe they are not. Changing limiting beliefs most readily occurs when people are confronted with evidence, over time, that consistently contradicts the belief, and which is perceived as real and meaningful. This is where stories can provide an effective means for initiating alternative ways of looking at the world (i.e., reframing). Reframing refers to looking at things in different contexts and, in doing so, give them different meanings. For example, in certain contexts, if I feel it may have relevance and meaningful learning impact for students, I have deliberately used personal stories relating to my own experiences at school (you will have noted this from Chap. 1). Has this led to some students actually significantly reframing aspects of their belief systems and then going on to make the necessary behavioural changes (e.g., put in personal effort, acquire key knowledge and skills, develop metacognitive capabilities, persist when things get tough and seek good feedback) and, as a result, becoming more successful (however defined)?

Typically, as teachers, unlike plastic surgeons, we rarely know how impactful we have been in some transformative way on our students' lives. For plastic surgeons, given that all as gone as planned, there is likely to be immediate grateful feedback from the patients concerned. However, for teachers, significant personal change in students is usually difficult to ascertain as it often takes time to develop and by then they have probably long left our tutorage. I am perhaps fortunate to have experienced one very striking example that occurred from a chance encounter. Some 20 years ago, I had just completed a consultancy assignment in Hong Kong and was sitting in the airport lounge waiting to catch the flight back to Singapore, when I heard someone call my name. Looking around, I recognized a student I had taught previously in the UK but had not seen since he graduated as an electrician from a further education college many years earlier. He recognized me in this far away location, which was pleasing as it meant I had not aged that much. We had a drink at the airport, and I discovered that since leaving the college he had completed a degree in electrical engineering and established a successful company. However, what was significant was his reference to my influence on him in terms of affecting his beliefs about himself through the stories I told of my experiences in growing up in a tough East London community. I was quite surprised at the time by his perception of my impact on his thinking and behaviour, but on reflection, it's not that surprising given the way the mind works.

Perhaps, even such a chance encounter, in a globalized world is also not so surprising. However, generally, and perhaps sadly, teaching is perhaps, as was once described to me, "The second most private act". How many people have been significantly influenced by good teachers, as Petty (2009) described in Chap. 1? Most of those teachers will never know and will remain 'unsung heroes. Equally, many may not care about such acclaim. That's because they chose to teach and they know what this means.

Also, it is important to bear in mind that stories do not need to be highly exceptional in terms of the human experience they communicate. They simply need to be authentic, relate to what is being taught and be meaningful to the learners involved. For example, I remember a teacher using a personal story to communicate the experience of dealing with a 'no' in the context of working as a salesperson. This was a business studies lesson focusing on selling. One of the big challenges that salespeople face in dealing with constant 'no's' to their offers of products or services, is a loss of personal feelings of esteem and confidence. His story involved asking a girl for a date over the phone (in the days of rotary dial phones). As he told the story, he modelled the dialing of each number, showing his nervousness visibly. He explained his fear about her saying no and made the powerful point that unless you ask in the first place, you already have a 'no' and if you get a 'no', then you have a simple choice—persist or give up. The story had a happy ending—she said yes. Stories represent a unique, novel and personal means to enhance the learning experience for students. The very best personal stories, used in context, are a key component of creative teaching as they provide both deep anchors for remembering the subject context and, even more importantly, a potentially positive transformative experience for many students. Table 5.1 provides an advance organizer for building a portfolio of useful stories.

Table 5.1 Summary framing questions for using stories

Stories	
What are the different ways in which stories can be used to creatively enhance student learning?	• Provide interesting and effective advance organizers for a new topic area or key concept • Create emotional anchors for enhancing learning impact and building rapport • Model good attitudes and dispositions, as stories will evoke mirror neurons which have strong subconscious impact on perception and, in the longer term, beliefs
What is important in telling stories?	• A clear and lively presentation style • Timing and emphasis of key learning point(s) in the story • Relevance to the specific topic to be learned and/or to the process of learning • Involve students where possible in terms of eye contact, posture and gestures • Draw out and analyse the relevance of the story to the topic (if necessary)
Where can I get useful stories to make lessons more interesting and creative?	• Personal experience • Colleagues, family members, friends—almost anybody • Media sources such as books, journals, newspapers, television, films, internet, etc.

5.7.2 Humour

The importance of humour and its uses to enhance learning were identified in Core Principle 9: *A psychological climate is created which is both success-orientated and fun.* The ability to create and use humour productively for the benefit of others is a rare and highly sought-after skill, which may explain why professional comedians earn significantly more than teachers. Useful specific definitions of humour have proved problematic, as it takes many forms and is always situated to persons and context. What some people may find extremely funny, others may simply find deeply offensive. Earleywine (2011), in this context, frames humour generically to focus on its interpersonal and outcome features:

Humour is an intricate interaction between the perceiver and the perceived…

…humour is anything that somebody deems funny. (p. 21)

Similarly, Tamblyn (2003) sees humor as "a state or quality", which has a number of aspects:

…humor is openness, optimism – a sort of yes-saying to life. Humor is creativity. Humor is, above all, play. Humor, creativity, and play are the same thing because they all involve the same act: *Finding new connections between things.* (pp. 9–10)

While there are many genres or types of humour (e.g., jokes, anecdotes, wise-cracks, witticisms, banter, wordplay), they all typically play out in an interpersonal context. If the humour is perceived by some as funny, it will result in smiles or laughter in some form. You may also have noticed that when some people laugh, it often has a contagion effect causing others to join in, often unknowing of the exact source of the humour. It is as though laughter of others, just as listening to sad songs or watching emotional scenes in a film, evokes our mirror neurons (Rizzolatti and Sinigaglia 2008). These are neurons that fire automatically when observing someone else having an experience, creating an inner feeling of having the experience oneself and we respond with sad emotion, often with tears, almost instinctively. Mirror neurons may also be an important consideration for how we interact with students in other ways. For example, showing enthusiasm and displaying good equity in dealing with students may have similar productive unconscious influences at the neural level.

In a similar vein, Martin (2007), drawing from brain imaging research, reports that exposure to humorous cartoons activates the well/known reward network in the limbic system of the brain (e.g., Mobbs et al. 2003). The funnier the particular cartoon is rated by a participant, the more strongly these parts of the brain are activated. As he summarizes:

> This explains why humour is so enjoyable and why people go to such lengths to experience it as often as they can; whenever we laugh at something funny, we are experiencing an emotional high that is rooted in the biochemistry of our brains. (p. 7)

Humour has a particularly powerful effect on human motivation as it affects all motivational dimensions. Humour creates pleasure (typically manifesting in laughter), reduces pain (as it distracts attention away from the object or perception of pain, if only fleetingly) and is typically novel in some way. Humour and creativity may well be fundamentally linked in terms of shaping aspects of our subjective experience. As Earleywine (2011) noted:

> Creativity and humour appear to go hand in hand. Some researchers view humour as another form of innovative, inspired flair…A few minutes of comedy, if it leads to genuine guffaws, can make folks happy and innovative. A good mood enhances creativity anyway, at least up to a point. (p. 137)

Similarly, Morrison (2008) argued that:

> The creative process flourishes when accompanied by a sense of humour. (p. 3)

Perhaps the most critical aspect of humour for learning is its impact on attention and perception in the learning situation. Firstly, humour, by definition is typically unpredictable and often is a surprise element in human experience. Hence, it has both a Von Restorff effect as well as creating a sense of immediacy (e.g., Anderson 1979). Martin (2007) summarizing the research on the impact of immediacy, notes:

> Past research has indicated that greater levels of immediacy are associated with more positive student attitudes towards the class and instructor, greater enjoyment and motivation, and greater perceived learning. (p. 353)

However, a new joke that we find funny, is funny because of its novelty. If we hear someone tell it a second time, it may be boring. My wife, for example, will berate me for repeating a joke or funny story with "We've heard that one before." Now, of course, when told to a new audience, it will be novel and if their mindset is 'in sync' with that type of humour it will get the usual positive response. If something is new to a particular person, it is creative in their eyes. The essential point is that humour typically catches the attention of the brain as it creates a strong Von Restorff effect, as we explored in Chap. 1 and as Morrison (2008) explained:

> The surprising aspect of humour affects the attentional centre of the brain and increases the likelihood of memory storage and long-term retrieval. Humour has the potential to hook easily bored and inattentive students. As brain food, humour can't be beaten. (pp. 2–3).

Note, in teaching, while a good piece of humour will work well in getting attention and helping to build a positive perception of you in the students' minds, but it has no or limited positive impact once repeated. However, the good news is that once one can use humour, there is an almost unlimited supply of resources. It's not the genre that becomes habituated to, it's only the specific example, whether joke, cartoon or story.

Humour is not just an attention grabber; it is also an experience shaper. I have a good friend who has a far greater creative capability for humour generation than anyone I know. He can tell the funniest of stories, display spontaneous wit, and typically gets folk laughing almost at will. When he is not present at an event, people notice immediately and ask, "Where is Tom?" If he is not coming, the groans of disappointment can be audibly heard. Tom's presence creates pleasure and novelty and people feel comfortable in talking (and laughing) openly with him. He creates that type of rapport.

In the context of teaching, the same patterns of human attention and perception play out, again much of this is unconscious. Being liked is a big factor in positive student evaluation of their teachers and humour shapes this perception helping to foster a positive psychological climate and facilitates the building of rapport. There are several subtle interacting components at play in this experience formation. Firstly, as Morrison highlighted:

> Humour thrives in an environment of trust and is a major factor that contributes to building trust. (p. 6)

Similarly, as Liston and Zeichner (1990), drawing on the work of Macmillan, suggest:

> Honesty and trust are inherent in the activity of teaching, irrespective of context or time... (p. 236)

Collectively, the experience works towards engaging students more emotionally in the learning experience. It's not rocket science to think of ways in which people, teachers or otherwise, can evoke such emotional states in others. In situations in which negative emotions are evoked (e.g., fear, disgust, anger), the only attention given is one of 'how best to exit or put up with the existing reality'. The converse is also true, enthusiastic, humorous and fair-minded teachers are more likely to be perceived as 'nice people'. Morisson (2008) goes as far as arguing that:

> Humour is a key element in building positive relationships with students that will make classroom management an invisible element. (p. 59)

Furthermore, humour is more than a strategic technique to generate attention or refresh the brain but is fundamental to our basic need structure, enabling us to experience both the joyful emotions as well as deal more effectively with some of the more negative aspects of human experience (e.g., stress and personal loss). The case for humour as a powerful resource for creative teachers is strong from an evidence-based perspective, both in terms of its outcomes on student learning and attainment, and as a key aspect of the creative design process and ability to teach creatively in situ. However, for many teaching professionals, this may seem a daunting challenge. Many concerned teachers have asked me, particularly in the context of students who appear increasingly distractible in class (so-called Gen Y students), "Do I have to entertain the student now?" and "Am I supposed to be a comedian also?" Well, there's some bad news and some good news. The bad news is that students may well be more distracted in formal classroom situations, in the sense that they won't just sit there and be bored as many might have done in yesteryear. Instead, they may simply access their mobile devices and indulge in more pleasurable activities, disrupting the classroom in some non-productive way. The good news is that we now have *much* more knowledge on how to teach better, much better and creatively. Of course, we are not going to motivate all students all of the time, that's a silly notion. It's like asking medical professionals to cure all illnesses for all people. The reality is that teachers do not need to be highly capable in the comedy stakes, but they should recognize the benefits of humour in creatively enhancing student learning and attainment. As a teacher, you need little competence in delivering humour, just the intent to foster it. As students don't generally expect their teachers to be humorous, when this is the case, it is typically experienced as a Von Restorff effect (though students are unlikely to frame it in these terms). I remember vividly an interesting scenario when observing a business studies lecturer teaching 'optional pricing' to a large group of young adult students in a lecture theatre format. He used his honeymoon in New Zealand as the context for some humorous ways to engage and entertain the students, as well as teach the concept in varied and authentic contexts. He explained that on his honeymoon in New Zealand there were many tours with options, at additional costs, of which one was 'bungee jumping'. He added quizzically that he did not choose this option in case an alligator was waiting below and ate him. You probably did not find this particularly funny, neither did I. However, the widespread laughter from the students was quite striking. What is significant is that a teacher does not need much skill in being humorous, only the intent to encourage lighter moments. Students react to what the teacher is trying to do, which is to make the learning experience a bit more fun than is often the case in many dry classrooms. This teacher simply used a bit of humour to lighten up a morning lecture while effectively teaching the designated content knowledge in the syllabus outcomes. His use of humour, as well as stories, had helped to create a positive psychological climate that worked very well in terms of student learning. This teacher did win an excellence in teaching award, and it was not difficult to see why. Apart from the ability to use humour, there were several other aspects of his practice that, as a total experience, made perfect sense from an evidence-based perspective. He also had the creative component—using his honeymoon experience (a Von Restorff effect) as the foundation for teaching optional pricing. That's everyday *Little c* creativity, and it's not that difficult to do when one understands the underpinning syntax and heuristics.

Humour then, like other aspects of human capability can be understood, and, while difficult to define, can be described in very specific terms and therefore, as Morisson (2008) argued:

Humor is a procedural skill that can be learned. (p. 58)

For example, when telling a joke, as when telling a story, there are key aspects of an effective presentation. These include keeping it moving fairly quickly, using movement and expressive voice tone when modelling a conversation between people, and a quick pause before the punchline. This is not difficult to model, understand and, with some deliberate practice, show a reasonable proficiency in delivery. When people tell me they can't do this, I can usually change their minds in around 1–2 h, through modelling of the story/joke and getting them to do the necessary practise with appropriate feedback. They are not quite ready for a professional career as a comedian but are good to go in terms of adding this humour component to their teaching skill repertoire of resources. Even easier is the use of materials', such as audio or video clips. The important point is that you don't have to be funny, just the communication of this intent is a rapport builder and communicates your humanness. I recall one teacher who, every week, shared either an amusing story about his dog or a cartoon with his students. It was hardly highly sophisticated humour, but it always lifted the mood in the classroom and communicated to the students that he was making an extra effort to make learning more interesting (or at least, less painful).

Quick wit is a really powerful humour resource but requires more skill in terms of recognizing when and how to use it. I have seen this used by skilful teachers to manage a wide range of potentially disruptive behaviours. Quick wit can break up situations of potential conflict, as it's hard to build up an aggressive psychological state once interrupted by something funny. This often works simply by slowing down the negative response long enough for the evoked emotions to settle at the neurological level and a more rational state of mind to take prominence in consciousness. I often use wit to 'call the room', an old comedian's term for being straight and honest with the audience. For example, in a long session on a dry subject, I might say to the students something like, "I know you would rather be on an exotic Island with 'Person x' (I identify a local celebrity of noted glamour as exemplar) than here with me doing this on a Tuesday afternoon, but what can we do about it?" This needs to be done with a bit of 'playfulness' in terms of voice-tone and gesture but typically works as intended because it communicates your empathy and that you share some of their learning 'pain'. Quick wit is probably the most difficult of all humour to do consistently well, and the 'quick' component can lead to saying something that may not be intended and which may be perceived as offensive by some. However, I use this type of humour extensively and have yet to be slapped or reprimanded in any way, so it's not that risky—if one is thoughtful of context.

Even for the teacher who is genuinely self-conscious and chooses not to experiment with any form of humour, there is a fairly easy solution, and this is to allow the students to generate some humour and simply participate with a genuine smile. The teacher does not need to be, nor should be, the seat of all humour, it's much better when it's collaborative. Table 5.2 provides a guide for using humour as part of one's creative teaching repertoire of resources.

Finally, it is important to recognize that infusing some humour into everyday classroom interactions is not taking valuable time away from learning the subject. Quite the contrary, as Dewey argued:

> To be playful and serious at the same time is possible, and it defines the ideal mental condition.

Table 5.2 Summary framing questions for using humour

Humour	
What are the different types of humour that can be creatively used to enhance student learning?	• Jokes; riddles; anecdotes; stories; cartoons; witticisms; impersonations; humorous video clips, audio segments or objects—almost unlimited
What are the main purposes for using humour in teaching?	• Get good attention (humour is a great resource for creating von Restorff effects) • Create and maintain a positive psychological climate and build rapport • Creatively illustrate a fact, concept or principle with high impact • Icebreaker for new classes (if done well this is a good primacy effect)
What must we consider carefully before using humour?	• Sensitivity to the learning group and individuals In terms of political correctness issues, such as ethnicity, gender, sexual orientation, etc. • Confidence in using the particular humour genre/type (e.g., witticism is more difficult)
Where can I get resources of humour that will work for me?	• Personal experience • Colleagues, family members, friends—almost anybody • Media sources such as joke books, journals, newspapers, television, films, internet, etc. • Observe professional comedians (live or on video)—model what they do and how they do it

5.7.3 Activities

In designing an instructional strategy, one of the most impactful aspects of the learning experience is the activities that students are engaged in to facilitate key learning outcomes. Activities are usually used in unison with methods, and can sometimes refer to the same thing. For example, case studies are considered a method of instruction, but the actual case is an activity in itself. In most basic term's activities provide specific structures for students to engage their thinking skills with selected content knowledge and work towards understanding and subsequent application. Activities can take numerous forms, varying from a single question posed to a large project or dissertation, but they all share a common purpose which is to enhance the learning process towards designated learning outcomes.

In practice, all instructional strategies are a sequence of planned activities. Some are predominantly teacher-centred, such as explanation (typically framed as lecturing or previously 'chalk and talk'); others involve greater participation and student autonomy in terms of choice and management. There is much talk about the need to make learning more student-centred, rather than teacher-centred and, as noted in Chap. 1, 'The role of the teacher being changed from the "sage on the stage" to "the guide on the side". Frankly, I find this quite disturbing as all instructional strategies should be student-centred, whether they are more teacher-directed or student-managed. For example, lecturing, while typically involving more teacher talk than that of students, does not inevitably mean a lack of student engagement and thinking. Invariably, long periods of teacher talk, especially if it lacks organization and presentation style is likely to be both boring and ineffective. However, where lectures are delivered in ways consistent with an evidence-based approach (e.g., appropriate chunking, a variation of presentation medium, focus on key concepts, questions to encourage thinking) and good presentation skills, they can be highly effective and creative in terms of enhancing student learning and attainment. It is a serious misconception to perceive lectures as a passive experience for learners. As Hattie and Yates (2014) make clear:

> Within the world of psychology, there is no such thing as passive learning, unless this term implies learning to do nothing, in a manner akin to learned helplessness. When we are learning from listening or watching, our minds are highly active...People will often learn more effectively from watching a model perform than from doing and performing that same action in the flesh. Although we note that learners need to be active, this does not mean being active in the physical sense of having to respond overtly. (p. 47)

In contrast, activity for activity's sake is both un-motivating in many cases, and not useful for enhancing learning and attainment. The important point is that an effective instructional strategy is typically an organized series of well-chosen methods and activities designed to meet the learning goal/outcomes for a given student profile; the main decisions should relate to the design and facilitation of the overall strategy, rather than a generic framing of it being teacher-centred-versus student-centred. The aim, over time, is to enable students to develop the necessary learning-to-learn skills and sufficient content knowledge to increasingly initiate, direct and manage their learning, becoming self-directed learners. Of course, well designed and appropriately calibrated activities (i.e. sufficiently challenging and achievable with effort for the student group) are fundamental to effective learning. Chickering and Gamson (1987) pointed out:

> Learning is not a spectator sport. Students do not learn much just by sitting in class listening to teachers, memorizing pre-packaged assignments, and spitting out answers. They must talk about what they are learning, write about it, relate it to past experiences, apply it to their daily lives. They must make what they learn part of themselves. (p. 3)

The most basic form of activity is the *question*, which powerfully impacts all aspects of the learning process. Robbins (2001) went as far as arguing that:

> Thinking itself is nothing but the process of asking and answering questions. (pp. 179–8)

The very nature of posing a question suggests some gap in knowledge in long term memory, hence the question. Once the answer is not found in long term memory, other possible sources of information are then identified and sought to provide the necessary information. If we ask the right questions about what we need to learn, find appropriate resources and persist in building the necessary understanding and competence, we should be well equipped to meet necessary attainment targets.

Questions can take many forms and serve different purposes. For example, they can be closed, and focused on the memory of key factual knowledge, such as "What is the currency used in Brazil?" This will provide immediate specific feedback on whether or not students know this particular fact. However, helping students to build a solid understanding of the subject content—that is memorizing relevant information and making the necessary connections through good thinking—requires the use of open-ended questions that specifically cue the relevant types of thinking documented in Chaps. 2 and 3. Questions that can effectively promote these types of thinking are typically *what* and *how* questions. The following are some examples relating to aspects of this book:

- What is the relationship between deliberate practice and expertise?
- What are the similarities and differences between whole-class interactive teaching and problem-based learning, and how does this affect the role of the teacher?
- What inferences and interpretations can be drawn from Hattie's meta-analysis of the research on the impact of learning styles on student attainment?
- How might we evaluate the effectiveness of our evidence-based teaching in terms of enhancing student attainment?
- What other ways might we encourage our colleagues to take on the challenge of being more creative in their teaching?

You will have noticed that the first four questions focus on critical thinking skills (e.g., analysis, compare and contrast, inference and interpretation, evaluation) and the fifth on creative thinking (generating possibilities). Of particular importance, students need to clearly understand what good thinking entails (e.g., the cognitive heuristics identified and outlined in Chap. 2), have opportunities for deliberate practice to apply these in authentic real-world contexts, as well as receive clear and useful feedback from expert professionals.

The creative challenge with activities is to produce authentic learning tasks that are sufficiently challenging as well as to systematically infuse key concept knowledge and appropriate thinking skills, tailored to desired learning outcomes. While thinking may not be a desirable activity in all situations, for reasons outlined earlier, the brain responds well to interesting mental challenges, and in some way novel. Hence, we may be naturally curious and take pleasure in solving problems. For whatever other reason would people do crosswords and other puzzles on underground railway system and buses, apart from relieving boredom? However, apart from personal dispositions, other factors influence our responses to activities, whether school-based or otherwise. Firstly, what is sufficiently challenging will vary depending on the student group and even for individuals within it. Secondly, as Willingham (2009) pointed out:

> ...curiosity prompts people to explore new ideas and problems, but when we do, we quickly evaluate how much mental work it will take to solve the problem. If it is too much or too little, we stop working on the problem if we can. (p. 10)

Activities that facilitate a range of differentiation (e.g., where all students can be successful, but the activity enables more competent or motivated students to go further in terms of depth or breadth of the knowledge and skill areas involved), is particularly challenging and requires creativity on the part of the teacher. Such activities can take various formats (e.g., cases, projects, problem-based learning, simulations and experiments) and are based on real work applications of the subject content How to design of real world tasks was explained and illustrated in Chap. 3.

Table 5.3 provides a guide for enhancing student learning through the creative use of activities.

It must be emphasized that activities need to be appropriately integrated within the overall instructional strategy which may also include other methods (e.g., cooperative learning, reciprocal teaching, peer assessment). Also, just as a funny story can always be ruined by poor presentation skills, a potentially high impact activity can fall relatively flat with poor facilitation skills.

5.7.4 Presentation Style

The importance of good presentation skills is obvious in the case of teaching. Unclear voice, disorganized sentences, monotonous tone, dull body language, irritating mannerisms (however defined) etc., all contribute to a quick loss of attention, boredom, disengagement and, for those wanting to learn, frustration. Increasing research evidence is highlighting very specifically how aspects of our presentation style impact on others' perception of who we are and what we are like. The powerful impact of voice tone and smile were identified prior, and it is certainly my experience that smiling, often supported with appropriate humour, provides important feedback to students about your mood and approachability, both at conscious and subconscious

Table 5.3 Summary framing questions for using activities

Activities	
What are the different types of activities that can be used to creatively enhance student learning?	• Specific learning tasks; quizzes; competitions; projects; visits; forums; simulations; cases; work experience; brain gym/puzzles; experiments; role-play-almost unlimited
What is important in designing and managing activities?	• Relevant to the learning goal(s) and specific learning outcomes • Tasks are real life focused, challenging but achievable, based on students prior knowledge and skills, and in the context of resource and time allocation • Logistic/resource support is available for successful completion • Clear and sufficient notes of guidance are provided for students in terms of task requirements and assessment components and criteria • The overall activity provides opportunities for differentiated learning • Good facilitation skills are maintained throughout the duration of the activity
Where can I get relevant activities that will be challenging and achievable for the students I teach?	• Produce these yourself—that's a real creative challenge. But a great way to fully customize activities to the student group and learning outcomes • Industry personnel, colleagues • Media sources such as books, journals, commercial packages, Internet, etc.

levels, and has contagion effects. For example, if you smile at a student, he or she is likely to smile back at you and this can quickly spread to his or her classmates. This has such face validity—excuse the pun—but it is not so easy for many people to do this and, even more importantly, to do it well. In my first year in Singapore, there was a National Smile Campaign, as it was felt that local people did not smile much, and this would be a good thing to encourage in the community. The intention was well-founded, and it provided me, a former teacher of social psychology, to conduct one of those 'strange experiments' that social psychologists often partake in. The basic experimental design was simple, I smile at everybody I walk past in the local 'hawker centre' (this is a Singaporean term for an area comprising many vendors providing food outlets, usually but not always outdoors). Social psychologists have thick skins and can deal with embarrassment. What other occupational group goes around with bold faces asking people to give up their seats in crowded underground stations, to investigate the impact of social norms on people's response behaviour? (Milgram 1977). Many did get up, without asking for reasons. It might on the surface seem unlikely that few people would ask why, but would you like a response such as, "My colostomy bag has just broken and…."? Anyway, I conducted the smiling experiment and held firm despite the great majority of local people, for the first day or two, looking at me as though I was from another world. Persistence is another trait of social psychologists, so I continued the daily smiling routine, and guess what? Some started smiling back to me and, for one week (my designated time for the coverage of the experiment), I had several brief conversations with local Singaporeans. A few even congratulated me on my bravery. Of interest, no one during this week initiated smiling at me. That experience may support the view that smiling is not particularly easy for many people.

Furthermore, there is a skill in smiling, as in most aspects of interpersonal communication. In the context of classroom teaching, it should be brief, natural and unforced, involves scanning the whole group with quick friendly eye contact. Yes, it's a skill requiring deliberate practice, but one well worth developing both for effective teaching and social interaction generally. Of course, smiling is not the content of one's interaction with others but facilitates setting the climate or context of the interactions, especially towards building rapport. And, as the famous success coach, Anthony Robbins (2001) once wrote:

> Rapport is the ultimate tool for getting results with other people. (p. 231)

Rapport is very much bound up with positive feelings towards someone you like and results from your perceptions of what they do and how. There is a saying that goes something like this, "People like people like themselves." It also includes liking people, who you might like to be like. This is often an unconscious bonding process and reflects the human tendency for identification with desired social models. I like Molden's (2001) frame:

> Rapport is long-lasting, elegant, respectful, and acknowledging in nature. Rapport connects emotional centres and creates enjoyable bonds between people. Rapport is the intelligent approach to influencing, regardless of positional power, whereas power and authority are defaults for people in positions of power who have poor interpersonal skills and little flexibility. (p. 72)

Presentation style is not just the ability to use one's voice and specific aspects of body language to maintain positive attention and build rapport, but also to quickly recognize the reactions of other people, and modify one's communication style accordingly to encourage desired changes in their behaviour. This ability to quickly notice, monitor, and to make sense of the external cues from other people, through careful observation and empathic listening, is often referred to as sensory acuity. These skills, while typically associated with professionals working in various fields of applied psychology and detective or military work are now increasingly becoming part of the training of salespeople and customer service personnel. The field of neuro-linguistic programming (NLP) which focuses on influencing other people's behaviour through the use of language, voice tone and a range of non-verbal communication strategies has much to offer teaching professionals in terms of enhancing presentation style. It involves skill in recognising patterns in body language and voice characteristics to understand the states of mind of other people and to be able to make accurate inferences about their perception and feelings. In this way, it is then possible to communicate more effectively and get better results in terms of building mutual understanding and rapport Ultimately, perception and judgements about other people, accurate or otherwise, are the product of their behaviour, and of course, our pre-existing beliefs. The use of NLP is explored further in Chap. 9, in the context of developing teacher expertise.

In teaching, as in other human interaction situations, the processes of perception, apprehension and response play out and lead participants to construct their realities, and these can be favourable or not for the teachers concerned. For example, Wadd (1973) warned:

> In establishing the order he has decided upon, the teacher must be fully aware that what happens in the first few encounters with the pupils is likely to establish the relationship which he will have to live with for the rest of his contact with that particular class. (p. 87)

Like on a first date, for a teacher encountering a new class for the first time, it can be the first step toward a long-lasting positive relationship, or it can be a disaster. In the language of psychology, as introduced prior, this is a Primacy effect, which has its roots in our evolution as a survival mechanism. For example, it probably wasn't helpful to take too long in ascertaining that the large sabre tooth tiger at the entrance to our cave may not be there offering us a meaty barbeque item. Similarly, the Recency effect also discussed earlier, is impactful in these ways. What you say or do at the end of a teaching session, or on your first date, may have a similarly strong impact on the other's perception, feelings and subsequent response to you. Hence, good Primacy and Recency effects can put one in a potentially strong position in terms of the 'person perception stakes'. To make this even more impactful, introduce a novel and interesting aspect—a Von Restorff effect—to the communication encounter, and you may well be on your way to achieving positive framing by others, in many (not all) situations.

For example, enthusiasm is typically considered a positive feature of a teacher's presentation style. However, the inference and interpretation of a person's enthusiasm can only be made by others based on their observation of behaviour, not the person's intent. Hence, if you want to be perceived as an enthusiastic teacher you need to behave like one (e.g., displaying the range of behaviours that people, within the cultural and normative context, typically frame as enthusiasm). We know that variations in movement and proximity, voice and tone, certain types of posture and gestures and eye contact, all contribute to positive perceptions of enthusiasm. We also know that this needs to be done skillfully for it to work in practice. In other words, the best-contrived performance works best, but it must not be perceived as contrived—sounds kind of strange, but true. That's why great actors earn such big bucks.

What this means is that the way we structure and conduct our communication behaviours to other people is crucial for influencing their perception of us and the kind and amount of attention, if any, they are likely to give. The impact of good presentation style is particularly significant in influencing students' perception of you as a teacher, especially in the area of being seen as a 'nice person'.

Many factors and contingencies will affect how others perceive our behaviour (e.g., personality configuration, cultural values and norms, prior experience, mood and situation) but it is far from a random process. People with high proficiency in emotional and social intelligence typically have a deep understanding of how interpersonal communication works, and, perhaps more importantly, why it often does not. As a result, they can, through good thinking and careful structuring of their communication strategies; mitigate the likelihood of ambiguity and misperception. Note, I say mitigate not eliminate. I remember a quote, but cannot trace the reference (apologies), but it captures the context so nicely:

> Life is a matter of perception, though more often than not a misperception.

While presentation style is influenced by the personality and biographical aspects of the presenter and is a holistic performance, it is understandable in terms of the key behavioural components involved and how they work to influence people's perception and behaviour. Skilled observation, listening, voice modulation and calibration of one's body language to the situation and audience are learnable through modelling how they work and through deliberate practice. The ability to use one's presentation style to specifically engage and motivate learners is a high-level skill, akin to that of great orators or other persuasive political figures. In the classroom it is less one of high verbal and visual rhetoric, but equally powerful in terms of creating impact than can be used productively for learning. I often use the analogy of Presentation Style being the *Pedagogic Glue* that binds together all components of the instructional strategy and is a major aspect of creative teaching competence. Table 5.4 identifies some key practical ways to develop greater creativity in this important area of practice.

5.7.5 Examples

In teaching students over three decades and across many educational and cultural contexts, a generic response that sticks in my mind is students asking for examples. It's as though an example will provide that special key to open the door to understanding the concept or principle being taught. Examples are a representation of a class or a group of things and, as understanding requires classifying things to generalize and differentiate, they play a key role in this organizing process. More complex conceptual understanding is analogous (this is also a kind of an example) in many ways to completing a large jigsaw puzzle. I can remember back to my childhood years when opening the box of a large jigsaw puzzle containing many hundreds of pieces, that feeling of both challenge and anxiety at the thought of the task ahead. However, by using certain strategies (e.g., finding and assembling all the straight edge pieces to form the border; putting together pieces of specific items in the picture) the task becomes more manageable and eventually, towards completion, easy, as the last several pieces can be quickly put in. Building complex understanding is a bit like this. In terms of the jig-saw analogy, examples are key 'instructional pieces' to facilitate this process, enabling the mind to organize information and build a clear and accurate

Table 5.4 Summary framing questions for developing presentation style

Presentation style	
What are the various aspects of presentation style that make it most effective in creating and maintaining attention and interest?	• Clarity, pace and tone of voice (expressive voice being most effective) • Calibration of body language (e.g., posture, gestures, eye contact, appropriate use of smile) to voice presentation • Appropriate variation and ability to creatively use stories, humour and examples (appropriate to the learning group)—remember this is your *Pedagogic Glue*
How can I develop a presentation style that is effective, creative and fits my personality?	• Apply the relevant Core Principles of Learning to oneself. Remember you must understand how it works (key concepts and principles, etc.) and be able to do it well across a range of situations (hence good deliberate practice with feedback is essential) • Observe directly (or from videos) effective presenters and motivational speakers and model what they do and how they do it—(remember NLP?) • Ask colleagues who have very good presentation style to observe you teaching and provide relevant feedback • Practice, evaluate and modify, etc. from an evidence-base—I might have said this prior!

picture of reality (mental schemata) for the topic being learned. The use of a range of worked examples (Effect Size of 0.57, Hattie 2009), in which students can clearly see the full process or procedure for completing an activity, enables them to fully connect their prior knowledge with new knowledge, build understanding and be able to transfer learning to other relevant contexts in which it has useful applications. According to Hattie and Yates (2014):

> The worked example effect now stands as one of the most robust findings from applied psychology research. Worked examples provide a form of modelling through demonstrations of successful procedures or products. (p. 151)

Examples also include such things as analogy and metaphor. These liken one thing with another for simplification and making things mean in terms of the particular prior experience of the student involved. For example, I often use the analogy of creative teaching as akin to good cake-making. A good cake requires a solid base and an attractive topping. Creative teaching requires an evidence-based instructional strategy, which has great SHAPE. It's a very simple analogy, but easy to remember. The reason I have selected examples as a key area for creative teaching is that there is almost unlimited potential for finding or creating powerful examples, analogies and metaphors to support the learning process. Like stories, examples cover all aspects of

Table 5.5 Summary framing questions for using examples

Examples	
What makes an example a good example?	• Relevance to a key concept, principle, skill being taught • Students can relate to it through their own experiences • It has a strong real life current impact
When is it most effective to use examples?	• Before or immediately following the teaching of a concept, principle, procedure or skill • When concepts are abstract or difficult to visualise
Where can I get good examples for the topics I teach?	• Personal experience • Colleagues, family members, friends—almost anybody • Media sources such as books, journals, newspapers, television, films, Internet, etc.—almost limitless

human experience, and we can use them to creatively communicate difficult concepts in simplified but authentic ways. Good analogies and metaphors are also very likely to result in a strong Von Restorff effect, stimulating attention and interest. Like stories, we can never find or create all the possible relevant examples for the topics we teach, and this always provides us with a challenging and creative avenue for practice. However, over time, with diligence and some creativity, it is not too difficult to build an extensive and varied portfolio of these valuable pedagogic tools. Table 5.5 provides an advance organizer for this.

5.8 Creative Activity and Resource Blending: The *Art* of Teaching

SHAPE is a metaphor for a range of activities and resources that can be utilized in the creative design and facilitation of instructional strategies. The selection, blending and enactment of such activities and resources determines, in large part, the experiences of the learning group. Better the blend and facilitation, and its alignment to the learning outcomes, student profile and situated context, the more likelihood of better quality in terms of the learning experience and its outcomes (e.g., effectiveness, efficiency and engagement). To use an analogy, it is like the very best chefs, who can use a range and blend of ingredients to getting that special taste that customers respond to, that can differentiate them from other expert chefs. Creative teachers use their 'ingredients' (personal and professional knowledge and skills) to design and facilitate lessons to better connect with student needs and interests than those less creatively competent.

Creativity in lesson planning, in most basic terms, is the ability to combine methods, activities and resources in novel and useful ways that can significantly heighten the learning process for students in terms of attainment and engagement. Often some of the methods, activities and resources may seem to have little connectivity in themselves, but when creatively combined and contextualized to the subject content, they make the learning of key concepts almost easy and fun. Again, another analogy, this time from the field of environmental engineering: one may ask, "What is a mirror got to do with solving a problem of tenants complaining about long elevator waiting-times?" There seems no immediate connection. However, there is a well-told story involving a multi-storey office building in New York, where many occupants complained about the slowness of elevators at peak hours. Several of the tenants threatened to break their leases and move out of the building because of this. In response, the management authorized a study to determine what would be the best solution. The study revealed that because of the age of the building no engineering solution could be justified economically. The engineers said that management would just have to live with the problem permanently. However, a young psychologist who took on the challenge of solving this problem reframed it differently and concluded that the complaints were as much a consequence of boredom as slowness. Therefore, he took the problem to be one of giving those waiting for something to occupy their time pleasantly. He suggested installing mirrors in the elevator boarding areas so that those waiting could look at each other or themselves without appearing to do so. The management took up his suggestion. The installation of mirrors was made quickly and at a relatively low cost. The complaints about waiting stopped. Today, mirrors in elevator lobbies and even on elevators in tall buildings are commonplace.

In the context of teaching, here's an example of making creative connections between what would appear to be unconnected aspects of reality to produce an effective learning experience for a group of students. Many years ago, mentoring and coaching a teacher who had received consistently low feedback scores, I remember him lamenting on how students found his teaching of Newton's Second Law of Motion particularly boring and difficult to grasp. He agreed to me observing his lesson on this topic, which began with a typical technical verbatim definition of the law, which went something like this:

> Newton's second law of motion can be formally stated as follows: The acceleration of an object as produced by a net force is directly proportional to the magnitude of the net force, in the same direction as the net force, and inversely proportional to the mass of the object.

The definition was then followed by around 40 min of exposition and the writing of formulae on the whiteboard. I was confused and bored, but no more than the students, based on my observations. In our post-lesson discussion, he acknowledged the students lack student engagement and interest, but could not see how it was possible to make this topic area interesting or meaningful for students to learn effectively. Having explored with him exactly what Newton's Second Law of Motion entailed, we designed the following strategy as a way of attracting increased attention and putting the law into a more practical perspective for the students. In summary, this involved showing a picture of the famous soccer player, David Beckham, then a

Manchester United player, taking his trademark free-kick. As soccer is very popular in Singapore and David Beckham is considered to be particularly good looking, this seemed both a good Primacy and Von Restorff effect combined to get good initial attention from both the male and female students, albeit for different reasons. The students were then asked to consider the following two scenarios and the impact they might have in terms of the acceleration of the soccer ball once struck by David:

Scenario 1   One of the opposition players changes the soccer ball before the free kick has been taken with a ball that is 20% heavier than the original ball.

Scenario 2   The ball remains the same, but David has been doing extra fitness training and can now strike the ball with around 10% more power.

I'm sure you have worked this out, so I won't need to bore you with my limited display of physics jargon. While not a perfect analogy, it was sufficient to get the students attention, create some interest and make the psychological climate a bit more fun than usual. The strategy also included a lively and humorous presentation style in which the scenarios were simulated by the teacher (e.g., putting a real soccer ball down and asking the students if he looked like David Beckham). As the teacher concerned lacked certain skills in terms of voice and gesture it was necessary to provide some measure of coaching here before the teaching session. In summary, the strategy worked in that student attendance was high and they quickly saw the relationship between mass, force and acceleration. Also, it lifted the mood of the class in an afternoon session noted for low student attention. From this basis, the teacher then made connections between the free-kick analogy and other real engineering contexts, inviting and answering questions, before proceeding with the mathematical formula and how it worked. He also chunked up the session and conducted short quizzes and activities to check to understand and provide feedback. This was creative teaching (remember *Little c*) as it was novel and produced useful results in terms of student learning.

How many teachers have previously introduced Newton's Second Law of Motion through a simulation of David Beckham's free-kick? Some may have, but that would have been creative teaching also. The connections are not readily apparent until you see them, as is the case with visual illusions, such as the famous 'old' and 'young' woman visual illusion. This is an example of lateral thinking in the context of teaching, as one is unlikely to automatically connect a David Beckham free-kick with Isaac Newton's Second Law of Motion. As outlined previously, creativity, as for good thinking generally, it is not necessarily something the human brain likes doing naturally. For this reason, we need to provoke it into action and make a conscious effort to think laterally.

The lesson incorporated several core principles of learning (e.g., focus on key concepts, good thinking, psychological climate) and it also had a creative spin, which made the experience more attention-grabbing and impactful (e.g., David Beckham was a fairly powerful Von Restorff effect in this situated context). Also, when attempting creative activities in class with students, it communicates an important latent message that you are genuinely interested in their learning and this is fundamental in determining their perceptions of you as a person.

The teacher in question went on to receive significant improvements in terms of feedback over the following two semesters. Invariably, this was not based on one session but his realization that good teaching is much more than positive intentions in a teacher's mind, but the actual behavioural performances in class over time. Most importantly, he was keen to improve his teaching and obtain better results, both in terms of student learning and, of course, his feedback scores. Of particular significance was his reframing of himself as a teacher, which was the result of the direct experience of feedback from students. As he taught better and responded to them more positively, they were more responsive and positive towards him, and good rapport was developed over time.

Creative teaching can also occur in the here and now teaching situation in response to the teacher's perception and subsequent reframing of the learning situation. No matter how well we try to plan an effective (evidence-based) instructional strategy, there are occasions when the methods or activities do not work out as expected. We may have made some incorrect inferences and interpretations about the prior knowledge of the student group, not delivered the lesson activities as well as we can, or it may simply be that some of the students are not in a good mood on that occasion, for whatever reasons.

Even the weather can influence people's behaviour. Mlodinow (2012) quoted research by Cunningham (1979) in which waitresses in a shopping centre in Chicago kept track of their tips and the weather over thirteen randomly chosen spring days. Customers were probably unaware that the weather influenced them, but when it was sunny outside, they were significantly more generous. It is therefore not surprising that student attention and behaviour can vary so much, even with the same teachers. Hence, don't take it too personally, if things don't always work out well; sometimes they just don't.

In situations where the planned strategy is not working the teacher, faced with little by way of positive response, maybe at his/her most creative. As the saying goes, "Necessity is the mother of invention". For example, several years ago I was teaching an elective module on learning strategies and skills at 2 p.m. on a Wednesday. These electives were compulsory though students could choose which ones they took. However, they were noted to be challenging in terms of getting student attention and participation as students did not receive much by way of academic credit and many thought this constituted unnecessary work. In my first lesson, before I had even spoken a word, I quickly noticed the look of disinterest on the faces of many students. It was apparent that if I just went ahead with the planned lesson, there might be little value to their learning and tedious experience for all, including myself. I was acutely aware of the need to change the students' perception of the situation and was seeking a strategy. Here's the summary story in context. Firstly, of note, the students were Singaporeans. In Singapore, education is highly valued and very well-funded, which means that all students have good access to learning opportunities. A few years prior, I was involved in a consultancy project in Kolkata, India. What's the connection you might ask? While working there I usually went for a walk after breakfast, just for some short exercise and mentally revising what I needed to do that day. On one occasion I was approached by a teenage boy of around 13 years of age who asked me if I was a businessman. Intrigued, I asked him why he was interested in that. In summary, he pointed out that he was living on the streets, wanted to avoid getting into trouble and was looking for an opportunity to get a job and learn some useful skills. He thought I might be able to find him employment. Somewhat sadly, I explained that I was not a businessman and could not provide him with any employment opportunities (though I wish I could have done this). He left and that was that. My only significant reflection at that time was that he did not ask me for any money.

Going back to the classroom situation in Singapore, a strategy flitted into my mind. I walked around and looked at the students, one and all, and told them that they were so lucky. One immediately perked up and said, "Why are we lucky?" I replied, "Well, you are young, healthy and Singaporean". They, of course, could not dispute the first two assertions, but there was a quick response to the third, "What's so lucky about being Singaporean?" I told them the Kolkata story and while walking around the classroom with a fairly serious expression on my face, making quick eye contact with the students. On completion of the story, I asked them to discuss in pairs what made them different from the boy in Kolkata. I used a verbal emphasis on the word *different* by slightly raising the tone and slowing pace. It was not long before they identified their situation of excellent learning opportunities and good job prospects, which were lacking for the boy in Kolkata. I then said something like, "Ok, well let's not waste our time being negative" and started the lesson. To my surprise, they settled down and the lesson seemed to progress quite well, especially in the context of the earlier scenario.

I subsequently gave this little thought but was quite astonished by the response of the students at the beginning of the next session. On my arrival, I was greeted by the students with words akin to, "It's ok, Cher (Singaporean slang for Teacher) we get the message, no need to tell us the Kolkata story again". They were quite good fun to teach for the next 14 weeks and many gave feedback that they had learned some useful stuff out of the elective. Was the story that impactful, or was I just lucky? Sure, I exploited what is often referred to as a 'teachable moment'; a situated unplanned activity that I grasped as an opportunity to create an impactful learning experience for the group at that specific time. In this case, the learning purpose was to change the present negative attitude into one more conducive to learning and to do this I needed to get some reframing of their present situation, by changing their perception of it. In terms of lucky, who knows? On another day, I may not have thought about this Kolkata story and I would have had to deal with the situation in a different way, which may or may not have been as successful. However, without the story encoded in my long-term memory, it could never have been part of my instructional strategy, albeit constructed in situ. Hence, creativity requires both resources in long-term memory as well as the creative competency to be able to see new combinations of methods, activities and resources to structure a novel and effective instructional strategy for a particular group of learners.

Similarly, while the core principles of learning have universality in terms of how humans learn, learners come to the learning event with different biographies, personality configurations and prior knowledge which will influence their perception and initial motivational status. Therefore, in planning the learning experiences, it is really useful to ascertain as much as possible about the learners, both collectively and individually. Invariably it is not possible to do this as thoroughly as one might like, as it can be time-consuming. Equally, one must be cautious in making inferences and interpretations from prior information about learners from secondary sources (e.g., attainment reports, other teachers' framing) as objective or fixed. Often, I have found that prior descriptions have been quite different and even at variance to what I experienced. Teachers construct their realities, through their teaching practices and human interactions with groups of students, as do the students themselves. I once inherited a class of students where their prior teacher referred to them as, "That bunch of animals". For the first two weeks, I could understand the basis of that teacher's framing. The students showed no interest in anything I tried to do, with many using a range of negative responses to try to 'wind me up'. It was obvious that many of them did not like school or teachers. In this situation, there's little point in trying to persuade them verbally to see meaning in any aspect of the formal curriculum when they do not. Furthermore, in my experience, there's no point in doing anything that might be perceived as confrontational as this will go nowhere useful for all concerned. In such situations, my response, based on a strategy that has worked previously on most occasions, starts with not showing fear or stress (of course, more easily said than done) and maintaining a positive stance towards them. What this means at the behavioural level is maintaining a friendly voice tone, smile and calibrated body language. This will typically, over time, result in even the more vociferous of the students losing interest in the activity of 'winding up teacher'. At the psychological level, a type of cognitive dissonance (Festinger 1957) will come into play. Cognitive dissonance theory suggests that we have an inner drive to seek consistency in our beliefs, perceptions and attitudes, and will experience inner conflict in situations where two cognitions are inconsistent. For example, if students believe that teachers are not particularly interested in them but are then consistently presented with one who seems to be showing genuine interest, such dissonance may occur. In this situation, the person may either retain the existing belief (e.g., rationalization; denial) or change the belief in some way. However, it is often not so clear-cut in terms of perceptual change, and I would not expect students to suddenly completely reframe and start liking teachers. I am only realistically looking for a slight shift in perception towards 'Dennis is ok, for a teacher'. Once this has been attained, I am usually able to engage in some informal non-confrontational chat with them and gradually build a workable rapport. This is how most relationships develop over time, and it is as much an unconscious as a conscious process. The key outcome is that the reality of this situated context (e.g., negative confrontational student responses) will change for the better. I remember hearing stories about how some people survived the horrors of concentration camps by 'being nice' (at the behavioural level) to the guards. It seems that it may be harder to kill someone who gives you a friendly word and a well-calibrated smile. Whatever one's views on this as a survival strategy, if it sometimes works in such situations, what can the genuine behaviour achieve in most classrooms?

Having achieved a level of rapport, defined in terms of friendly banter with at least a few individual students (this usually has a contagion effect over time), I am then in a position to explore areas of possible interest and collaboratively identify school-based activities that have at least a minimal buy-in from their perspective. This is what happened with this particular class. By the end of the year, they were quite responsive to learning and fun to teach. At the beginning of each session, I had to run the gauntlet of jokes for several minutes, but they would always settle down enough to do some 'useful schoolwork'. The main significance of this story is that at the beginning of the following semester something really interesting happened. I was not timetabled for this group of students as another teaching faculty had been was. I approached this colleague and asked if I could take them on, and he could choose any one of my classes in exchange. He was somewhat surprised but readily agreed. On arrival in class on their first session, the students were surprised to see me, though visibly pleased. When I explained I had exchanged another class to teach them, one stated that this had never happened before in their school life. Over the next two years, they choose to do a City & Guilds qualification in Communication Skills (with all passing and many getting distinctions) and the 'O' level English Language (in which around 50% passed). Even the principal of the institution was shocked by such results and congratulated me. This was not what seemed the likely outcomes after the first few lessons, which were largely encounters of sarcasm and nihilism. Had I tamed 'that bunch of animals', or were they not that bad in the first place? Life is a matter of perception, and at any point time, it's the reality. All I can say is that I preferred the latter reality, and I think the students did. And I also think they benefited in terms of learning and well-being—I certainly did.

When I have been asked, "What was your most significant achievement as a teacher", I often tell this story. After establishing a positive learning relationship with these students, and seeing them develop a real sense of personal belief as able learners, this fully reinforced my perception of the potential value of teaching and the impact it can have on student attainment and well-being. It also taught me that it was damn hard work.

5.9 The Magic of Expertise: Getting into Great Shape

A few years past my wife persuaded me to accompany her to see the magician David Blaine perform a live show in Singapore. I rarely go to such events and have never before seen a world-renowned magician perform live. I don't believe in magic in the metaphysical sense but can appreciate the illusion of the experience. David Blaine did not disappoint on this count. However, I was a bit disappointed not to have been selected to participate in one of his magic segments, not for reasons of 'being on stage' but to get that close up view of how he does things. As a psychologist, who should possess a good level of sensory acuity in terms of observation skills (he says hopefully), I thought I might be able to work out how he performed the particular piece of 'magic', at least in theory. Even though I did not get this opportunity of a close-up view, I was impressed with David's expertise—it was surely magic for us mere novices. I can, in contrast, remember my father doing card tricks and other bits of magic such as separating his thumb in two when I was a kid. It seemed quite awesome when I was five years old, but by eight years old I had worked it all out. The card tricks had a planned sequence (arranged beforehand) and the separating thumb was the thumb of the other hand, disguised by two fingers.

How might we explain the apparent magic of such expertise? The answer lies in the earlier discussion of Core Principle 7: *The development of expertise requires deliberate practice* in Chap. 2. David has developed a level of expertise in his magic acts that it has become part of his physiology (cognitively, effectively, and in terms of neural wiring) that makes him different from the rest of us. It's amazing, but it's not magic. For example, in his final act in Singapore, David immersed himself in a tank of water for over 10 min. I would have died within a minute, but in 2012 Stig Severinsen was awarded the record of "Longest time breath held voluntarily (male)" by Guinness World Records for holding his breath for 22 min. This makes David's performance almost routine for such experts. Stig has a doctorate in medicine and started experimenting with holding his breath as a child at the bottom of his parents' pool. Hence, while his performance is exceptional and world-class, it is explainable. Experts can do things far better than the general population because they are different in significant ways and, therefore, it feels like magic to the rest of us.

There is a saying, "One swallow does not a summer make". This essentially means that seeing one swallow (swallows are birds that typically migrate toward warmer weather) is not sufficient evidence that the summer is, in fact, on its way. I have no idea how many swallows one must see to feel confident about the impending arrival of summer in a given environment. The point in this context is that one creative act of teaching does not make an effective learning experience for students, nor define a creative teacher. There needs to be consistency in the overall evidence-based design of instructional strategies over time. This is the case with all areas of expertise in any field. However, while we would expect expert teachers to be able to teach to consistently high standards in terms of their ability to maximize student learning opportunity and achievement scores, it is unlikely that even the most creative can come up with creative aspects (e.g., original components of SHAPE) every lesson. Furthermore, creativity can arise in so many ways in teaching, as SHAPE illustrates. For example, some teachers may be creative in terms of the activities and examples they create for students, others in their presentation style and humour. Very few are likely to be able to weave highly creative SHAPE, incorporating all components, in most lessons they teach. That would constitute creative teaching competence at the highest level of proficiency.

However, creative teaching competence, like other forms of competence, is based on the same core principles of learning and involves the teacher developing from novice to varying levels of proficiency towards expertise, and ultimately to *adaptive expertise*. This may throw better light on differing conceptions of teaching as 'art', 'craft' or 'science' that have appeared in the research literature (e.g., Eisner 1995).

Creative teaching is science, art and craft combined. We now understand, in large part, how this works in terms of the underlying syntax and heuristics involved. There is no real dichotomy between the science and art of teaching, as both are underpinned by strong evidence-bases from diverse fields in the human sciences. The art is the capability for creative weaving of methods, activities and resources into high impact instructional strategies. They can be seen in terms of great SHAPE, and just as David Blaine creates the aura of magic in his performance, the most creative teachers create similar experiences in their classrooms. It has the illusion of magic, but it is expertise and can be learned by motivated teaching professionals over time. Once attained it is exactly as Intrator (2003) depicts in his description of excellent teachers:

> A potent teacher will skillfully and gracefully create conditions and stage activities that inspire students to have a sustained and meaningful encounter with a subject – because they can. (p. 7)

Furthermore, the development of powerful and easy-to-use information-communication technologies (EdTech) are increasingly providing useful resource capabilities for teachers to become even more creative in their professional work. Utilizing the affordances of EdTech for enhancing student learning (e.g., attainment, engagement, differentiation) is now a fertile field for creative teachers. This will be the focus in the following chapter.

5.10 Teaching Can Be Improved with Reflection and Scholarship

The notion of reflective practice has long been a buzzword in teacher education in terms of how teachers can go about improving aspects of their practice. However, reflection like thinking is a very general term and asking somebody to do good thinking (or reflection) is making some very big assumptions about prior learning. If teachers are as confused on what constitutes critical thinking, as Wagner (2010) suggested in Chap. 2, we may similarly question the extent and quality of their critical thinking when reflecting on aspects of professional practice.

However, let's not ascribe blame to teachers for gaps in knowledge relating to current research on human learning or even a lack of application of evidence-based practices. Our earlier tour into Educational Jurassic Park in Chap. 1 provides ample explanation for teachers' reticence to buy into new initiatives. Furthermore, given their busy schedules and the increasing plethora of demands placed on them, it's a wonder that many function as effectively as they do.

Scholarship, which involves research and sustained interaction with ongoing developments and new knowledge relating to a field, is foundational to improvement in any professional arena. Reflection, when underpinned by good thinking and scholarship go 'hand and glove' in enhancing understanding and improving aspects of practice. How teachers can *thoughtfully* use an evidence-based approach to improve teaching, both at individual and collective levels will be further illustrated in Chap. 9.

5.11 Summary

This chapter, using the process and features of good design and building on the pedagogic framework outlined in previous chapters have sought to unpack what constitutes creativity in the context of teaching (e.g., the key underlying processes and how they work in terms of producing creative outcomes). We no longer need to view creativity as some mystical or ephemeral activity, limited to a few exceptionally talented people. The creative process can be understood in terms of the underlying cognitive processes involved, then modelled and applied to the design and facilitation of learning. Therefore, Creative Teaching Competence can be learned by any teaching/training professional through the acquisition of EBT knowledge and practices, creative thinking as outlined here, and a strong volition to achieve such capability.

References

Amabile TAM (1996) The meaning and measurement of creativity, chap 2. In: Creativity in context. Westview Press, Colorado, pp 19–40

Anderson JF (1979) Teacher immediacy as a predictor of teacher effectiveness. In: Nimmo D (ed) Communication yearbook, vol 3. Transaction Books, New Brunswick, NJ, pp 543–559

Andrews J et al (1996) The teaching and learning transaction in higher education: a study of excellent professors and their students. Teach Higher Educ 1(1)

Apple W et al (1979) Effects of speech rate on Personal attributions. J Pers Soc Psychol 37(5):715–727

Bain K (2004) What the best college teachers do. Harvard University Press, Cambridge

Blank WE (1982) Handbook for developing competency-based training programs. Prentice-Hall, Englewood Cliffs, NJ

Beetham H, Sharpe R (2007) Rethinking pedagogy for a digital age: designing and delivering e-Learning. Routledge, London

Bloom BS (ed) (1985) Developing talent in young people. Ballantine Books, New York

Bouchard T et al (1990) Sources of human psychological differences: the Minnesota study of twins reared apart. Science 250(4978):223–228. https://doi.org/10.1126/science.2218526

Bransford J et al (1999) Brain, mind, experience & school. National Academy Press, Washington, D.C

Brophy JE, Evertson CM (1976) Learning from teaching: a developmental perspective. Allyn and Bacon, Boston

Carr W (1989) Quality in teaching: arguments for a reflective profession. The Falmer Press, London

Chickering AW, Gamson ZF (1987) Seven principles for good practice in undergraduate education. Am Assoc High Educ Bull, 3–7

Claxton G (1998) Hare brain tortoise mind: why Intelligence increases when you think less. Fourth Estate Limited, London

Costa A (1991) The search for intelligent life. In: Costa A (ed) Developing minds: a resource book for teaching thinking. Association for Supervision and Curriculum Development, Alexandria, VA

Cunningham MR (1979) Weather, mood and helping behaviour: quasi experiments with sunshine Samaritan. J Pers Soc Psychol 37(11):1947–1956

De Bono E (1992) Serious creativity: using the power of lateral thinking to create new ideas. Ebury Publishing, London, U.K.

De Bono E (2003) Serious creativity 2. Allscript Establishment and Harper Business, Singapore and New York

Dweck CS (2006) Mindset: the new psychology of success. Ballantine, New York

Earleywine M (2011) Humour 101. Springer, New York

Edison T, Quote. Available at: https://www.brainyquote.com. Last accessed 6 Nov 2019

Einstein A, Quote. Available at: https://brainyquote.com. Last accessed 6 Nov 2019

Eisner EW (1995) The art and craft of teaching. In Ornstein AC, Behar LS (eds) Contemporary issues in curriculum. Allyn & Bacon, Massachusetts

Elliot J (1991) Action research for educational change. Open University, Milton Keynes

Fasco D Jr (2006) Creative thinking and reasoning. In Kaufman JC, Baer J (eds) Creativity and reason in cognitive development. Cambridge University Press, New York, pp 159–176

Gardner H (1983) Frames of mind: the theory of multiple intelligences. Ingram, New York

Festinger L (1957) A theory of cognitive dissonance. Stanford University Press, Stanford, CA

Harris S (2010) The moral landscape: how science can determine human values. Free Press, New York

Hatano G, Inagaki K (1986) Two courses of expertise. Child Dev Edu Japan, 262–272

Hattie J (2009) Visible learning. Routledge, New York

Hattie J, Yates GCR (2014) Visible learning and the science of how we learn. Routledge, New York

Intrator SM (2003) Tuned in and fired up: how teaching Can inspire real learning in the classroom. Yale University Press, London

Katz LG (1993) Dispositions: definitions and implications for early childhood practices. Perspectives from ERIC/EECE: A monograph series, No. 4

Liston DP, Zeichner KM (1990) Reflective teaching and action research in preservice teacher education. J Educ Teach 16(3):235–254

Martin RA (2007) The psychology of humor: an integrative approach. Elsevier Academic Press. ISBN 978-0-12-372564-6

Martin R (2009) The design of business. Harvard Business Press, Massachusetts

Marton F (1981) Phenomenography—describing conceptions of the world around us. Instr Sci 10:177–200

Mauzy J, Harriman R (2003) Creativity Inc: building an innovative organization. Harvard Business School Press, Boston, Massachusetts

Milgram S (1977) The Individual in a social world: essays and experiments. Addison-Wesley, Reading, MA

Mittler P (1971) The study of twins. Penguin Books, London

Mobbs D et al (2003) Humor modulates the mesolimbic reward centers. Neuron 40:1041–1048

Mlodinow L (2012) Subliminal: how your unconscious mind rules your behaviour. Vintage Books, New York

Molden D (2001) NLP business masterclass. FT-Press, New Jersey

Morrison MK (2008) Using humour to maximize learning. Rowman and Littlefield, Lanham, Maryland, USA

Nucci IP (2001) Education in the moral domain. Cambridge University Press, Cambridge

O'Connor MC, Paunonen SV (2007) Big Five personality predictors of post-secondary academic performance. Pers Individ Differ 43(5):971–990. https://doi.org/10.1016/j.paid.2007.03.017

Petty G (1997) How to be better at…creativity. Kogan Page, London

Petty G (2009) Evidence-based teaching: a practical approach. Nelson Thornes, Cheltenham

Pinker S (2003) The blank slate: the modern denial of human nature. Penguin, London

Rizzolatti G, Sinigaglia C (2008) Mirrors in the mind: how we share our actions and emotions. Oxford University Press, New York

Robbins A (2001) Unlimited power. Pocket books, London

Sale D (2015) Creative teaching: an evidence-based Approach. Springer, New York

Schank R (2011) Teaching minds: how cognitive science can save our schools. Teacher College Press, New York

Schon DA (1987) Educating the reflective practitioner. Jossey Bass, San Francisco

Schon DA (1991) The reflective practitioner. Avery, Aldershot

Spearman C (1950) Human ability. Macmillan, London

Swaab D (2015) We are our brains: from the womb to Alzheimer's. Penguin, London

Tamblyn D (2003) Laugh and learn: 95 ways to use humor for more effective teaching and training. AMACON, New York

Tuxworth E (1982) Competency in teaching. HMSO, London Further Education Unit

Wadd K (1973) Classroom power. In: Turner B (ed) Discipline in schools. Ward Lock Educational, London

Wagner T (2010) The global achievement gap. Basic Books, New York

Willingham DT (2009) Why don't students like school: a cognitive scientist answers questions about how the mind works and what it means for the classroom. Jossey-Bass, San Francisco

Winget L (2007) It's called work for a reason. Gotham, New York

Chapter 6
Creative Teaching Competence and EdTech: Total Pedagogy

Abstract This chapter, based on the evidence-based creative teaching framework established in the previous chapters, provides a design model for utilizing the affordances of EdTech to positively impact specific aspects of the learning process, learning delivery arrangements, and student attainment and engagement. While there are emerging new technologies (e.g., virtual reality, augmented reality, and mixed reality), which have specific learning affordances (though often high cost in the present context), it is now possible to use low tech (in terms of user learning), low cost (mostly free) and user-friendly (essential) in effective, efficient and creative ways to enhance student learning outcomes. The design frame provides the practical heuristics for selecting, blending, and employing appropriate EdTech tools into the design and facilitation of learning for all delivery modes (e.g., face-to-face, blended/flipped classroom, and fully online).

6.1 My Early Scepticism Was *Not* Unfounded

I must confess to previously being very sceptical concerning the early euphoria relating to the supposed benefits of information-communication technologies (now summarized as EdTech) in enhancing learning effectiveness, at least in the short term. Like many others, I regularly experienced frustration when using technology-based databases and software, often questioning, "Why is it that such a simple process seems like the Mars mission?" I particularly remember attending an education conference, waiting to listen to a keynote talk on the benefits of using technology in teaching, only to see the speaker struggle with the applications and not able to even get his PowerPoint slides up on the screen. After some 15 or so minutes he aborted (or postponed) the presentation. I did not even bother to check. Anyway, such experiences did little to inspire us non-techie folk to embrace technology for learning in any sustained manner. Also, it does not seem many years past that e-learning was being touted as, to use an old English metaphor, "The best thing since sliced bread." However, such early overhyping soon waned and it was not that long before a significant evaluation of the use of e-learning in education was referred to as a "Thwarted Innovation" (Zemsky and Massy 2004). Similarly, Oliver (2007), commenting on the

© Springer Nature Singapore Pte Ltd. 2020
D. Sale, *Creative Teachers*, Cognitive Science and Technology,
https://doi.org/10.1007/978-981-15-3469-0_6

lack of EdTech's widespread application in educational settings to create engaging and effective learning experiences noted that:

> What appears to be still missing for teachers is appropriate guidance on the effective pedagogical practice needed to support such activities. (p. 64)

Robinson and Schraw (2008), in reviewing the literature on e-learning research, further supported this overall perception:

> Unfortunately, empirical research informing decisions regarding "what works" ranges from sparse at best, to non-existent at worse. This is because e-learning has focused on the delivery of information rather than the learning of that information. (p. 1)

More recently, Gallagher-Mackay and Steinhauer (2017) pointed out:

> Despite our high hopes and high financial investment, however, most evidence suggests that, so far, technology impact on schools and on achievement is relatively limited. In 2015, the OECD released *Students, Computers and Learning,* a report analysing tests of literacy, numeracy, science, digital reading and computer-assisted mathematics taken by fifteen-year-olds in thirty-one countries or economies around the world. Despite the fact that all the countries surveyed have invested mightily in information and computer technologies over the past decade, there has been no discernible improvement in literacy, mathematics or science test scores over the same period. Moreover, in countries where it is less common for students to use computers at school, student's performance in reading improved more rapidly, on average, than in countries where it is more common. (p. 123)

Given a relative lack of widespread application of evidence-based practice in mainstream teaching and training, the adding of technology tools that were often far from user-friendly, in contexts of variable operability and bandwidth capability, these findings were not surprising. The reflections by Shea-Schultz and Fogarty (2002) provide a poignant insight into this apparent failure of EdTech to make the expected significant far-reaching positive impacts on teaching and student learning:

> One thing is certain – e-learning will evolve into something so simple, so elegant yet all persuasive and natural, that our grandchildren will wonder in dismay why we didn't see it coming. (p. 165)

> Truly human-friendly technological design won't appear anytime soon. Computer, networking and software engineers cast the die five decades ago. (p. 89)

However, much is now changing for the better concerning the use of EdTech in teaching and learning. Several factors are contributing to this, and overall, their impact is to merge pedagogy and technology into one seamless enterprise to offer the increasing capability for highly effective, efficient and differentiated learning experiences. This will become the arena in which professionals can fully display their pedagogic knowledge (i.e., Pedagogic Literacy) and CTC. I would also like to think that it may move teaching further towards the profession it has only fleetingly threatened to become: one noted for its wide range of knowledge bases relating to human learning, high skills in learning design and creative competence in practices that significantly enhance learner attainment, engagement and well-being.

Firstly, and most significant, there is the recognition that technology tools alone do not constitute anything near a learning revolution. No matter what we can create in terms of computer-generated resources, human brains are still little or no different from those of our distant ancestors. As Mlodinow (2012) explained:

> Our genus, *Homo*, has been evolving for a couple million years. Brain evolution happens over many thousands or millions of years, but we've lived in civilized society for less than 1% of that time. That means that while we may pack our brains with twenty-first-century knowledge, the organ inside our skull is still a Stone Age brain. (pp. 129–130)

Hence, no matter how much information we have in terms of gigabytes and terabytes, it is not going to get quickly assimilated and nicely integrated into long-term memory. The same Core Principles of Learning apply irrespective of mode or medium. In most basic terms, looking into a computer screen does not change how our memory systems work.

Secondly, there has been a significant reframing of the use of EdTech towards pedagogic considerations and how they might enhance specific aspects of the learning process, rather than the technologies per se. For many years I sat, and frowned, in meetings on the use of technology tools. I listened to enthusiasts who showed that with several clicks on fuzzily conceived icons one could read other people's opinions as well as offer one's own opinion on an online discussion board. It seemed to them that this technology affordance would exceed the frustration and inconvenience of its complex technical use. It did not for most of us. The pain exceeded the relatively small pleasure, and it was far from novel. From a pedagogic focus, and using an evidence-based approach, we can now analyse and evaluate the use of various EdTech tools in terms of how they can enhance aspects of the learning process. In this way, we can select and creatively combine those e-tools which are most effective and efficient in promoting learning, and not use technology simply because we have it.

Finally, the technologies are becoming more stable, much faster and, most importantly, user-friendly. For many years, apart from a lack of good pedagogic design, there has been much criticism of online learning in terms of its ease of usability. Shea-Schultz and Fogarty (2002) observed that very basic design failure is common in e-learning environments:

> When most learners complain about e-learning, it's often not the training they object to but the confusing menus, unclear buttons, or illogical links. (p. 117)

Similarly, as Shank and Sitze (2004) have pointed out:

> Your success as a designer and developer of online learning is directly tied to your ability to build instructional materials that don't leave users frustrated. (p. 138)

Taken together, while the initial hype of EdTech was premature and exaggerated, there is now a strong evidence base for their potential to significantly enhance learning and attainment opportunities for a wider range of learners. EdTech tools are not going to change brain capability and functioning in highly significant ways but driven by a strong pedagogic literacy, they now provide an exceptional resource capability for teachers to design learning experiences and teach in ways that are more effective, efficient and engaging. Miller (2016), from a cognitive psychology perspective, summarizes the present scenario:

> I believe that technology gives us many advantages over and above traditional face-to-face classroom techniques, but clarification is in order. I don't believe that instructional technology promotes learning by its mere presence. Nor does it let us evade some of the immutable truths about how we learn – especially the fact that learning requires focused attention, effortful practice, and motivation. Rather, what technology allows us to do is amplify and expand the repertoire of techniques that effective teachers use to elicit attention, effort, and engagement that are the basis for learning.
>
> In a short space of time, technology in higher education has gone from a smattering of fully online distance-only programmes and self-created web resources of a few individuals to near-ubiquitous. (p. 1)

We can see, therefore, that many factors have led to more optimistic and viable use of technology in education—but another significant factor is the increasing demand for high-quality flexible access learning opportunities with reduced costs. As Miller (2016) summarizes:

> …one reason that interest in technology is exploding is that it is easier and cheaper than before. (p. 12)

6.2   Framing EdTech Genres

Even though this is a rapidly developing and changing arena, with new applications and e-tools emerging almost daily, there is an underlying set of generic learning affordances that remain relatively stable. All are related to enhancing some aspect(s) of the learning process, whether the focus is more on providing subject content knowledge, facilitating the building of understanding or skill acquisition. Furthermore, while there may be many applications and specific e-tools available, they will inevitably fall into a limited number of key genres or categories, relating to these broad learning areas. For example, while tablets come in many formats and have different features, they are essentially similar in terms of being compact mobile personal computers. The same is true of the wide range of smartphones, and social media platforms such as Facebook and Twitter.

In terms of EdTech genres, several broad categorizations have been suggested. For example, Pacansky-Brock (2013) offered the following four main categories:

- Cloud-based applications—accessible from anywhere
- Web 2.0 tools, that make the creation and sharing of multimedia content simple
- Social media, technologies that enable communication and sharing
- Mobile apps.

This captures the essential range of EdTech options, which collectively provide a rich resource base for teaching/training professionals to exploit in designing learning experiences and facilitating their practices in effective, efficient, and creative ways. I will be primarily focusing on e-tools that are user-friendly and can effectively facilitate:

1. the production and delivery of differentiated multi-media content and hyperlinked resources
2. communication, sharing and collaboration of learners.

In analysing and evaluating EdTech, especially specific e-tools, the *Core Principles of Learning* will be used as the guiding heuristics. In this way, it becomes possible to identify the key learning affordances for each technology genre and specific e-tool. In summary form from Chap. 2, learning and motivation are enhanced when there is:

- clarity of learning goals, objectives and expectations
- activation of learner's prior knowledge
- a focus on key concepts and principles
- the facilitation of good thinking
- variation in the methods and mediums of presentation
- teaching that works in accordance with memory systems and processes
- ongoing quality feedback from formative assessment
- a success orientated and fun psychological climate.

The following sections explore the various affordances of EdTech, and how specific e-tools can contribute to enhancing aspects of the learning process. This will be framed in the context of designing and facilitating Blended Learning and Flipped Classroom Learning, which are becoming the most used formats in the structuring and delivery of learning using EdTech.

6.3 The What, the Why and How of Blended Learning

As with most things relating to teaching and learning, there are different perceptions of blended learning and any one particular definition will lead to some contestation. Indeed, it could be argued that most learning designs are blended in the sense that different methods and resources are typically combined in the creation of teaching and learning strategies. As Littlejohn and Pegler (2007) pointed out:

> Blending is an art that has been practised by inspirational teachers for centuries. It centres on the integration of different types of resources and activities within a range of learning environments where learners can interact and build ideas. (p. 1)

Effective teachers typically blend (or weave) methods, activities, and other resources into pedagogically sound instructional strategies to meet desired learning outcomes for the students they teach. Furthermore, as we saw in the previous chapter, the generation and blending of appropriate methods, activities and resources are at the root of creativity in teaching. Here, the focus is specifically on the pedagogically driven blending of EdTech into learning design and teaching practices. Over time, we will most likely talk less about the use of technology in teaching and learning, as it will just be part of *Creative Teaching Competence*.

Even in most lectures today, it's rare not to see at least some use of EdTech, such as PowerPoint slides or a video, blended into this traditional teacher centred format. At the other end of the spectrum, we are seeing the growth of fully online courses, in which there is no traditional face-to-face contact time. Invariably, discussions on blended learning raise questions as to what content areas are best delivered online, and on what basis, as compared to the face-to-face mode; as well as what percentage of a programme should be delivered in these different modes. Such questions will be addressed later in the chapter. In terms of operationally useful definitions of blended learning, I find the following conceptions capture the essential framing:

> Blended learning is the combination of different training "media" (technologies, activities, and types of events) to create an optimum training for a specific audience. (Bersin 2004, xv)

> In the best-blended learning design, the selection and organization of learning activities and assessments support desired learning outcomes while maximizing the strengths and minimizing the weaknesses - of both online and onsite environments. (Stein and Graham 2014, p. 28)

Most importantly, as Picciano et al. (2014) summarized:

> Maximizing success in a blended learning initiative requires a planned and well-supported approach that includes a theory-based instructional model, high-quality faculty development, course development assistance, [and] learner support. (p. 3)

The concept of blended learning is attractive for several reasons. Firstly, from a pragmatic point of view, there are affordances in terms of cost, time and convenience. As the demand for higher education increases, in the face of public funding and personal finances decreasing, high-cost long-duration face-to-face instruction may become an option only for the wealthier minority. It's not a high-level prediction to forecast an explosion of low cost, even free, online or blended learning, as MOOC (Massive Open Online Courses) have demonstrated. How this will eventually position itself in the market context is open to anyone's guess at present, but there's little doubt that blended learning, in whatever format (e.g., Flipped Classroom Learning) will be a major curriculum option. The challenge will be to make it as effective, efficient, and engaging as possible for increasingly diverse groups of learners (e.g., differentiation). This is where creative teachers will be most impactful and needed.

There is evidence that blended learning can enhance learner attainment. For example, The U.S. Department of Education, in a meta-analysis of online research, reported that students in online courses performed modestly better, on average than those in face-to-face courses, with blended students performing the best (e.g., Means et al. 2010). Furthermore, according to the Educause Horizon Report (2019) Higher Education Edition:

> Students report a preference for blended learning, citing flexibility, ease of access and the integration of sophisticated multimedia. (p. 12)

6.4 Designing and Facilitating Blended Learning

The EdTech design frame presented here can be applied both at the macro-curriculum level (e.g., a module or unit of study), as well as for individual sessions or lessons. Invariably, as in the face-to-face situation, even the best pedagogic design and practices will not engage all learners, and certainly not all of the time. As we know, when dealing with humans, you will not, as the saying goes, "Please all the people all of the time". There are *just* better heuristics, but it's an important 'just'. From an evidence-based approach, I use the following broad heuristics in the creative pedagogic design and facilitation of blended learning:

1. Good learning design is based on Evidence-Based Practice (e.g., methods that work best and consistent with how humans learn, i.e., embodied in the Core Principles of Learning)
2. EdTech (e.g., internet, e-tools, multimedia) are used strategically and creatively to enhance specific aspects of the learning process across the design, planning, facilitation and evaluation of learning events
3. The completed blended learning design maximizes the affordances of a range of learning modes, mediums and methods to enhance learning effectiveness and efficiency (e.g., attainment levels, interest/engagement, differentiation, access and flexibility).

For example, the selection and use of technology platforms and tools become one of identifying the affordances of different categories of e-tools (e.g., content creation & delivery tools; communication & collaboration tools; assessment & feedback tools) and how they can be used to positively impact aspects of the learning process. The following EBT guiding heuristics apply:

- In principle, if an EdTech facility (e.g., e-tool or an e-tool combination) enhances any aspect(s) of the learning process (e.g., taking in, processing and applying information in practice) for a group of learners, then there is potential use in terms of infusion into the instructional strategy
- EdTech combinations that enhance a number of the core principles of learning simultaneously are more likely to have a greater synergistic impact in terms of enhancing student attainment opportunities and the experience of learning
- The creative blending of high effect size methods and cognitive scientific principles (e.g., the core principles of learning) with appropriate e-tools is where teachers can be most impactful, both on the motivational and attainment stakes. Such expertise, both in planning and when the situation demands (e.g., when the planned strategy is not working with a group of learners), can re-invent a new more effective strategy in situ—is ***Creative Teaching Competence***. In terms of Hattie's (2009) 'Russian Doll' analogy, the dolls are getting an added EdTech 'makeover', so to speak. We would then be using the best method combination in terms of pedagogic design and e-tools integration to support such methods.

This guide can be systematically worked through in the learning design process. Key focal areas for creative thinking and application will be in the strategic enhancement of the learning process and maximization of the blend. You will often find that while working through this design process in practice, new ideas or potential resource blends will come into mind, making this as much an iterative process as a linear one, and that's where much of the creative connections will incubate and hopefully flit into conscious thought—as we explored previously/That's the way the brain typically works in terms of creativity in any field.

6.4.1   Good Learning Design Is Always Grounded on Evidence-Based Practice, Incorporating Core Principles of Learning

The 'brain is the brain', whether it's in a face-to-face situation or processing stuff online. Do we need to adapt the teaching strategy, customizing what we know about human learning and teaching methods to the online environment? Of course, we do, just as there is always customization of instructional strategies in different face-to-face contexts. However, we now have to effectively negotiate an added customization and adapt it to the particular nature, affordances and limitations of this different instructional mode. Clarke and Lyons' (2005) analysis, in the context of human learning, remains relevant for the foreseeable future:

> The most robust instructional principles are those based on a model of human psychological learning processes …Any given instructional method will be effective or ineffective depending on the extent to which it supports or disrupts basic-learning psychological processes regardless of the delivery media. (p. 594)

Hence, no matter how much information we have in terms of gigabytes and terabytes, it is not going to get quickly assimilated and nicely integrated into long-term memory. As Moroder (2013) discovered from her experience:

> Technology does not make learning more engaging or meaningful. A great lesson does this…technology can make it more effective and efficient.

However, as summarized earlier, the situation is now changing for the better concerning the use of technology in teaching and learning. Firstly, the technologies are becoming more stable, much faster and, most importantly, user-friendly (and many are free). Secondly, and most significantly, there is a reframing of the use of technology towards pedagogic design considerations and how they might enhance specific aspects of the learning process, rather than the technologies per se. Hence, Horton (2006) makes the summative point in this context:

> At its best, e-learning is as good as the best classroom learning. At its worst, it is as bad as the worst classroom learning. The difference is design. (p. 3)

Similarly, Olbrish Pagano (2013) pointed out:

> Technology will change, but good design is constant. (p. 8)

Quite simply, disorganized and over complex content in the online environment is no less disruptive than in the face-to-face context—perhaps even more so. Similarly, dull is dull, wherever, whenever; and we know how this works in terms of brain processes. The framework developed in the previous chapters equally applies here. EdTech simply provides a resource capability which, if thoughtfully used, has the almost unlimited creative capability for enhancing learning opportunities for a wider differentiated range of learners. Treadwell (2017) summarizes the present scenario accurately:

> The focus of technology is not to make the learner's work look pretty or create far more 'stuff', but to give greater agency to the learner and drive their learning capacity deeper by focusing on the new end-point of building conceptual frameworks of understanding that learners can apply creatively to be innovative…

> The key here is to leverage technology to make that happen, but the technology must be simple to use, requiring no significant training. If ongoing training courses of any length are required, then it is unlikely that in the long term, the technology will be used by educators as it was intended. (p. 156)

6.4.2 EdTech Is Used to Strategically and Creatively Enhance Aspects of the Learning Process

Pacansky-Brocks' (2013) position on the use of e-tools is particularly pertinent in this context:

> The tools here are merely colours in a palette. As an artist, your task is to select a tool and align it to your creative vision to construct relevant, engaging learning activities for your students. (p. 130)

EdTech does not change the fundamental ways in which the brain works and therefore our focus must be on what the different e-tools can specifically do to enhance aspects of the learning process. For example, at the most generic level, we know that EdTech provides anytime and anyplace access to online resources. Also, computers do not suffer from mental fatigue and we can, therefore, expect consistency in performance, if the technical architecture is good. Hence, this is a potentially good affordance for those who cannot attend class at designated times.

In understanding more fully the specific range of learning affordances that EdTech offers, it is useful to consider what technology and human brains are disposed to do well with information processing. This provides evidence-based guidance as to the learning contexts in which technologies may be most effective. For example, we know EdTech tools are much better than the human brain at:

- retrieving information from vast resource banks of data
- rapidly, accurately and effectively processing complex sequences of clearly defined facts
- reconstructing and representing large amounts of information.

In contrast, the human brain is better (at present anyway) than the computer at:

- conceptualising ambiguous problems
- exploring concepts
- formulating and communicating ideas.

Sylwester's (1995) summary, from an extensive review of the literature, pulls the present discussion together nicely:

> Our brain is better than a computer at conceptualising ambiguous problems. Conversely, a computer is better at rapidly, accurately and effectively processing complex sequences of clearly defined facts. (p. 120)

Based on the brief comparison and contrast above, certain inferences and interpretations of what types of e-tools offer significant learning affordances for different aspects of the learning process are readily apparent. For example, cloud-based applications such as Google Drive and Dropbox enable the storing, organizing, sharing and collaboration of a wide range of content and applications. This enables teachers to present extensive content resources in various formats and mediums, catering to a wide range of learners' needs and competency levels. The capacity to decentralise the structure of knowledge bases and reconstruct them in dynamic customised digestible bits (knowledge warehousing) makes knowledge even more directly accessible and manageable. Similarly, and perhaps the most significant single learning affordance of the online learning environment is the hyperlink which, at the click of the mouse, can bring together a wide range of text-based, multimedia and personnel resources way beyond what is possible in the traditional classroom. Hamilton and Zimmerman (2002) illustrate this vividly when they wrote:

> …the hyperlink, which is practicable without counterpart in the physical world of traditional academics. Within an internet document, hyperlinks are used to bring multi-sourced information into the primary text or to give the reader a path to alternative media. In essence, this eliminates the physical separation of material messages that are logically connected. In addition to text, hyperlinked messages may be pictures, sound files, animations, or video clips. External links can refer students to other information-rich Internet sites, including personal Web pages, specialized bibliographies, and professional specialists. (p. 270)

This provides the capability of creating networked resources that enable both faculty and students to create, share and continually develop an extensive and varied range of resources that can support the desired learning outcomes. These enable the capability to:

- centralize key resources relating to a module syllabus (e.g., learning guides, module maps, advanced organizers, annotated bibliographies of key resources, guidance on how to negotiate potentially difficult topic areas)

- select prepared resources to support learning (e.g., notes, cases, videos, animations, activities)
- select web links to provide a networked architecture of extended and dynamic resources
- access, where appropriate, to other digital learning exchange portals (e.g., libraries, specific learning communities).

6.4.3 The Completed Blended Learning Design Maximizes the Affordances of a Range of Learning Modes and Mediums

This concerns determining what curriculum components and specific learning outcomes can be effectively and efficiently met in the online environment, and what can be better facilitated in a face-to-face context, the 'balance of the blend' so to speak. From an evidence-based approach, the answer is primarily pedagogic and situated rather than numeric. It is not a question of how much online learning versus how much face-to-face learning; rather about how the face-to-face learning context can be enhanced through ICTs and vice-versa. The real indicator of effective blended learning is not the amount of face-to-face or online learning but their effective integration within a programme (Garrison and Kanuka 2004).

Therefore, if the previous two stages of the design process have been appropriately negotiated, this final stage is essentially one of practicality and creativity. For example, while we may have an 'ideal' blend in our mind, in practice the 'right blend' may depend on some other factors, which typically include the following:

- Programme type and focus (e.g., cost reduction, high impact on attainment)
- Learning group (e.g., prior competence, motivational level, cultural factors)
- Resources (e.g., budget and technology infrastructure)
- Content stability (e.g., enduring, relevance to key outcomes).

As the creative combination of methods, activities and resources underpin creative teaching in the face-to-face context, the same design principles equally apply in the online environment. We now have an increasing range and variety of e-tools that provide affordances for different aspects of the learning process, hence the increasing potential for more and more creative combinations. Furthermore, as we create and develop effective and efficient method and e-tool combinations (blends), the creative process will, over time, lead to highly synergistic embedded learning experiences that will move us towards the ideal of maximising learning opportunities and attainment for all students. This is creative teaching competence at the level of adaptive expertise.

The following section outlines the Flipped Classroom Learning approach (a variant of blended learning) and illustrates the EBT Blended Learning Design Model with examples from a 3-year research project conducted at Singapore Polytechnic

(SP), which systematically implemented Flipped Classroom Learning in a series of 15-week module programs. (Sale et al. 2017).

Flipped Classroom Learning

The Flipped Classroom is a blended learning format, in which online work on key underpinning knowledge is completed before more application orientated work is facilitated in the face-to-face context. It was highly popularized by Bergmann and Sams (2012), and since then, there had been many publications attesting to the usefulness of flipped classroom in improving various aspects of students learning (e.g. Herreid and Schiller 2013; Berrett 2012), and many resources offering guidelines and tips for implementing flipped classroom (e.g. Moffett 2014; Margulieux et al. 2013). Strayer (2012), in the teaching of statistics, noted that students in an inverted classroom become more aware of their learning process than those in more traditional settings; and increases in student cooperation, innovation and task orientation. Based on analysis of self-determination theory and cognitive load theory, Abeysekera and Dawson (2015) suggested that flipped classroom can improve student motivation and help manage cognitive load. Other benefits include increased student's engagement and satisfaction with the course of study (e.g. Strayer 2012; Wilson 2014), and improvement in meta-cognition (van Vliet et al. 2015).

Equally, flipped classroom learning has received a significant share of critical review. For example, Mason et al. (2013) reported that the flipped classroom "at best-improved students' understanding of engineering concepts" and that at worst "did not harm". Similarly, McLaughlin et al. (2013) compared final exam scores of students and found no significant differences. In a recent publication, Jensen et al. (2015) reported that flipped classroom does not result in higher learning gains or better attitudes compared to a non-flipped classroom when both utilized an active learning, constructivist approach. These authors proposed that such gains are most likely a result of the active learning style of instruction rather than the use of flipped classroom per se. One of the strongest objections to a flipped classroom is a view that it is "simply a time-shifting tool that is grounded in the same didactic, lecture-based philosophy. It's a better version of a bad thing" (Ash 2012). Pienta (2016) noted that the literature is clear that flipped classroom, just like other forms of active learning, requires engaged students. However, not all students are motivated to put in the required effort to learn lesson materials on their own before coming to class.

Most importantly, in this context, according to Abeysekera and Dawson's (2015, p. 3) review of research on flipped classroom learning, it remains "…under evaluated, under-theorized, and under-researched in general." Recent research is still showing that much of design research is still focusing on surface features, or physical attributes (e.g., online, face-to-face), of the design without articulating clearly the core pedagogical attributes. As Picciano et al. (2014) point out:

> …the heavy focus in existing models on physical or surface-level characteristics rather than pedagogical or psychological characteristics are impeding progress. (p.29).

The following extracts are from two of the research-based cases on implementing *Creative Flipped Classroom Learning: An Evidence-Based Approach* (Sale et al. 2017).

Supported Experiment 1: Digital Electronics

The Digital Electronics module in the School of Electrical and Electronic Engineering in SP has been flipped for 3 years. Apart from its impact generically, we have become interested in how it specifically impacts weaker students. The learning design process incorporated appropriate high 'Effect Size' teaching methods (Hattie 2009) and cognitive scientific principles (e.g., Core Principles of Learning, Sale 2015). Furthermore, the selection of EdTech tools was based on their predictive capability to enhance specific aspects of the learning process. For example, the EdTech tool Kahoot was chosen for its capability, when used effectively (especially creatively) to get good student attention and engagement with the content learning, activate prior knowledge and check conceptual understanding. In the language of cognitive science, this had an excellent *Von Restorff effect* and hit the 'sweet spot' in terms of an appropriate motivation strategy for many students.

Data from 'Student Co-participants' (e.g., Lincoln 1990) was collected to ascertain how the students experienced the different components of the flipped classroom innovation These were student volunteers who actively participated in the research, providing regular feedback to the research team of their experience in learning throughout the modules. The focus group of student co-participants comprised 11 of the class members. Based on agreed areas by the majority of students, the following inferences and interpretations were recorded:

- The anytime, anyplace and opportunities for repeated exposure received a strong majority affordance (which was to be expected).
- The use of the messenger app for smartphones like WhatsApp and EdTech tools such as Socrative Exit Poll and Kahoot was widely noted as supporting learning. Students found the provision of the short videos both supporting content understanding and interesting. They also identified and confirmed the learning benefits of some of the key explicit EBT strategies employed (e.g. activation of prior knowledge, checking understanding, timely and quality feedback, and the creation of humour and fun as part of the learning experience).
- Of note, while the design of the instructional strategy is important, much of the feedback seemed to be contextualized to how individual teachers facilitated the learning process and interacted with students. In most basic terms the teachers' style, personality and competence are key components determining the success or otherwise of a flipped-classroom approach.

Student Performance

Apart from what the students told us about their learning experience, we also wanted to find out the actual attainment levels in terms of the meeting stated outcomes. We used both formative and summative assessment strategies to facilitate our learning goals and objectives throughout the semester.

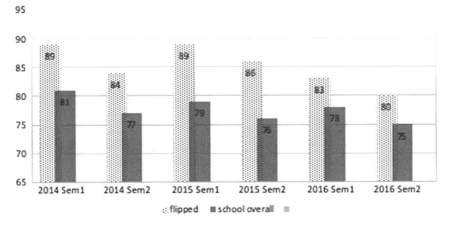

Fig. 6.1 Comparison of student performance

The comparison was made between the flipped classes and the entire cohort of students taking the same module, with overall results computed using marks from general participation, continual assessment, laboratory tests, mid-semester test and final exam. Figure 6.1 shows the comparison of student performance for over 3 years. The results show better student attainment for the flipped classes as compared to school overall. Furthermore, there is a consistent improvement in student attainment for consecutive 3 years.

Supported Experiment 2: Plant Safety and Loss Prevention

Flipped Classroom Learning has been implemented for the Year 3 core module *Plant Safety & Loss Prevention* in the Diploma in Chemical Engineering since Academic Year (AY) 2015/2016.

The motivation for adopting flipped classroom stemmed from the observation (and frustration) of the lecturer that students taking this module are often unable to integrate knowledge and skills gained from earlier studies and developed a sufficient understanding of the subject content. Students also lacked real-world experience in the chemical process industries to fully appreciate the application of loss prevention principles in the chemical plant. This, plus the 'fact' that safety itself is a relatively dry topic due to its 'common-sense-on-hindsight' nature, makes teaching the module especially challenging.

Like the Digital Electronics module, the learning design process utilized Hattie's high effect size strategies (2009) and Sale's (2015) Core Principles of Learning. We specifically focused on the effectiveness of the following high effect size interventions and using the 'Russian Doll strategy' (Hattie 2009) outlined earlier:

- Learning Outcomes: Advance Organizer (start-of-class) and Checklist (mid-point and end-of-class) (effect size 0.40)
- Whole-class Interactive Teaching: Challenging Goals (effect size 0.56), Classroom Discussion (effect size 0.82) with Peer Influences (effect size 0.53)

- Facilitation: Questioning (effect size 0.46), Scaffolding (effect size 0.53)
- Assessment for Learning: Feedback (effect size 0.73), Formative Evaluation (effect size 0.68)
- Deliberate Practice: Worked Examples (effect size 0.37), Mass Practice (effect size 0.60).

A designated website was created to host all the resources students needed for the flipped classroom lessons. Bite-sized pre-class mini-lectures were organized using EBT practices and principles to ensure activating student's prior knowledge, facilitate the acquiring and assimilation of new context knowledge, as well as managing cognitive load (e.g., consistent with effective memory processing). EdTech tools were used effectively and efficiently (e.g. where it was felt that they could positively impact one or more of the core principles of learning) throughout the pre-class learning experience. For example, all mini-lectures were created in MP4 format and made available in a dedicated YouTube Channel via links in the module website. Similarly, animations of process flow were created using PowerPoint exported to YouTube and drawings of the process control loop and safety interlock were created using Video-Scribe. The animated piping and instrumentation diagrams were created using Thinglink; while instructions to students are delivered via cartoons created using PowToon. Other pre-class activities are introduced as and when appropriate, e.g., reading of industry journal, or visit a laboratory. At the end of the pre-class lessons, students get to evaluate their understanding using Socrative, usually via a combination of True/False and multiple-choice questions. Students can also get a "sneak preview" of how the knowledge gained from the pre-class learning will be used in the classroom.

In the face-to-face learning context, calibrated to the prior online learning experience, real-world case incidents (e.g., Bhopal Gas Tragedy; Piper Alpha Explosion) are used to demonstrate that failure to adhere to safety principles often is a leading cause of these events. Understanding is further reinforced through worked examples and transfer of learning encouraged through the application of key concepts to other case study scenarios. To facilitate a conducive learning environment, one that is success orientated and fun, the use of positive language, tone of voice, calibrated body language and smiling (e.g., Hattie and Yates 2014) is consciously employed to facilitate rapport building.

Some class activities involve all groups working on the same problem, and results from different groups are then compared. To maximize learning via collaboration, two key strategies are used. One requires that each group of students work on different aspects of the same problem. For example, one classroom activity requires students to assess the adequacy of a reactor systems protective system from an over-pressure condition. One group can look into excessive pressure scenarios due to uncontrolled chemical reaction, while another group can investigate potential over-pressures due to failure in the reactor's cooling jacket, and yet another group can study the failure of its pressure control system. Each group can then share their findings with the rest of the class. Another strategy requires that each group of students work on a different case study and share the results with the whole class. In this way, groups are purposefully active towards meaningful goals and have variety in the learning experience. There are also more challenging questions for any group to tackle if they finish the assigned questions earlier.

Google Doc, Google Slide and Padlet are used for student work submission in a class, enabling the lecturer to review and give appropriate feedback in real-time. Google Doc is used most frequently, predominant for the posting of group answers for the activities mentioned above. Google Slide is used for activities requiring students to arrange events in a particular sequence, for example, how a series of plant modifications made out of good intentions, resulted in an unintended consequence that proved fatal. The use of Padlet enabled individual students to pose questions to lecturers on topics that they did not understand, as well as for posting individual work in response to concept-type review questions posed by the lecturer.

The opportunity to view students' work in real-time is perhaps one of the most significant benefits of flipped classroom learning, as it afforded both the lecturer and students to analyse and evaluate the learning experience in situ—so to speak. This benefits the students as they get to clarify any doubts or uncertainty in understanding on the spot. For the lecturer this means that he/she can get feedback on the effectiveness of a particular classroom strategy employed, and depending on the situation, can make adjustments in real-time as well.

To ascertain the effectiveness of the different high effect size strategies used, survey questionnaires, focus group discussion, and student co-participants were employed (as in the previous Supported Experiment). Overall, students find the learning experience enriching. On the more specific strategies, the students reported that the use of advance organizers is useful in helping them to keep track of their learning progress and how the various topics are connected to form the big picture, although some found it difficult to understand at first. They also found that the use of self-evaluation exercises after every topic (multiple choice and/or true/false questions in Socrative) are useful. Likewise, students reported that the use of Google Doc and Google Slide in the classroom is very useful in helping them to learn, especially from each other. Students also informed that they found the use of mock assignments and marking with rubrics is useful to help them understand the lecturer's expectations and how to improve on their work. They, however, did not find the use of checklist very useful, partly due to the way the lecture used it – which is seen as more for the lecturers own check rather than for the students to track their learning.

Besides feedback from student co-participants, the lecturer also engaged in Evidence-Based Reflective Practice (Sale 2015). This involves a structured thinking process (e.g. analysis and evaluation) using evidence-based teaching (EBT) principles to all valid evidence sources (e.g. students, peers, peer observers), rather than personal reflections in isolation. As a holistic process, it enables a better understanding of the reality of classroom learning (e.g. what is happening, and how this is affecting the learning process). Through this process, the lecturer can pinpoint areas where strategy/EdTech tool blending can be improved, and also identifying one's blind spot in assuming that students can retain key learning points from earlier modules. Furthermore, via this process, the lecturer can draw similarities in one classroom activity that posed difficulty and adjust the delivery approach in a later classroom activity, e.g. making better alignment of topics and making connections to topics covered earlier.

Table 6.1 summarizes the findings on the impact of each strategy employed from the students' perspective (using broad categories of "Very Useful', 'Useful' etc., based on students' responses to the survey questions) alongside with the lecturer's own reflection based on his classroom observation.

In the following subsections, I will illustrate how selected EdTech tools can significantly enhance different aspects of the learning process. This will involve a consideration of e-tools from the genres identified prior, focusing mainly on content development and delivery, knowledge building for understanding, and supporting skill acquisition. This provides the key underpinning components for the development of competence. The e-tools selected are not meant to be comprehensive coverage of the genres or prescriptive in any way, as many others can serve similar pedagogic purposes. Also, I am focusing on those tools that I find user-friendly, both in terms of the teacher/developers' perspective as well as from the learner's experience. Most of the tools discussed below are either freeware or 'affordable', recognizing the latter is always relative.

Table 6.1 Effectiveness of various strategies employed

Strategy employed		Students' reported usefulness	Lecturer's own class observation
Learning outcomes	Advance organizer	Useful	Students initially have some difficulty
	Learning checklist	Not useful	Agreed—largely due to the way it was used!
Whole-class interactive teaching	Challenging goals classroom discussion and peer influences	Very useful	Allowing students to form own group is preferred
Facilitation	Questioning and scaffolding (activation of Prior knowledge)	Useful	Students initially need encouragement to ask questions. Once rapport was built (after about 3 weeks) this process is going on well. They are generally open to feedback, and are now not hesitant to ask questions. However, getting students to do extra exercises on their own still proved challenging
Assessment for learning	Feedback and formative evaluation	Useful	
Deliberate practice	Worked examples and mass practice	Very useful	

6.4.3.1   Rapid Content Development Software Tools

The importance of good subject content knowledge, the essential information that needs to eventually end up as well as formulated mental schemata in the learner's long-term memory is a crucial element of effective learning. Hence, the selection, organization and presentation of content are important considerations in all learning contexts, and especially so in the online context, which usually lacks immediate opportunities for clarification and feedback. It's important also to be highly sceptical of the notion that content is all out there in cyberspace, just waiting to be downloaded by a few clicks of the mouse. This is wildly over-optimistic. While there are numerous and varied resources on the Internet, much may lack validity and usefulness. Keen (2007), for example, makes a damning criticism of so-called internet expertise knowledge contained in such sites as Wikipedia:

> ...the real consequence of the Web 2.0 revolution is less culture, less reliable news, and chaos of useless information. One chilling reality in this brave new digital epoch is the blurring, obfuscation, and even disappearance of truth. (p. 16)

Invariably, this problem can be mitigated to some extent by helping learners to be more critical, and apply good thinking, to what they are reading or looking at, often referred to as *Media literacy* or *Digital Literacy*. The ability to do this effectively and efficiently will constitute an important competence for learning from such resources. It is noteworthy that this is not a new human faculty or competence but generically applies to all aspects of human communication. Newspapers, books and other media have been around a long time, which need such critique; it's just that we now have more and more varied information sources to validate.

From an evidence-based approach, in the real world of teaching and learning, well-prepared resources, tailored to the desired learning goals and outcomes, with multi-modal and differentiated learning experiences, are a key affordance to support student attainment in most learning contexts. To illustrate this, let's go back to my GCE 'A' level experience of yesteryear. The lessons typically had the following format for the full 2-year duration of the programme:

- The teacher dictates notes for the whole duration of the lesson
- The students copy these down verbatim or in personal shorthand form
- The students write these out neatly and store them safely after the lesson
- The students memorize these notes for tests and exams.

There were probably some opportunities for questions but I cannot recall these as a significant part of the instructional strategy. It was not pleasurable, and it was not novel, but a dull and boring 'slog' to pass the exams. Such practice was forgivable some 40 years back, but it is not now. We are in a position to use EdTech to create, organize and present our subject content knowledge in highly organized, dynamic and interesting ways. This is where rapid content development tools are particularly useful. In most basic terms these are technologies that enable the production of e-learning content and learning experiences which have the following key elements:

- User-friendly functionality requiring only a short learning curve (in some cases only a few hours) to master
- Teaching and learning resources can be produced and updated rapidly. This depends on the number of resources produced but is significantly quicker than the previous e-learning development software
- The content mix can include text, graphics, embedded videos and podcasts, hyperlinks to more detailed and differentiated content, activities and assessments.

This enables faculty to quickly get up to speed in being able to produce and integrate a variety of media-rich and interactive learning resources tailored to programme learning outcomes and accommodating a range of student learning capabilities. When guided by a strong pedagogic literacy and creative teaching competence, these tools provide an enormous capability for enhancing the student learning experience at the level of exposure to the content knowledge to be learned; these specifically include:

- Content structuring that ensures good chunking to reduce cognitive overload
- Multimodal presentation to enhance interest and application for the content to be learned
- Ongoing formative assessment, enabling retrieval practice of key concepts through short quizzes and immediate feedback.

From my experience, apart from PowerPoint which has been around for a long while, I have found the following rapid content development software tools to be particularly useful:

SoftChalk LessonBuilder enables the creation of interactive web pages for e-learning courses. The software is easy to use (really) and it enables the quick production of interactive lessons that have a professional look to them. Specific features include pop-up text annotations, self-assessment quizzes, and interactive learning games. After production, you can package the lessons for delivery via CD-ROM, Intranet, Internet, or integrate with a Learning Management System. As their homepage states:

> If you can use a word-processing program, you can use LessonBuilder. Designed for teachers and content-experts that don't have time to learn complex software, LessonBuilder is simple, yet powerful, with only the features you need to create exciting, interactive content for your online course. It claims to enable you to:

> Create custom lessons by combining your materials with interactive learning content. The mixture of personalized content, embedded assessment, and interactivity will increase student engagement and improve learning outcomes.

They offer free trial downloads and the software is well priced in terms of comparative products on the market. The current website is https://softchalk.com/.

Camtasia Studio enables the creation of packaged lessons within a self-contained video format that can be web-enabled. Its screen recording system that will capture a prepared lesson (e.g., PowerPoint presentation) as well as your voice-over during the recording. Using good pedagogic design, high-quality teaching videos can be shared with students on the Web, CD-ROM, as well as on portable media players such as the iPod. The live-action video component adds the human touch to the presentation material and enables both technical professionalisms as well as the use of informal narrative and humour. The current website is: https://www.techsmith.com/store.

VoiceThread is a web-based application tool that facilitates the presentation of an environment of integrated learning resources (e.g., images, video, documents) in which participants can interact and contribute (e.g., voice, video upload) both synchronously and asynchronously as part of collaborative discussion. It is a more interactive collaboration tool than the other tools outlined. The application is easy to use, provides a versatile learning environment that is easily modifiable and reasonably priced. The current website is: https://voicethread.com/.

VideoScribe is an easy to use tool that enables the production of content, incorporating text, graphics and audio into a visually powerful and animated video format. In terms of learning affordances, it enhances presentation impact, can highlight key concepts, and works particularly well in terms of our memory systems. For creative teaching and the development of one's creative teaching competence, it's a real playground for such activity. The opportunities for creating novelty—those powerful Von Restorff effects—into the content structuring are probably unlimited. It is both fun and challenging to use creatively and, from what we know about human learning, it will have positive impacts on learning, especially student attention and engagement. The current website is: https://www.videoscribe.co/.

While these applications have quite different affordances, all have significant capability to enhance learning effectiveness and efficiency in terms of content provision. Given the user-friendly nature of these e-tools, it will not be a time consuming or frustrating experience to experiment with them or view some good exemplars. From that basis, it should then be readily apparent which applications (and you can use more than one) are best suited for particular parts of your curriculum and for the students you teach.

6.4.3.2   Communication and Collaboration Tools

The ability to engage with a wide range of relevant content in the context of open communication and collaboration with peers and experts, where there is ease in posing and answering questions to facilitate building understanding, has much by way of learning affordance and cuts across many of the core principles of learning. This is another area in which EdTech can have an increasing creative impact on how learning is facilitated but requires, of course, creative teachers. Students themselves will likely find creative ways in which to use these technologies as they are supposedly the 'Digital Natives' (Prensky 2001). However, are they? There is much to challenge such popular generalizations, as Hattie and Yates (2014), from reviewing the research, pointed out:

> The central problem with the digital native theory is that it is advanced in the absence of any known database…In its raw form, the digital native theory has to be seen as considerably overstated and incorrect. Human capabilities are not as flexible or tied to experience as this theory might suggest. (p. 197)

They also pointed out:

> The same is true of students today being able to multitask – again the evidence is negative. The notion that the brain can genuinely do two things at one is widely recognized to have no serious validity. (p. 188)

This is fully supported by Miller (2016) who notes that:

> There's little reason to believe that ubiquitous computing has transferred the younger generation of students into tech-savvy, tech-dependent individuals who think in ways that are distinct from those of their older counterparts. (p. 5)

The whole notion of students' today being significantly different in terms of learning from previous generations makes no sense in terms of human evolution. It is unlikely that brain functioning, despite the arguments for brain neuroplasticity, could have undergone such systemic neurological change so quickly. Hence, the significant changes in student behaviour are most likely to be both phenotypical and memetic. It is, however, worrying, if Winget's (2017) assertion that many of the new generations are true 'snowflakes' (I don't think I need to explain the analogy) has validity, and accredits this to:

> Our values are slipping and it shows up in every area of our society, and it terrifies me that people have convinced themselves this isn't the case. (p. 7)

He goes on to argue:

> Because of the collapse of the core values of kindness, charity, love, being-nice, and respect, people resort to meanness as a way of dealing with each other. (p. 76)
>
> Most people are cursed with a combination of too much want and too much won't. (p. 82)

How accurate Winget's framing is, and how generalizable it is across countries and cultures, is likely to be contested by many, and I am not going to pursue the debate here. However, if there are increasing numbers of students who are lacking in such values, this will make the job of teachers even more challenging in promoting self-directed lifelong learning, which is based as much on values and volition as it is on cognition. Such concerns are not just pertinent to this analysis and evaluation on teaching with technology, nor of teaching per se, but of wider societal concern.

Anyway, back to the specific issue of technology for learning in this context. Of all the EdTech features, perhaps the most prolific in terms of impact on young people's engagement is the ever-increasing range of Web 2.0 and social media e-tools that enable communication, content sharing and collaboration. There is some confusion between what exactly are the differences between Web 2.0 tools and social media tools? It seems that it is as Beattie (2011) wrote:

> It would be difficult to find two popular buzzwords that are in more of a quagmire than social media and Web 2.0. (technopedia)

He goes on to argue that while social media is a Web 2.0 innovation:

> …referring to Web 2.0 as social media is incorrect because it ignores all its less social aspects, such as blogs, YouTube, and so on.

I am not too concerned about these fine differentiations, though I note the qualitative difference of social media being more focused on the ease and simplicity of user-generated content sharing and reviewing. Collectively, these tools provide an extensive platform for both asynchronous and synchronous communication, sharing and collaboration. Learners can engage in multiple platforms of subject content knowledge, share and collaboratively work with these knowledge sources. Such activity has the potential to help learners to build understanding, join and participate in learning environments and communities that focus on their specific interests and learning goals, whilst in a psychological climate that works for them. Furthermore, related to the capability for extensive resource connectivity, there is the specific connecting (both synchronously and asynchronously) of people globally. Learners can now connect with their tutors, peers and other experts who give their time to enthusiastic learners, as well as pretty much anyone prepared to communicate with them. In today's internet society, we can readily go beyond this physical local community of learners, to a global community of learners. In this context, there is the emergence of "Communities of Inquiry", which Garrison and Vaughan (2008) defined as:

> …a formally constituted group of individuals whose connection is that of academic purpose and interest who work collaboratively toward intended learning goals and outcomes. (p. 6)

However, some words of caution, as popular notions of students being able to build new knowledge and deep understanding through their inherent creative capabilities with various social media and Web 2.0 tools such as blogs and Wikis may also be somewhat exaggerated. Indeed, Willingham (2009) argued that getting students to create new knowledge should not be the main goal, rather it's better to focus on developing a deep understanding of existing knowledge. He noted:

> …posing students challenges that demand the creation of something new is a task beyond their reach - but that does not mean you should never pose such tasks. (p. 109)

However, communication, collaboration and the opportunity to be creative is motivating, can facilitate differentiation of learning, and provides multiple feedback sources. Hence, from an evidence-based approach, there is much potential for enhancing learning opportunities. There are many e-tools in this genre, and certainly some I have probably not even heard of. Here's my experience of a few.

Facebook, while often not seen specifically in the context of teaching and formal education, offers many good affordances in terms of supporting different aspects of the learning process Firstly, it's free and easy to use. Secondly, though now perhaps less so, our students are on it and are comfortable with the format. They may be on it less now, as we are on it more. Nothing changes when it comes to intergenerational interactions, does it? Ok, what's the pedagogic affordances? I find it easy to keep students updated on key aspects of the programme, get feedback from them on my teaching, generate some lively chat around topics of interest, quickly add or update bite-sized resources and quickly hyperlink to more in-depth content of interest. The students can do this also. Once a good psychological climate is created with a bit of humour, it can be a fun and very effective collaborative learning experience. There's much that can be done with the good pedagogic application and a little creativity.

The blog is a Web 2.0 e-tool I find particularly useful. It is easy to use and young people are very familiar with it. Blogs are now an everyday part of communication channels for most students, albeit with a more social rather than educational orientation at present. However, a blogs capability to enhance learning and attainment is high if we look at its potential impact through the lenses of the Core Principles of Learning. For example, to learn a subject effectively, students need to connect new information with what they already know (Core Principle 2: *Learners' prior knowledge is activated and connected to new knowledge*). The very nature and design of a blog facilitate this principle very well, as blog posts typically appear in reverse chronological order with the most recent post appearing at the top of the web browser. If the blog is regularly updated, students will be able to see a progressive update of the material covered each week and will be able to link what they are currently learning with prior knowledge. Similarly, as different media types can enhance the learning experience (Core Principle 5: *Instructional methods and presentation mediums engage the range of human of senses*), blogs can enable the publishing of a range of multimedia content on the web (e.g., video, audio, animation). They can also make explicit and clarify learning outcomes, encourage good thinking, provide rapid, clear and constructive feedback, as well as create a psychological climate that is success-oriented and fun.

A further affordance of the online learning environment for promoting student thinking is that the use of asynchronous text can provide certain significant advantages over the typical face-to-face situation. In face-to-face learning, there is often too much information to absorb and too little time for critical and creative thinking around the content. As a result, knowledge may not be fully understood or even effectively transferred into long-term memory. In contrast, the provision of enduring text, which enables students to spend time revisiting this content, posing and answering questions around its application, helps to build a solid understanding of topics over time. As Hamilton and Zimmerman (2002) argued:

> The medium supports iterative exchanges of information and opinions over an extended period, so ideas are not merely "hatched" and delivered but rather allowed to evolve and be refined in a manner that makes information more convincing, narrative deliveries richer in detail, and learning more thorough. (p. 265)

The blog is certainly, from my experience, a high leverage e-tool. It is easy to use, low cost and has a range of learning affordances that can be utilized with a sound pedagogic literacy and some creativity.

Kahoot is a student response e-tool that works in an interactive game-based quiz environment. It is free (at the time of writing) as user-friendly as an e-tool can get and provides a platform in which student prior knowledge and understanding can be readily assessed in a non-threatening and fun environment. The creation of questions is simple, and the question types can be easily varied (i.e., the typical multiple-choice question format). Also, when used creatively, it can do all the above, plus be used as a catalyst for other parts of the instructional strategy (e.g., extended questioning and discussion post the quiz questions; an advance organizer for new input). It's also a good Von Restorff effect—but don't over-use this. Kahoot is, in terms of a previous analogy, an elegant ICT enhanced "Russian Doll" and, with good *Presentation Style*, open to much creativity in the classroom. The current website is https://getkahoot. com/.

While numerous e-tools are offering a wide range of communication options, it is important to recognize, as with most things, that more is not necessarily better. There is much online chat and sharing in these social media cyber-places, but we may also question just how effective such online collaboration is in real learning outcome terms. For example, Brown (2009) concluded that:

> The internet helps move information around but has done little to bring people together. Creative teams need to be able to share their thoughts not only verbally but visually and physically as well. I am not at my best writing memos…I haven't heard of a remote collaboration tool that can substitute for the give-and-take of sharing ideas in real-time. (p. 30)

Similarly, as Melchior et al. (1997) rightly pointed out:

> One pervasive myth is that the technologies themselves teach important complex skills…they need to be identified, taught, modelled, and reinforced by capable teachers. (p. 91)

Indeed, the development of good thinking is a major challenge in any medium and we explored the reasons for this in some detail prior. Furthermore, we also explored strategies which firmly established that student thinking can be developed through:

- The explicit modelling of the cognitive processes that are involved in good thinking
- Involvement in questioning processes that cues and reinforces specific types of thinking
- Engagement in authentic real-world meaningful tasks (e.g., projects, case studies) that are challenging but achievable and necessitate the use of the main types of thinking (e.g., critical, creative, and metacognitive).

There are now many EdTech applications that can be used to facilitate and enhance thinking, provided that they are employed thoughtfully within the context of the pedagogic considerations and practices. Such applications include:

- Online tutorials involving active problem solving with feedback
- Hypermedia software integrating knowledge, multimedia, activities and feedback
- A range of communication tools (e.g., email, blogs, bulletin boards, forums)
- Constructing software (e.g., desktop publishing, spreadsheets, etc.) where learners can produce, manipulate and change information
- Simulations and virtual reality programmes.

For example, in a chemical engineering module, in which students had to solve a range of problems in a chemical reaction plant simulator it was noted that certain key factors in the instructional strategy combined to enhance the quality of students' thinking. Most important was that the design of activities was a challenge but achievable in the time allowed, coupled with appropriate questioning strategies such as cueing such critical thinking skills as analysis, comparison & contrast, evaluation and making inferences & interpretations (Sale and Cheah 2011). Where activities were not experienced as challenging by students, the simulator lost this capability for enhancing the development of such critical thinking skills. It was also noted that in situations where both faculty and students had shared notions of what constitutes good thinking, there was evidence of further enhancements in the quality of student thinking.

Interactive Videos

We are primarily visual learners, and our visual system is so advanced as compared with our other senses. While we struggle to keep several (at best) pieces of new information in our working memory, we can immediately apprehend (not necessarily understand) a new complex visual experience. That's not to say we pay attention to, or will remember, everything in that visual field, as other mental systems come in to play, including memory. However, we don't suffer from visual overload in the same way we do with cognitive overload. It was long popular, but now largely refuted (e.g. Hattie 2009), that people had distinct learning styles (e.g., visual, auditory, kinaesthetic). I had long challenged such notions, based on reason rather than the research findings that have surfaced since then. Visual dominance seemed obvious from experience, and captured in the saying, "a picture paints a thousand words". Words and auditory learning are of course also important, and these enhance visual learning. Equally, kinaesthetic learning is important, as much learning, especially skills, involves touching stuff—so that's obvious. However, the implication of the importance of our senses is not to seek to understand student's learning styles (quite the opposite); rather be creative in designing learning experiences that engage the necessary range of senses relevant to supporting the learning outcome(s). This is also the case for engaging emotions, as these can aid attention and memory encoding. However, we may not need to engage smelling and tasting—or emotions—in everything we teach. The same generic heuristics, as in the selection and use of e-tools, applies—can it/they enhance an aspect(s) of the learning process (in cost-effective ways)?

I have used video extensively as a key learning resource in teacher education, long before they had the newer technological affordances of interactivity and augmentation in the form of quiz pop-ups, reflective cues, hyperlinks and other embedded features that enable increasing differentiation. It makes sound pedagogic sense in teacher education to use video of actual teachers, doing their work in authentic contexts and then exploring aspects of practice with them through skillful mediation. Furthermore, short, focused videos can provide key conceptual understanding and illustration of specific teaching methods and tools, which can be logically packaged and customized (basic curriculum design stuff) and hey presto—you have some good teaching and learning resources.

Previously, in micro-teaching or post teaching video-recorded sessions, I would manually stop the video at those times that I felt were relevant to exploring a specific teaching practice or principles of learning (professional judgment here) and facilitate accordingly (e.g., open sharing of experience and connections with evidence-based principles). I may also capture key points and/or questions on the whiteboard for further or later reference. In many ways, pedagogically, I was doing then what now can be done more efficiently and in varied ways with technology.

Interactive videos offer many affordances in terms of supporting all the core principles of learning as well as motivation generically. It is not surprising that Treadwell (2017) offers such a bold assertion:

> Over the next 25 years, we will experience the rise of video as the primary information source that learners will use for research and inquiry. It will also increasingly become the medium that educators use to demonstrate their comprehension and understanding. (p. 4)

Greenberg and Zanetis (2012) in a report commissioned by Cisco Systems Inc, identified three key concepts that summarize the pedagogical impact of videos:

1. Interactivity with content (the learner relates to visual content, whether verbally, by note-taking or thinking, or by applying concepts)
2. Engagement (the learner connects to the visual content, becoming drawn in by video, whether on-demand or real-time)
3. Knowledge transfer and memory (the learner may remember and retain concepts better than with other instructional media).

In terms of human learning, a video produced and facilitated with EBT pedagogic design principles is likely to get good attention, engagement and be intrinsically motivating for learners.

Furthermore, in today's context, with easy to use production tools, teachers can produce customized videos, tailored to key concepts in their subject areas, and the context (linguistic or otherwise) of their student groups. There is little need for the video production to be of 'National Geographic' standard, though the key EBT pedagogic and communication principles need to be fully employed. Much research has been conducted that is extensively validated and can be easily integrated into the design and facilitation process. For example, Colvin Clark and Mayer (2011) recommend that E-learning (which includes video formats) include both words and graphics and provide evidence to support the importance of delivering information in the correct audio and visual mix, to create balance in the visual and audio channels of the student. The key principles of good instructional design for audio-visual presentations were identified in Chap. 2, Core Principle 5: Instructional methods and presentation mediums engage the range of human senses. Also, Mayer reinforces what has been explained prior in terms of what constitutes active learning, as in the false dichotomy that students listening, looking and hearing is inevitably passive learning while doing stuff is invariably active and more effective. He argues that:

> well-designed multimedia instructional messages can promote active cognitive processing in students, even when learners seem to be behaviourally inactive. (p. 9)

Miller (2016), argues that one of the major directives from Mayer's research is that:

> ...narrated animations work better than animations paired with text, most likely because the latter arrangement requires the learner to shift visual attention back and forth between these two visually presented sources of information. Auditory input can be processed somewhat independently from visual input, a claim that makes sense given what we know about separate brain mechanisms used for visual and auditory processing. It follows that the two can complement one another when presented in pair fashion.

> Furthermore, narration works best when it uses conversational everyday language, compared to when it's formal and academic in tone. (p. 151)

The video is also exceptionally versatile in providing variation in the learning experience, as it can combine multiple methods and techniques, such as stories, questions, examples, presentational style and humour. I have found that a two-person informal interview in a video is effective, especially in terms of engagement. In new teacher induction programs, I employ an instructional strategy in which a colleague and myself, through a series of role-play videos in which I play the experienced teacher (I am, I think) and the other teacher plays the novice (new teacher), and we explore practical ways to apply EBT to different aspects of planning, facilitation, and using EdTech for enhancing learning—as outlined above. In the first video, for example, the new teacher asks me what he should do in the first lesson, as he informs me that colleagues have said to him, "just go in and teach", which he finds confusing and of little value. The video then captures a dialogue in which I ask him what he thinks he might do, and based on his responses, I offer some contextualized options, and so it proceeds. The basic aim is to introduce the concept of an *Instructional Strategy* and its key components, and how one can plan a lesson. I gradually introduce EBT practices and principles, noting to work situationally and not create cognitive overload. I focus on key concepts and work from there. Subsequent videos build on this, but always incorporate some variation, often a bit of a plot with some humour. Based on much feedback, new teachers consistently find these videos both engaging and useful for preparing and teaching their lessons. The videos are typically short (around 5–7 min), informal in style, reinforce the key concepts, incorporate illustrative examples and stories, facilitate retrieval practice and feedback, and maintain a positive and fun presence. Hence, the overall programme seeks to utilize the affordances of EdTech, especially those of the video, in the context of the EBCT framework.

Even changing location in itself helps on the attentional front (e.g., don't have that white wall or same pot plant in the background all the time). There's much scope for creativity in the overall design and use of video, and it is cost-effective. For example, I make all my videos,—you can view them on YouTube. Most are one take, and I don't worry about them not being pristine; the focus is on content and learning. Hence, making your video, as long as you are not copying, avoids plagiarism, and students respond well to their teachers' effort to engage them more personally—it sends positive subliminal messages of your commitment—remember those mirror-neurons? In my experience, it's a good rapport builder, and often creates a bit of fun, which always works well. Finally, students can also learn in meaningful and motivating ways from making their own videos content—again an area for creative collaborative learning.

Mixed Reality and Artificial Intelligence

Mixed reality (MR) is an umbrella term for a range of technologies, though drawing on Virtual reality (VR) which most people are now familiar with (where the user puts on a helmet and is immersed into an entirely computer-generated environment) and Augmented Reality (AR) (which may use a headset or a smartphone) overlays images or other content onto the physical world. From AR, MR deploys overlays, but like VR, these are interactive and can be manipulated.

The potential learning affordances of VR, AR and MR need little by way of explanation in that they provide immersive experiential learning events. Invariably, their design and connectivity to learning outcomes are of central importance as the cost can be high in production. It may be that their widespread use in mainstream education is unlikely in the immediate future as it can be expensive, with the time for adoption forecasted by the Educause Horizon Report (2019) as between 2 and 3 years. One could argue that AR is not a new concept. I remember using overlays in the days of overhead transparencies, to gradually reveal more of a process or object. Of course, technology-based AR makes possible the immediate capability of delving both very wide and deep into, and across, areas of learning—as was suggested in the movie 'Minority Report', a 2002 film in which such technologies were being used to synergise databases to build models of crime scenes before they happened.

In most basic terms, Artificial Intelligence (AI) uses computer systems to accomplish tasks and activities that have historically relied on human cognition (Educause Horizon Report 2019, pp. 25–27). AI has attracted much attention as its impact across society is potentially life-changing and often referred to as "the next electricity" (e.g., Ilkka 2018, p. 2). There is no doubt that AI will impact many occupational structures, as outlined prior. How much AI will impact teaching and learning, and in what ways, is open to debate. Certainly, key issues revolve around the capability of technology to fully model all aspects of human intelligence and maybe surpass it. For example, Gee (2017), sees this as unlikely in the foreseeable future, and uses the following logic:

> Since computers cannot have experience, they cannot learn humanly. They start with facts and generalizations as strings of symbols they cannot understand. We humans do not generally start with facts or generalizations; we start with embodied experiences. In turn, these embody experiences give deep meaning to the facts and generalizations we eventually derive from them or learn from others. (p. 10)

Ilkka (2018), from a similar analytical standpoint, sees the current AI systems as severely limited, suggesting that there are technical, social, scientific and conceptual limits to what they can do (p. 3) and concurs with the view of Luckin (2018) that at present AI lacks most of the human's metacognitive regulatory capabilities.

In contrast, Harari (2018) paints a more sinister and worrying picture of the future of AI. He suggests:

> AI not only stands poised to hack humans and outperform them in what were hitherto uniquely human skills. It also enjoys uniquely non-human abilities, which make the difference between an AI and a human worker one of kind rather than merely of degree. Two particularly important non-human abilities that AI possesses are connectivity and updateability. (pp. 22–23)

Harari's inference and interpretation of this scenario are not one of a replacement of millions of human workers by millions of individual robots and computers—but being replaced by an integrated network, that is immensely more powerful. What we can infer and interpret from the consequences of this are unknown, and we may easily fall into the realms of science fiction, envisioning something akin to that of the Terminator film series, in which machines take control of the world. Is it possible that an AI integrated network could possess a superordinate form of consciousness that could turn on humans?—we simply don't know. As Harari (2016) states:

> The rise of AI and biotechnology will certainly transform the world but it does not mandate a single deterministic outcome. (p. 461)

In terms of impacting specific aspects of teaching and learning in educational institutions, there are some clear indicative possibilities, including intelligent tutoring customized to individual learning needs and the process of assessment. The two can be related in many ways to provide fully customized personalized learning. It is likely that neural AI will enhance the areas of learning diagnostics, analytics and data mining. As the Educause Horizon Report (2009) summarizes (with a warning):

> It is possible to imagine many exciting possibilities for AI in teaching. Without clear pedagogic principles, it is, however, probable that AI vendors will provide products and services that affect key-decision-makers' perceived immediate problems, instead of more fundamental social and economic challenges.
>
> AI may, therefore, mechanize and reinvent outmoded teaching practices and make them increasingly difficult to change. (p. 32)

The report highlights several challenges, the most salient being a redefining of what educational aims and goals educational institutions should best focus on; it highlights:

> As AI will be used to automate production processes, we may need to reinvent current educational institutions. It is, for example, possible that formal educational institutions will play a diminishing role in creating job-related competencies. This could mean that the future role of education will increasingly be in supporting human development. (p. 34)

For example, Gallagher-Mackay and Steinhauer (2017) make the case for increasing social and emotional skills in the school curriculum. They point out that:

> Strong social-emotional skills, like self-regulation, growth mindset and peaceful problem-solving, give students the tools to manage themselves and their relationships better and clear the way for improved learning. (p. 103)
>
> SEL programs do not merely impact behaviour; they are also correlated to higher academic performance. (p. 75)

Such a viewpoint is very much within my earlier framing of Metacognitive Capability as the Superordinate twenty-first century competence. Certainly, as EdTech makes the organization, delivery and assessment of learning more effective and efficient, this enables a greater emphasis of teaching—indeed educational—to focus on how best to learn, how we might enhance human well-being, as well as derive a value system most conducive to such aims (e.g., Harris 2010).

6.4.3.3 Presence in the Online Learning Environment

Creating presence online is not specifically related to a particular EdTech genre or e-tool, but is an essential aspect of using a blended or fully online learning format. In the previous chapter, the importance of *Presentation Style* was explored in detail as the **pedagogic glue** that creates and structures the learning experience for students. This is where the quality of teaching ultimately plays out at the level of subjective experience for the participating learners, creating communication features and behaviours that have high impact in the building of rapport and facilitating a psychological climate that is success-orientated and fun. The importance of voice, pace and modulation, the use of eye contact and smile were considered from an evidence-based approach. These subtle but powerful aspects of a teacher's presentation style are key determiners of the way students (both consciously and unconsciously) make meaning of the learning experience and orientate their level of participation accordingly. In the face-to-face situation, the teacher has the advantage, if practices are well executed, to quickly establish, monitor and evaluate the psychological climate of the classroom, and make modifications instantly. However, in the context of the online environment, where the direct visceral aspects of a positive presence are lost, or at least significantly dulled, this is much more challenging for the teacher.

To create and maintain an effective presence online, many researchers have looked at online presence in terms of interrelated role functions. For example, Hodges and Saba (2002) suggest that there are three role dimensions for online tutors to negotiate if they are to be effective in online tutoring:

- Organisational Role: This involves creating the agenda for the online programme, establishing objectives of the forum discussion, timetabling, creating procedural rules, and decision-making norms
- Social Role: This involves creating a friendly social environment for learning. It will involve a frequent and lively presence, as well as a sense of humour
- Intellectual Role: This is essentially about educational facilitation. As in any kind of teaching, the moderator should focus discussions on crucial points, ask pertinent questions, and probe responses to encourage critical thinking (pp. 399–401).

Similarly, Garrison and Vaughan (2008) refer to 'social', 'teaching' and 'cognitive' presence in an online community, as summarized below:

- Social presence is focused on open communication and building group cohesion
- Cognitive presence is focused on the process of inquiry and encouraging thinking and discourse among participants
- Teaching presence is concerned with the management of curriculum and instruction, guiding activities and providing structure.

These dimensions of online role functions or aspects of presence, however, framed, essentially relate to facilitating the learning experience online, especially the specific ways in which the teacher (facilitator or tutor—as these terms are now increasingly used) can best utilize the resources available in optimizing participants learning. There are many e-tools available for creating and managing this online presence, which range from the most common but highly effective asynchronous tool, the email, to more interactive synchronous virtual classrooms, such as Lync. The choice of particular e-tools is often a matter of preference, cost and access. Many can do the same things, which are to provide communication through the various mediums of text, audio and visual. What is most important is how communication is conducted, and this is essentially a good *Presentation Style*. The creative challenge is how best to apply this contextually for effective presence in the online environment. From an evidence-based approach, we have a set of heuristics to ascertain what is likely to work well, how, when and on what basis. As in the face-to-face situation, the initial experience, the *Primacy effect* is very important. For example, faced with a new group of students, the disorganized teacher with an uninspiring presentational style is likely to experience a very quick downturn in levels of attention and engagement. In the online environment, this negative experience is likely to be even more heightened for participants, often leading to early attrition. In my online tutorials, I typically start with a short video, which will have some carefully crafted and positioned supporting text to make the best initial contact, without cognitive overload. A major goal at the onset is to communicate my approachability and commitment to supporting the learning group. I am very mindful of my voice tone and body language, and try to work as much on the unconscious mind as the conscious. I am also seeking to convey the best possible clarity on what the purpose of the programme is, how it works, what to expect, and how to deal with any questions and concerns. I then focus on establishing an open and trusting base for ongoing two-way feedback.

There is much that can be modelled and customized from the field of customer service practices. Customer service professionals are particularly aware of important *touchpoints* in shaping the relationship between the customer service provider (e.g., in this case the online tutor) and the customer (e.g., in this case the online learners). For example, first impressions are significant as we have outlined previously, but these can quickly fade, if not maintained and developed to learner expectations. In the hotel industry, other touchpoints include the contact with customers in their coming and going from the hotel, making requests—no matter how small—and creating nice surprises (delighters) such as leaving a favourite magazine on the table in the customer's room (previous researched by hotel staff). Of course, we expect a high level of customer service when staying at the so-called top hotels, and usually (not always) receive it. However, this can also be the experience anywhere, as it's not that difficult to do, when you know how and, most importantly, want to do it really well. In my many travels, I have had some of the best customer service experiences in the most modest of places in terms of pricing or ascribed status. The same touchpoints apply in the online environment. For example, students will need to be given information and assignments at different points, there will be times when they need clarification and other support, and there will simply be times when they get a bit fed up of doing the work. Hence, try to make these touchpoints less painful than they could otherwise be. Even better, and this is where creativity can come into play, introduce novelty and pleasure into the mix, a humorous caption of the present situation that provides a reframe in which the 'funny side' can be seen or introduce a fun activity. Anyway, here's a few guiding frames that often (not always) work well:

- Ensure clarity, access and ease of use of all designated feedback channels. People don't like being left in 'limbo', so to speak. It's much worse than being given a 'no' in many cases
- Avoid overburdening learners with too much information at any one time, it'll cause cognitive overload and strain. Use the announcement board, and any other programme organizer to provide a clear structure and bite-sized guidance on what needs to be done, how and when
- Maintain regular contact, but don't overdo this. Too much communication can become boring and eat up participants' valued time. Most importantly, identify and deal with concerns quickly
- Work towards an informal communication style that fits the comfort zone of your learners. You can find this out through experimenting with your use of language and tone, and some safe humour. The more you can work in a friendly informal manner, the better is the likelihood of rapport. Once you have this, coupled with sound pedagogy, everything (ok, most things with most people) will work better, especially retention rates and student attainment.

To be more creative, look for opportunities to incorporate an appropriate *Von Restorff effect*, in the context of the learner group. I like to use humour and the occasional poignant story to achieve this, as it supports rapport building and creating a positive psychological climate—"Russian Doll" stuff. Invariably, be careful not to overuse these strategies, and ensure contextualization to the learner profile. Also, you may remember the *Recency effect*. At the end of any specific period of learning (e.g., transitions and endpoints), check key understanding and provide supportive feedback, key summaries of what's been achieved and what's coming next.

As the online tutor, do not contribute unnecessary confusion and complexity to your learners' already busy lives. Do the opposite—provide structure, be predictable in supporting their learning, and enhance their lives with a bit of fun and humour. And, finally, to the point of repetition, work not only with the conscious aspects of the mind, but pay good service to the unconscious aspects of human psychological functioning. You will find that this works well.

6.5 Summary

There is now a convergence or 'singularity' of pedagogy and technology in terms of learning design, and there is little doubt of the potential affordances of EdTech for content development, deployment and management as well as multiple global communication and collaboration mediums to support the learning process. Apart from technical, access and administrative issues, the main differentiator between the good and poor usage of EdTech is one of pedagogy. Hattie and Yates (2014), made the summative point (as for now) on the impact of computers on teaching:

What became apparent, through a careful reading of the extensive literature, was the reali-
sation that such positive effects are achieved through applications of the same principles of
learning that apply in all other areas of human learning. (p. 199)

It is now viable for most teaching professionals to be able to produce effective,
efficient and creative blended learning experiences. It is for this reason that blended
learning is unlikely to be just another *creature of fashion* in the educational landscape
for our foreseeable future. There is also little doubt that interactive videos with
augmented reality will significantly transform the shape of teaching and the context
of learning.

The day when we can ask a device that activates an AI agent which, at the blink
of an eye, produces customized resources, video and other relevant features, and
then facilitates our learning in the communication style of our preference, is not yet
available for most of us. However, 'the train has left the station'—so to speak—and
the digital revolution, with sound pedagogy, is on its way. It's not a quantum learning
revolution, as we still have a stone age brain—but teaching will need to change quite
radically.

On a positive note, if teachers of yesteryear, armed with only a blackboard and
a set of coloured chalks, could create interesting and effective lessons, (and some
could do this) then what's the potential for the highly creative teacher in the present
context?

References

Abeysekera L, Dawson P (2015) Motivation and cognitive load in the flipped classroom: definition,
rationale and a call for research. In: Higher education research and development, vol 34(1)

Ash K (2012) Educators view 'flipped' model with a more critical eye, Education Week, S6–S7

Beattie A (2011) What is the difference between social media and web 2.0? Technopedia.com, Nov
29. https://www.techopedia.com/2/27884/internet/social-media/what-is-the-difference-between-
social-media-and-web-20

Bergmann J, Sams A (2012) Flipped your classroom: reach every student in every class every day.
International society for technology in education

Berrett D (2012) How 'flipping' the classroom can improve the traditional lecture. The chronicle
of higher education

Bersin J (2004) The blended learning book. Wiley, San Francisco

Brown T (2009) Change by design. HarperCollins, New York

Clark R, Lyons C (2005) Graphics for learning: proven guidelines for planning, designing, and
evaluating visuals in training materials. Pfeiffer, San Francisco

Colvin Clark R, Mayer RE (2011) E-learning and the science of instruction; proven guidelines for
consumers and designers of multimedia learning (3rd edn). Wiley, San Francisco. Retrieved from
https://formulasi.googlecode.com/files/e-Learning.pdf

Educause Horizon Report (2019) Higher education edition. Louisville, CO

Gallagher-Mackay K, Steinhauer N (2017). Pushing the limits, how schools can prepare our children
today for the challenges of tomorrow. Penguin Doubleday, Canada

Garrison DR, Kanuka H (2004) Blended learning: uncovering its transformative potential in higher
education. Internet High Educ 7(20):95–105

Garrison DR, Vaughan ND (2008) Blended learning in higher education: framework, principles, and guideline. Jossey-Bass, San Francisco

Gee JP (2017) Teaching, learning, literacy in our high-risk high-tech world: a framework for becoming human. Teachers College Press, New York

Greenberg AD, Zanetis J (2012) The impact of broadcast and streaming video in education. Wainhouse Research, LLC, Cisco Systems Inc.

Hamilton S, Zimmerman S (2002) Breaking through zero-sum academics: two students' perspectives on computer-mediated learning environment. In: Rudestam KE, Schoenholtz J (eds) Handbook of online learning: innovations in higher education and corporate training. Sage, London

Harari YN (2016) Homo deus: a brief history of tomorrow. Penguin, UK

Harari YN (2018) 21 lessons for the 21st century. Jonathan Cape, London

Harris S (2010) The moral landscape: how science can determine human values. Free Press, New York

Hattie J (2009) Visible learning. Routledge, New York

Hattie J, Yates GCR (2014) Visible learning and the science of how we learn. Routledge, New York

Herreid CF, Schiller NA (2013) Case studies and the flipped classroom. J. of College Science Teaching, National Science Teachers Association, pp 62–66

Hodges P, Saba L (2002) Teaching statistics online. In: Rudestam KE, Schoenholtz (eds) Handbook of online learning: innovations in higher education and corporate training chap. 18. Sage, London, pp 389–404

Horton W (2006) e-learning by design. Pfeiffer, San Francisco

Ilkka T (2018) JRC science for policy report: the impact of artificial intelligence on learning, teaching, and education. Publications Office of the European Union, Luxembourg

Jensen JL, Kummer TA, Godoy PD (2015) Improvements from a flipped classroom may simply be the fruits of active learning. CBE Life Sci Educ 14:1–12

Keen A (2007) The cult of the amateur: how today's internet is killing our culture and assaulting our economy. Nicholas Brealey Publishing, London

Lincoln YS (1990) The making of a constructivist: a remembrance of transformations past. In: Guba EG (ed) The paradigm dialog. Sage, London

Littlejohn A, Pegler C (2007) Preparing for blended e-learning. Routledge, London

Luckin R (2018) Machine learning and human intelligence: the future of education for the 21st century. UCL Institute of Education Press, London

Margulieux L, Majerich D, McCraken M (2013) Guide to flipping your classroom. Center for 21st century universities, Georgia Tech

Mason GS, Shuman TR, Cook KE (2013) Comparing the effectiveness of an inverted classroom to a traditional classroom in an upper-division engineering course. IEEE Trans Edu 56(4):430–434

McLaughlin JE et al (2013) Pharmacy student engagement, performance, and perception in a flipped satellite classroom. Am J Phar Edu 77(9):1–8

Means B et al (2010) Evaluation of evidence-based practices. In: Online learning a meta-analysis and review of online learning. U.S. Department of Education Office of Planning, Evaluation, and Policy Development Policy and Program Studies Service, Washington, DC

Melchior T et al (1997) New technologies. In: Costa A, Liebmann RM (eds) Supporting the spirit of learning: when process is content. Corwin Press, Thousand Oaks, California

Miller MD (2016) Minds online: teaching effectively with technology. Harvard University Press, Cambridge, MA

Mlodinow L (2012) Subliminal: how your unconscious mind rules your behaviour. Vintage Books, New York

Moffett J (2014) Twelve tips for "flipping" the classroom. Med Teach 37(4):331–336

Moroder K (2013) You tube video. Learning that technology is not a silver bullet: Ed tech challenge intro (1/4). Available at https://www.youtube.com/watch?v=pXkkoAVrmvo&index=4&list=UUQHWPc4O7z68ZqZwMjgNznA. Last accessed 26 Nov 2019

Olbrish Pagano K (2013) Immersive learning: designing for authentic practice. ASTD Press, Alexandria, VA

Oliver R (2007) Describing ICT-based learning designs that promote quality learning outcomes. In: Beetham H, Sharpe R (eds) Rethinking pedagogy for a digital age: designing and delivering e-Learning. Routledge, London

Pacansky-Brock M (2013) Best practices for teaching with emerging technologies. Routledge, New York

Picciano AG et al (eds) (2014) Blended learning research perspective, vol 2. Routledge, New York

Pienta NJ (2016) A "flipped classroom" reality check. J Chem Educ 93:1–2

Prensky M (2001) Digital natives, digital immigrants. On the Horizon 9(5):1–6

Robinson DH, Schraw G (2008) A need for quality research in e-Learning. In: Robinson DH, Schraw G (eds) Recent innovations in educational technology that facilitates student learning. Information Age Publishing, Charlotte, NC

Sale D, Cheah SM (2011) Developing critical thinking skills through dynamic simulation using an explicit model of thinking. Paper presented at the 7th international CDIO conference at DTU Lyngby, Denmark, June 20–23

Sale D, Cheah SM, Wan M (2017) Symposia on creative flipped classroom learning: an evidence-based approach. In: Redesigning pedagogy international conference, National institute of education, Singapore, May 31–June 2 2017

Sale D (2015) Creative teaching: an evidence-based approach. Springer, New York

Shank P, Sitze A (2004) Making sense of online learning. Pfeiffer, San Francisco

Shea-Shultz H, Fogarty J (2002) Online Learning Today. Berret-Koehler Publishers Inc, San Francisco

Strayer JF (2012) How learning in an inverted classroom influences cooperation innovation and task orientation. Learn Environ Res 15:171–193

Stein J, Graham CR (2014) Essentials for blended learning a standards-based guide. Routledge, New York

Sylwester R (1995) A celebration of neurons: an educator's guide to the human brain. Association for Supervision and Curriculum Development, Alexandria, VA

Treadwell M (2017) The future of learning. The Global Curriculum Project, Mount Maunganui, NZ

Van Vliet EA, Winnips JC, Brouwer N (2015) Flipped-class pedagogy enhances student meta-cognition and collaborative-learning strategies higher education but effect does not persist. Life Sci Educ 14:1–10

Willingham DT (2009) Why don't students like school: a cognitive scientist answers questions about how the mind works and what it means for the classroom. Jossey-Bass, San Francisco

Wilson S (2014) The flipped class: a method to address the challenges of an undergraduate statistics course. Teach Psychol 40(3):193–199

Winget L (2017) What's wrong with damn near everything!: how the collapse of core values is destroying us and how to fix it. Wiley, New York

Zemsky P, Massey WF (2004) Thwarted innovation: what happened to e-learning and why. University of Pennsylvania, Learning Alliance Report

Chapter 7
Framing a Curriculum for the Twenty-First Century Competencies

Abstract This chapter explores and addresses the key considerations that need to be thoughtfully negotiated in curriculum development. It provides a critical analysis and evaluation of what constitutes twenty-first century competencies, how these are best derived from a cognitive science perspective, and the implications for framing educational aims and outcomes. A core valuation is that while we must frame and enact curriculum to meet the demands of industry and provide employability, there is also a need to accommodate competencies for wider issues of well-being and citizenship. There are concerns that technology—especially Artificial Intelligence— may make employability increasingly difficult for more people, and this will provide a systemic new challenge to twenty-first century curriculum planning and teaching.

7.1 Introduction

In Chap. 1, the short tour of Educational Jurassic Park illustrated the contestation about what constitutes the nature and practices of *good* teaching. Equally important, in terms of framing educational quality (however defined), there are similar parallel issues in terms of what is the *good* curriculum. A stark description of curriculum divergence over the decades (centuries) is aptly captured by Kelly (1989) who describes it as:

....the battlefield of many competing influences and ideologies. (p. 149)

Framing the essential curriculum aims and components for work and living in the twenty-first century may indeed be a battle; it will certainly constitute a major educational challenge. At present much of the curriculum focus is on defining, and of course teaching and assessing the so-called twenty-first century competencies. These competencies are deemed essential for success, progress and well-being—though it often feels more like survival—in the face of rapid technological and social change.

In this chapter, I unpack and explore the key curriculum issues in terms of an EBT approach, which has similarities with what Pinker (2019) refers to as *reason* and *science* in decision-making, concerning what we need/want our workers and citizens

© Springer Nature Singapore Pte Ltd. 2020
D. Sale, *Creative Teachers*, Cognitive Science and Technology,
https://doi.org/10.1007/978-981-15-3469-0_7

to learn, to be, and on what basis. This provides the framework for deciding how best to achieve such educational aims and goals, based on the pedagogic framework established in the previous chapters. It will involve analysing and evaluating prevalent conceptions of twenty-first century competencies, how they align to an evidence-based approach to learning and the implications for professional practice in teaching. Notions of curriculum innovation and change imply that certain forms of knowledge, skill sets and attributes (e.g., attitudes, dispositions) are more important than others in terms of facilitating a better society and/or in response to perceptions of a challenge to existing societal arrangements. In 'nuts and bolts' terms, it has become apparent, or it is perceived by significant interest groups (stakeholders) that significant changes need to be made in terms of what is taught in school curricula to meet changing societal needs; often those of industry, but also for issues of citizenship and well-being.

Curriculum change, whether now or in yesteryear, is always based on considerations of what is a *Good Society* and how do we prepare (socialize) people into achieving it and making it work—as best as is possible, given the existential conditions that prevail. The good society is typically framed, in Edu-speak, as *Educational Aims*; the means to bring this about is the *Curriculum*; and *Pedagogy* is what we do in the various delivery forms and arrangements (e.g., instructional approaches) to bring about that socialization.

7.2 Human Learning and Curriculum Framing

It is useful to anchor this discussion in evidence-based facts about learning and the concept of curriculum. Firstly, learning is foundational to the human species and the developmental of culture in its myriad of forms, and it appears that we are the only species, as far as we can ascertain, that possesses the unique capability for metacognition. Such capability is immensely powerful as we explored prior. Evolution, as documented by Darwin (1859) occurs across all species, including humans, and can be said to constitute a biological fact (irrespective of whatever, if any, metaphysical spins we may choose to put on it). However, it has been a slow process and has left us with an existential burden, as Pinker (2019) documents:

> ...our cognitive, emotional and moral faculties are adapted to individual survival and repro-
> duction in an archaic environment, not to universal thriving in a modern one. To appreciate
> this, one doesn't have to believe that we are cavemen out of time, only that evolution, with
> its speed limit measured in generations, could not possibly have adapted our brains to mod-
> ern technology and institutions. Humans today rely on cognitive faculties that worked well
> enough in traditional societies, but which we now see are infested with bugs. (p. 25)

No matter how much we try to wriggle out of this problem, with vague notions of 'accelerated learning', 'clicking on the creativity switch', and other quick-fix solutions to achieve success, we cannot get away from certain existential facts. Firstly, we are living in an age when the discrepancy between our brains' evolutionary capacity for processing information and the amounts of information hitting it, (albeit that most of it is probably of no real use) is becoming increasingly problematic and untenable. Cognitive overload, probably of little concern for our primitive ancestors, is now a nemesis and may underpin much of modern stress in human learning. Learning in primitive societies would have been little in terms of content knowledge, and the important bits probably taught (and learned) relatively quickly. Of course, skills (e.g., hunting, building shelter) may have taken a long time, depending on how long it took to become competent or expert at doing this. This fits squarely with our lazy brain (e.g., Kahneman 2012) and our desire to avoid thinking (e.g., Willingham 2009). Also, as documented in Chap. 3, research suggests that human brains were pretty much the same (morphologically) some 50,000 years ago (Neubauer et al. 2018). Hence, folk then probably had similar cognitive abilities and motivational dispositions to us now—but they just played out in different contexts, with different resources, and different contingencies.

The problem is stark and clear, we are stuck (in the present cognitive arrangements) with a stone-age brain, perfectly equipped for living in savannah in Africa some hundreds of thousands of years ago, but unequipped to deal with the exponential knowledge explosion of modern times.

Hence, we have an existential problem, though there are some credible futurists, such as Kurzweil (2005) who argues that *Radical Evolution* (e.g., the coming together of genetics, robotics, information-communication technologies and nanotechnology into a singularity) will, apart from significantly increasing the human lifespan, also increase human capability through bionic enhancement, transforming physical health and cognitive capability through the connectivity of IT and brain neural networks. Just as map applications have largely made redundant the painful task of years of learning (memorizing) the names and locations of all London's roads (which was the case to acquire a taxi license for driving the famous black taxi-cab in London), downloading a full language system straight into an organized set of neural networks in the brain would take the drudge out of learning a foreign language. The futurist writer, Harari (2016) makes an interesting assertion:

> In the twenty-first century, humans are likely to make a serious bid for immortality. Struggling against old age and death will merely carry on the time-honoured fight against famine and disease. (p. 24)

In terms of modern science and modern culture, he goes on to suggest that:

> They don't think of death as a metaphysical mystery, and they certainly don't view death as the source of life's meaning. Rather, for modern people, death is a technical problem that we can and should solve. (p. 25)

Well, what are the curriculum implications of such radical evolution events? Off the bat I don't think this will happen for the next 50 years or so and, in the meantime, we are living with a stone-age brain and dealing with a VUCA world. Hence, I will continue writing this book.

Animals do learn, and this has certain correlates with human learning. For example, there are similarities in terms of learning through the senses, especially visual, auditory and kinaesthetic. I had a budgerigar (who was called Joey) and he could consistently orate-well, "Who's a pretty boy then" (note: with an East London accent). There's nothing mystical about how Joey acquired such exquisite linguistic capability. I must have repeated this to him hundreds of times, and as he picked up bits of the sentence syntax, I would give him a tasty bit of millet (that's a food treat for budgies, apparently). He learned from hearing (we would not call this empathic listening), and inevitably from practice (though we would also not call this deliberate practice).

However, animals, lacking such metacognitive capability do not build culture as humans do; their knowledge production over time is limited and their basic survival mechanisms, deeply rooted in instinct, remain largely untouched. Invariably, as the environmental conditions around them change, they will evolve new responses and can communicate this information to each other (maybe more effectively than us humans). For example, foxes are now quite prevalent in some English cities, and they are not there for aesthetic reasons; they are seeking food and/or shelter—no more, no less. Without degradation to their previous environment, they would not be foraging around the cities.

While primitive men (and women) had the capability for extensive learning as their distant successors (us), there was little need for it. Curriculum, though it is unlikely that the word existed in their vocabulary, would constitute locally constructed knowledge and practices mainly relating to skills in finding food, building shelter, dealing with threats, and human conduct among themselves. They would also have notions about their existence and its meaning, as noted in the wide variety of beliefs in the supernatural among primitives. Such knowledge would largely be passed on orally, though many created cave drawings and other artefacts, from generation to generation, and that's how we know about them. Socialization into this cultural milieu would not have required decades of formal education and lifelong learning. The big difference for us today, for sure, is that its damn more complex now.

The unique human capability of metacognition may well have been relatively latent for primitive people, as they did not have to do much thinking, and if there is little by way of change, even less need for 'thinking about thinking'. However, in the present situation of exponential knowledge growth, as well as rapid social and technological change, the need for good thinking has become somewhat of a 'Pandora's Box'—in that once opened, the nemeses are let out and tough challenges ensue. Pandora's box is considered one of the most descriptive myths of human behaviour in Greek mythology in that Ancient Greeks used it not only to instruct themselves about the weaknesses of humans but also to explain certain human misfortunes. As the legend goes, Pandora was given a box by Gods who told her that it contained special gifts from them, but she was not allowed to open the box ever. Even if you are not familiar with the myth, you have probably guessed the rest—curiosity got the better of Pandora and she opened the box allowing all the illnesses and hardships that gods had hidden in the box to come out. While Pandora quickly closed the box, it was too late—only Hope was left inside. Perhaps this is the metaphor of these times. As the old saying goes "Hope springs eternal".

However, while today's world is more complex, rapidly changing and potentially more dangerous at the macro-level, humans of yesteryear still faced similar existential challenges to their survival, making meaning of life, and dealing with death. As mentioned prior, humans then still needed to solve difficult problems such as finding food, staving off intruders, etc. It's just that in modern culture, technology has scaled up certain problems exponentially—as threatening tribes can be world-wide and the weapons of today offer the potential for extinction, not just local genocide.

7.3 Framing Twenty-First Century Competencies from an Evidence-Based Approach

We must be thoughtful in framing new so-called twenty-first century competencies, as there needs to be a sound base for this in terms of the most important human capabilities needed for meeting the demands and challenges of living in the VUCA world of the twenty-first century—as best as we can predict them. Equally important, how best to organize these in terms of practical curricula and learning arrangements that are most viable for learning (and teaching), whether in educational institutions or otherwise?

The P21 Skills Framework, a U.S. organization composed of educational, governmental, and business leaders, has developed a comprehensive framework articulating the competencies that are required for successful employment and citizenship in the twenty-first century. The framework does not negate the importance of traditional curriculum forms of knowledge (e.g., school subjects such as history, geography, etc.) or the so-called "3Rs" (reading, writing, arithmetic). It argues that in this rapidly changing digital age, other skills are increasingly necessary, highlighting the importance of the "4Cs" (Creativity, Communication, Collaboration, and Critical Thinking). In this context, in Chap. 3, I made the case for Metacognitive Capability as the superordinate competence, as all others are ultimately subsumed in terms of their effective and efficient enactment through the executive self-regulatory aspects of metacognition.

The above discussion may have seemed a bit like a detective plot, it was—but there is a point to this. Firstly, it raises the question of what are twenty-first century competencies, and how similar or different are they from those of yesteryear, in what ways and, most importantly, what does this mean in terms educational aims? Such questions are central to curriculum design and typically involve considerations of the following:

1. What knowledge is most useful to attain and on what basis?
2. How is this knowledge most effectively attained, or created ?

1. What knowledge is most useful to attain and on what basis?

To provide an advance organizer here, it is worth reflecting on the following two quotes, one from a nineteenth century writer, and the other from a former prime minister of Singapore.

> It is not proposed that the children of the poor should be educated in a manner to elevate their minds above the rank they are destined to fill in society... Utopian schemes for an extensive diffusion of knowledge would be injurious and absurd.
>
> (*Colquhoun, 1806, Writing About Education in England*)

> We must get away from the idea that it is only the people at the top who should be thinking, and the job of everybody else is to do as told. Instead, we want to bring about a spirit of innovation, of learning by doing, of everybody each at his level all the time asking how he can do his job better.
>
> (*Mr Goh Chok Tong, 1997, the Prime Minister of Singapore, at the Opening of the 7th International Conference on Thinking*)

Different times, different cultures, different views on the human condition, different needs—one could go on with this—but it illustrates the culturally constructed nature of the curriculum, reflecting the power structures and aims of education at any particular time in human history. The same framing can also be made for primitive mankind, though it was simpler, less formalized (there were no books about it), and less contested in the localized context. All contexts were largely localized; hence, homogeneity would tend to persist until threatened by external agencies.

Fundamental to the challenges of the curriculum today, issues revolve around the framing of our educational aims, which in turn reflect our perception (grounded in beliefs) of the *good society*, as identified earlier. Wringe (1988) captured this nicely:

> Human beings have the potential for developing in many directions and the problem of educational aims is deciding which kinds of development should be fostered and which discouraged. (p. 43)

Questions of what constitutes the *Good Society* today is underpinned by one of the major problems facing mankind now. In yesteryear, tribes would inevitably develop different local cultures, and while they would have to face the same existential problems of survival, there would be indexicality in how they tackled them. Most significant they would evolve different norms of human conduct for themselves, as well as for how they would interact with other tribes. For example, the Sentinelese are an uncontacted tribe living on North Sentinel Island, one of the Andaman Islands in the Indian Ocean, and they vigorously reject all contact with outsiders. In contrast, the Kombai tribes in Indonesia are more friendly to outsiders. Such 'cultural dispositions' would remain for centuries, as there were no challenges to their world view. However, in an internet rich society, a Flat World (Friedman 2006), we are awash with knowledge about what is good, on what basis (and this seems to change constantly), what we should aspire to be, what is the purpose of life, etc. It goes on and on, and it is not surprising that many people are confused as to what to believe, and on what basis.

7.4 What's Our Best Package of Competencies in a VUCA World?

As identified earlier, the basic questions are not new; it has always been prevalent (and current) at any time since the formulation of knowledge into organized structures (i.e., curriculum) for deliberate transmission to a learner group (e.g., the whole society or segments within it). Hirst (1974) argued that socialization into these organized structures of knowledge—what he referred to as, "Forms of Knowledge"—are very important as there is a close relationship between their acquisition and the growth and development of the mind. Such an approach makes the following psychological assumptions about the nature of the mind:

- Knowledge is a quality of mind. Failure to receive certain forms of knowledge is a failure to achieve rational ways of thinking in those areas.
- The mind does not develop rationally—it needs organised forms of knowledge.

Hirst argued that there are seven distinct forms of knowledge, each with its unique concepts, distinctive logical structure, testability against experience and unique methods of testing. These seven forms of knowledge are mathematics, the physical sciences, the human sciences, history, religion, literature and the fine arts, and philosophy and moral knowledge. The framing of traditional school subjects is consonant with Hirst's model of the intelligent mind. Latin for example, which was central to elite education, was perceived to be a valuable tool for the development of the mind, and Lowe (2017) suggests that there are reasons and evidence to support such a view. She notes that Latin helps with SAT scores and makes learning a modern vocabulary easier. True, but Lowe also argues that there are more important objectives that Latin achieves better than any other subject: The first is mental development, and the second is an understanding of English grammar.

Latin, like math, provides students with the experience of studying one subject to a mastery level which, according to her, is missing in modern education where we try to teach everything, and we cover too many subjects superficially. There are few opportunities to use higher-order thinking skills when you are merely a novice. It is only when the student has studied a subject enough to have some depth that his/her mind can be stretched and challenged with higher-order thinking skills. This relates to Hirst's (1974) notion that in moving upwards through the structure of a form of knowledge increasingly develops rationality in that form of knowledge. Latin and math give students the invaluable experience of studying one systematic subject to a mastery level over a long period. From this perspective, such learning is essential for mental and character development and is the most valuable academic experience a child can have in school. Latin and math, when taught to a mastery level, according to Lowe (2017), requires perseverance, hard work, stamina, will, and grit. Students need to plan, adopt a never-give-up attitude, and display flexibility in learning. Sounds very much like good preparation for self-directed lifelong learning! No, I am not arguing that Latin becomes a universal part of the curriculum for the twenty-first century—but the 'thinking behind the thinking'—so to speak—has merit.

The extent to which these forms of knowledge are distinct has been open to criticism. Young et al. (1971), for example, argued that knowledge is less delineated at the experience level and is best learned in a more integrated and holistic context. From this perspective, a well-integrated curriculum is more consistent with how we learn, the nature of knowledge in the real world, as well as making learning more interesting for students (e.g., Fogarty 2009).

As cultures develop, and extended forms of knowledge become available, there are inevitable choices concerning what forms of knowledge are most useful. Questions concerning how much of each form, and to what level of competence are the *essentials* of curriculum content decision making. Neary (2014) summarized the perennial issue:

> The structure and methods of education must help to sustain the traditional values of society, but they must also respond adequately to current cultural, social, industrial and technological issues, and to future change.

Lawton's (1975) notion of *cultural analysis* identifies the need to consider the wider context of a society in curriculum decision-making, though recognizing that this was still value-laden as decisions need to be made based on current perceptions of relevance, and this must reflect the dominant forces shaping curriculum at any particular time. Many factors or forces shape the curriculum; these include dominant political ideology, dominant educational perspectives, industry, educational institutions, practitioners, and students. The present context is no different, but the problems are perhaps more pressing than in yesteryears. For example, Collins (2017) argues that:

> We educators can't go on adding things to the school curriculum as knowledge grows exponentially. We can't keep people in school longer and longer until everybody needs several advanced degrees just to deal with the complexity they face in their lives. We can't make our textbooks much fatter than they already are and cover more and more topics in the same amount of time…Our strategies for coping with the exponential growth of knowledge are hitting a wall. (xv)

> The school curriculum is filled with stuff that most people will never use and hence will forget as soon as they leave school or move on to the next grade. (p. 1)

> Our current model of universal schooling is an Industrial Age institution, and it is not at all clear how well it can adapt to the Information Age, where thinking and creativity are prized. (p. 3)

The message and the problems are clear. There is a significant need to reframe curriculum, and this has led to the growth of frameworks attempting to define twenty-first century competencies. However, the same fundamental curriculum questions remain in terms of what the competencies should be, what range and depth of knowledge, skills and attitudes are the key content constituents, how to structure them into viable educational offerings, and how best to deliver them pedagogically to meet desired educational aims and outcomes. Certainly, as detailed in Chap. 3, much is now focusing on the need to develop self-sufficiency in our students so that they can become Self-Directed Lifelong Learners. Collins (2017) sees this as a major curriculum aim in that:

Having a critical mindset is critical to navigating through today's complex world. (p. 340)

Planning, monitoring, and reflection are the basic elements of the learning cycle that pervades everything we do. (p. 35)

There is abundant evidence that self-regulatory skills are central to living a happy and successful life. Recent studies have shown that it is possible to learn these new skills, but they are seldom taught in school, except in extra-curricular activities. They need to become central to the school curriculum. (p. 35)

Finally, in this context, he suggests that:

The goal is to develop schooling that will have a major impact on student motivation and learning and will better prepare students for the complex world they are entering. (p. 9)

In meeting this challenge, it is useful to conduct a thorough and thoughtful pedagogic analysis of what competencies are most useful (as far as this is possible)—akin to Lawton's approach—but driven from the perspective of learning science as well as cultural memetics and stakeholder interests. As Reimers and Chung (2016) emphasize:

In spite of the obvious need for a theoretical underpinning to the development of curriculum, most conversations about "twenty-first-century-education" to date have failed to draw a connection between the proposed twenty-first-century competencies and any psychological theories of how those competencies are developed, in particular in relationship to one another, as a unified developmental process. (p. 9)

The Towards Defining 21st Century Competencies, The Foundation Document for Discussion, Ontario (2016) makes an anchoring statement:

Many international thought leaders and business leaders – and many young people today are increasingly asking educational systems to prepare students with "21st-century" competencies that will enable them to face complex challenges now and in the future. These competencies – knowledge, skills, and attributes that help children and youth to reach their full potential - are added to the important foundational skills of literacy and mathematics and the core learning in other subjects. (p. 5)

Various groups, such as the OECD, the European Commission, the Partnership for twenty-first century Skills, and the U.S. National Research Council have made significant contributions to framing these competencies. While there are differences in the approaches taken, there is broad agreement that twenty-first century competencies are associated with growth in the cognitive, interpersonal and intrapersonal dimensions. For example, The National Research Council report titled Education for Life and Work: Developing Transferable Knowledge and Skills in the twenty-first century synthesizes psychological and social science research evidence on skills that have demonstrated positive short or long-term consequences for individuals. The report summarizes those skills in the following framework:

1. **Cognitive Competencies**

   1.1 *Cognitive Process and Strategies*

   Critical thinking; problem-solving; analysis; reasoning and argumentation; interpretation; adaptive learning; executive function

1.2. *Knowledge*

Information literacy, including research using evidence and recognizing bias in sources; information and communication technology literacy; oral and written communication; active listening

1.3. *Creativity*

Creativity and innovation.

2. **Intrapersonal Competencies**

2.1. *Intellectual Openness*

Flexibility; adaptability; artistic and cultural appreciation; personal and social responsibility; cultural awareness and competence; appreciation for diversity; continuous learning; intellectual interest and curiosity

2.2. *Work Ethic/Conscientiousness*

Initiative; self-direction; responsibility; perseverance; grit; productivity; type 1 Self-regulation (metacognitive skills, including forethought, performance, and self-regulation); professionalism/ethics; integrity; citizenship; career orientation

2.3. *Positive Core Self-Evaluation*

Type 2 self-regulation (self-monitoring, self-evaluation, self-reinforcement); physical and psychological health.

3. **Interpersonal Competencies**

3.1. *Teamwork and Collaboration*

Communication.

To derive standards for these competencies that are sufficiently clear, valid, and practical for teaching and assessment, it is necessary to clarify certain conceptual confusions about what constitutes a competency, especially as the terms competency and skills are often used in interchanged ways. Competencies involve much more than skills as such. For example, in the game of tennis, and this applies to any sport, there are discrete skills in terms of hitting the ball with the racket to gain speed, spin and other ball behaviours that might be effective in winning points. We talk about serving techniques, which typically involve some skill combinations. For example, there are serving techniques that put spin and slice on the ball, but these are skill combinations. The top players are top players because of their combined high level of skills in these different techniques. However, one can be skilful, but not that effective, and that's because of other factors (e.g., aptitude, traits/disposition) impact performance in real-world contexts—whether at work or in play, as explained in Chap. 5. Without clear framing of what we want our students to learn in order to develop a deep understanding and skill in using competencies (e.g., the specific performance criteria, evidence of what constitutes competence and expectancy levels), teachers with be unsure on what to teach and assess, creating much by way of confusion for all concerned, especially the students. As identified in Chap. 2, many teachers are unsure of what critical thinking entails or how to teach it (e.g., Wagner 2010). While teachers may be experts in their subject domains, many may lack the necessary level of expertise in the emerging science of learning and what this means for pedagogic practices (e.g., EBT)—as outlined in Chap. 1.

7.5   The Competency-Based Approach to Education and Training

Defining competencies has its roots in the functional analysis of work roles. Many different methods of developing competency frameworks have evolved, but the most effective ones share certain characteristics. All of them follow McClelland's (1973) dictate to determine what leads to superior performance and to identify top performers and find out what they do. This can be broken down into two important principles:

1. focus on highly successful people without making assumptions about their role
2. pay attention to what they do.

From this, the range of behaviours can be identified from observation in various contexts, then explored, clarified and made meaning of through interviewing, which often involves skilled questioning to evoke latent or tacit knowledge (e.g., Polanyi 1966). They can then be validated through checking with high performers and other stakeholders (e.g., supervisors; other workers who are part of the role-set for that work area). In other words, if the high performers can relate well to the competence, supervisors agree that this constitutes good competence, and those who engage with persons performing this work competence also feel that this is the case, that's probably competence in this work function(s). To use an analogy: 'If it looks like a bird, sounds like a bird, behaves like a bird'—it's probably a bird.

Hence, to reiterate, there are many components to competence, and some (i.e., knowledge and skills) are more conducive to training than others. Aptitude, Traits and Dispositions certainly have a hereditary/neurological base and related to personality configuration. Discussions on how many hereditary aspects impact competence, and how much results from environmentally structured experiences, is largely irrelevant in the world of mainstream teaching and training. Currently, we are not going to impact genetics much in the classroom. However, we can through the systematic use of metacognition and self-regulatory strategies help people to monitor, evaluate and modify their behaviour more in line with the requirements of the competency areas (at least in work contexts). Therefore, these are also subject to curriculum and pedagogic interventions as explained previously. There is also evidence that for persons who use such strategies, especially volition, in changing their actual behaviours, this can change brain structuring at the neural level (i.e., neuroplasticity), which starts to 'rewire' the person, in this area—so to speak. However, in my experience, while there is validity in this framing and it should be pursued as a regulatory ideal (i.e., give everybody a chance and be inclusive), there are some individuals who display almost a complete non-alignment to certain competencies. In working with thousands of teachers, many in training, despite much mentoring and coaching, for whatever of the above reasons, some just don't seem to have enough of the 'competency pack' to be effective teachers. Keeping such persons in the profession is not useful for the students or the folk themselves to continue. There is an adage that seems tautological (but it isn't), that it is easier and better, in many cases, to 'change people than to change people'.

It is also necessary to remember that competence, knowledge and understanding are all constructs and not something we can observe directly. Of course, we can observe behaviour, which is time-consuming and thereby make inferences and interpretations about a person's competence. We also know that:

- Knowledge and understanding contribute to competence
- Knowledge and understanding are best learned 'in use'
- Competency, knowledge and understanding are highly contextualized.

A significant part of the present confusion about framing twenty-first-century competencies has stemmed from a lack of clarity in framing educational outcomes, particularly, though not exclusively, to higher education. Diamond (1998), from a wide range of sources, argues that:

> A serious problem that institutions of higher education face is the perception by business leaders, governmental leaders, and the public at large that they have enthusiastically avoided stating clearly what competencies graduates should have and that as a result, they have provided little evidence that they are successful at what they are expected to do. (p. 4)

In an increasingly competitive educational landscape, institutions of higher education need to be perceived as relevant by key stakeholders if they are to stay viable and thrive. It is not surprising that many are conducting major reviews and adopting what is referred to as *Outcomes-Based Education* (OBE). This has resulted, in no small part, from the industry's dissatisfaction with the present more theoretical content-based curriculum that is seen as not adequately preparing graduates with the necessary competencies needed in modern work environments. OBE is a curriculum approach that provides the following framework (e.g., heuristics) for curriculum design and development:

- The starting point for any curriculum offering is the identification of clear student learning outcomes, which describe the result of learning for that curriculum. In the context of twenty-first century competencies, this requires the definition, delineation and organization of the knowledge, skills and attitudes of the various competencies into measurable learning outcomes.
- Assessment supports the learning process (formative) and is more performance-based (authentic assessment). Teaching, learning and assessment are therefore systematically interlocked.
- The learning environment and instructional system are designed to facilitate the desired outcomes. Note: This has typically been the case in theory, though not always the practice.

The Conceive-Design-Implement-Operate (CDIO) engineering education framework is a global example of a large-scale OBE curriculum innovation, which has responded to the need to make the curriculum more relevant to the world of work, develop skills for the twenty-first century, as well as making the learning experience of students more interesting. CDIO is an international initiative, originally conceived at the Massachusetts Institute of Technology (MIT) in Cambridge, Massachusetts in the late 1990s. In 2000, in collaboration with the Swedish universities of the Wallenberg Foundation, Chalmers University of Technology, Linkoping University and the Royal Institute of Technology, the CDIO initiative was formed. It has now over 300 member institutions worldwide.

The initiative represented a response by engineers in industry, government and academia to a concern about the present state of engineering education. Essentially, feedback on engineering education from the various stakeholders revealed perceptions of it prioritizing the teaching of theory, especially mathematics and science, while not paying enough attention to the real world of engineering practice and the need for skills such as design, teamwork and communications. As Crawley et al. (2007) summarized:

> …we identified an underlying critical need – to educate students who can Conceive-Design-Implement-Operate complex, value-added engineering products, processes and systems in a modern, team-based environment. It is from this emphasis on the product, process, or system lifecycle that the initiative derives its name - CDIO. (p. 1)

The importance of Conceive-Design-Implement-Operate, as an organizing frame for engineering education, is further explained by Crawley et al. (2007):

> Modern engineers lead or are involved in all phases of a product, process, or system lifecycle. That is, they Conceive, Design, Implement, and Operate. (p. 8)

The creation of an educational framework that would encompass this wide range of competencies, enabling students on graduation to be "ready to engineer" (Crawley et al. 2007, p. 6) is the basis and rationale of CDIO. The aim is to ensure that graduates will leave universities and colleges with both relevant practical competences as well as a thorough understanding of the role of an engineer in the present and future work context, and what this entails.

In The Challenge of Reframing Engineering Education (2014) I documented the complete curriculum development process from an OBE perspective. This included:

- Customizing the general syllabus (very much in the genre of most twenty-first century skills), but lacking specificity (which is a major challenge at present) to more specific and measurable skill areas.
- Calibrating the assessment approaches and tools for assessing more complex real work performances (outlined and illustrated in this chapter).
- Developing an instructional strategy using EBT generically, though with a specific focus on identifying the most effective pedagogic methods (e.g., signature pedagogies, Shulman 2005) conducive to key skill development in the context of engineering education (e.g., project-based learning, case-based learning, simulated practice).

- Developing a Professional Development Framework to provide faculty training and support in their learning of new pedagogic skills, which was essential for the success of the initiative.
- Conducting a 3-year longitudinal evaluation of the student learning experience.

Competency-Based Education and Training encompasses OBE and is likely to be the dominant curriculum planning model for the time being. However, as we frame the twenty-first century competencies into more specific performance elements and criteria, it is necessary to recognize that competency-based approaches, like other curriculum models, have inherent limitations and need to be addressed, as explored in Chap. 5.

7.6   Evaluating the Worth of Twenty-First Century Competencies

The Towards Defining 21st century Competencies, Foundation Document for Discussion, Ontario (2016), emphasizes that key-criteria for evaluating the worth of twenty-first century competencies must have measurable benefits for multiple areas of life and therefore critical for all students. The most prominent twenty-first century competencies found in international frameworks that have shown measurable benefits are associated with critical thinking, communication, collaboration, and creativity and innovation (p. 12). Certainly, the area of thinking, most notably identified—critical and creative thinking—has been extensively framed as essential in a society that is rapidly changing from a workforce embedded in an industrial model to a technology-driven, interconnected globalized knowledge-based economy:

> Researchers acknowledge that the need to engage in problem-solving and critical and creative thinking has "always been at the core of learning and innovation" (Trilling and Fadel 2009, p. 50). What's new in the 21st century is the call for education systems to emphasize and develop these competencies in explicit and intentional ways through deliberate changes in curriculum design and pedagogical practice. The goal of these changes is to prepare students to solve messy, complex problems – including problems we don't yet know about – associated with living in a competitive, globally-connected, and technology-intensive world. (p. 3)

Similarly, competencies of teamwork, communication and collaboration, as well as relating to important dispositions are well-validated in terms of positive outcomes in the present work and life contexts. As they document:

> Employers value skills such as teamwork and leadership.

> Studies in health and well-being have found that characteristics such as perseverance, grit, and tenacity are a more accurate predictor of success than IQ scores. For example, among intrapersonal competencies, the characteristic of conscientiousness (a tendency to be organized, responsible, and hard-working) is "most highly correlated with desirable educational, career, and health outcomes". (e.g., Pellegrino and Hilton 2012, pp. 4–5) (p. 10)

Fullan and Langworthy (2014) describe metacognition or learning to learn, as a twenty-first century competency that enhances student's ability to acquire skills, knowledge, and attitudes that are relevant to new areas of learning. Similarly, The Towards Defining 21st Century Competencies, Foundation Document for Discussion, Ontario (2016) highlight:

> Researchers and thought leaders see that metacognition and a growth mindset (including self-regulation skills and ethical and emotional awareness), while always important, are much more so in a connected, global context that requires an ability to communicate, work, and learn with diverse groups of individuals and teams worldwide. (p. 16)

As explored extensively in Chap. 3, MC provides the superordinate and anchoring role in terms of twenty-first century competencies, as it constitutes the executive functioning system, which is capable of initiating, maintaining, and evaluating self-directedness. I use the term capable with some caution, as previous chapters have documented and illustrated that rational thought—good thinking—is far from common sense, and even further away from being easy. To what extent individuals possess autonomous free will and have the ability to understand and regulate oneself remains contested among neuroscientists and philosophers alike. For us educators, we have to assume that people can learn the skill sets that underpin MC, and in doing so—learn better, think better, communicate and work with others better, and achieve a better sense of well-being in their lives. The more that we can achieve better MC globally, especially among young people, the more we may achieve a meaningful framing and direction towards global citizenship—transcending prejudice, bigotry and selfishness—recognizing that we need to solve the big global issues collaboratively and thoughtfully.

In this context, Rychen and Salganik (2003) argue that twenty-first century competencies should make measurable contributions to specific valued areas of life such as:

- Educational attainment
- Relationships
- Employment
- Health and well-being (pp. 66–67).

Yes, twenty-first century competencies must support attainment opportunities, and we know that teachers make a massive difference here; better teachers result in better student learning outcomes. Furthermore, we are a long way towards knowing how to teach better, as we have much more evidence-based knowledge of how humans learn and how the brain works in such matters. Hattie's work (2009, 2012) provided a strong empirical base to the relative usefulness of different aspects of the learning landscape, especially the effectiveness of instructional methods and strategies, as explained in Chap. 1. Similarly, the usefulness of the curriculum—what is to be learned—is equally important.

Yes, relationships matter to human functioning, well-being and happiness—as explored prior, and it is fairly obvious on how this works. Good relationships are also important in working environments and for living in the local community. And, wouldn't it be nice, and better, if this was more prevalent globally also?

Yes, we need to create employability opportunities, especially as the growth of artificial intelligence and other aspects of the 'internet of things' impacts occupational structures, resulting in the need for learner adaptability. Employability is helped through education, but the curriculum must develop both relevant functional work-related competencies and, perhaps more importantly, skills and attitudes that facilitate better learning, self-regulation and well-being. The nemesis of having a stone-age brain in the modern world is one thing, but as Pinker (2019) also points out:

> People are by nature illiterate and innumerate, quantifying the world by "one, two, many" and by rough guesstimates. They understand physical things as having hidden essences that obey the laws of sympathetic magic of voodoo rather than physics and biology…They generalize from paltry samples, namely their own experience, and the reason by stereotype, projecting the typical traits of a group onto any individual that belongs to it…They overestimate their knowledge, understanding, rectitude and luck. (p. 26)

Fortunately, such natural limitations can be mitigated in large part by firstly, being aware of them, and secondly, through developing MC—not just in schools—being able to self-regulate in more positive and effective ways. As Pinker points out (which is positive):

> However, long it takes, we must not let the existence of cognitive and emotional biases or the spasms of irrationality in the political arena discourage us from the Enlightenment ideal of relentlessly pursuing reason and truth. If we can identify ways in which humans are irrational, we must know what rationality is. Since there's nothing special about us, our fellows must have at least some capacity for rationality as well. And it's in the very nature of rationality that reasoners can always step back, consider their shortcomings, and reason out ways to work around them. (p. 384)

Similarly, Harris (2010), recognizing the impact of systemic biases in human psychological functioning, points out:

> As one might expect, many of these errors decrease as cognitive ability increases. We also know that training, using both examples and formal rules, mitigates many of these problems and can improve a person's thinking. (p. 123)

Furthermore, competencies in the intrapersonal domain contribute significantly to students' well-being, character development, and success. There is a growing body of research (e.g., Dweck 2006; Tough 2014) demonstrating that intrapersonal competencies such as perseverance, grit and tenacity, with a growth mindset, has a strong relationship with individuals' capacity to overcome challenges, achieve long-term success, and experience well-being.

7.7 Learning for What Purpose for the Many

I grew up in East London in the 1960s, when gaining employment was not a difficult task; it was indeed the "Affluent Society", so aptly described by Galbraith (1958). On leaving university for the first time in 1974, companies contacted me to see if I would like to be considered for employment with them. This is a far cry from the situation facing many graduates today, where applicants are in thousands, and they must go through a rigorous multi-stage recruitment process. It seemed a different world then, a world of plenty, and England seemed a great place to live. England also won the soccer world cup in 1966, a feat yet to be replicated. Of course, things were not like that for everywhere in the world; I was simply in an affluent bubble, but it was a nice bubble. I was not from a wealthy family, my parents were solid working-class folk, and we lived in a council flat (i.e., a local government-owned apartment). However, as a kid (twelve-something), on a Saturday and Sunday mornings, I earned 'pocket-money' doing car cleaning, edge cutting, and helping a local trader deliver paraffin. All was good, and my needs (e.g., money for chocolate bars and the amusement arcade) were adequately met—not that I had any concept of what a need was. However, it's a different frame now for many people. Of course, the wealthy are largely immune from such considerations—as the famous pop group ABBA recited in the song "Money, Money, Money", "it's a rich man's world." Of course, this applies equally to women. The worrying feature now is as Ford (2015) warns:

> …it is becoming increasingly clear that many people will do all the right things in terms of pursuing advanced education, but fail to find a foothold in the economy of the future. (p. 27)

> We are running up against a fundamental limit both in terms of the capabilities of the people being herded into colleges, and the number of high-skill jobs that will be available for them if they manage to graduate. The problem is that the skill ladder is not a ladder at all; it is a pyramid, and there is only so much room at the top. (p. 52)

For those of us working in the higher educational sectors, the vulnerability should be obvious, as Ford reminds us:

> If the higher education industry ultimately succumbs to the digital onslaught, the transformation will very likely be a duel-edged sword. A college credential may well become less expensive and more accessible to many students, but at the same time, technology could devastate an industry that is itself a major nexus of employment for highly educated workers. (p. 143)

There is much debate surrounding the impact of new technologies, particularly Artificial Intelligence (AI). We are seeing the impact of AI on many jobs already. I remember a friend of mine, who spent over 2-years doing *the knowledge* as it was then called in the local argot. The knowledge was a slang term for learning the location of the streets in London (and there are many of them), which was a mandatory requirement for acquiring the competence and licence to own and work as a registered black cab taxi driver in London. He achieved and acquired a comfortable living standard for his whole working life—retiring at 58. According to the cab drivers I once knew, this seemed to be the case generally, and many of them retired before 60 to live in warmer climates such as Spain and Thailand—so I believe them. However, what's the 'knowledge' worth now with smart maps? It may not be long before there are no cabs, as cars will increasingly become driverless. There is a view that it will be mainly the algorithmic jobs (e.g., can be totally modelled and turned into computerized 'neural networks', often referred to as machine learning) that will be eliminated or massively reduced. Again, Ford argues:

> …the ongoing race between technology and education may well be approaching the endgame: the machines are coming for the higher-skill jobs as well. (p. 121)

His most pessimistic frame is worrying:

In a perverse process of creative destruction, the mass-market industries that currently power our economy would be replaced by new industries producing high-value products and services geared exclusively towards a super wealthy elite. The vast majority of humanity would effectively be disenfranchised. Economic mobility would be non-existent. (p. 219)

Not everyone shares Fords' more pessimist framing. For example, in contrast, Brynjolfson and McAfee (2014) take a less pessimistic frame. They argue that:

Computers are not useless, but there are still machines for generating answers, not posing interesting new questions. That ability still seems to be uniquely human, and still highly valuable. We predict that people who are good at idea creation will continue to have competitive advantage over digital labour for some time to come and will find themselves in demand. (p. 192)

The impact of technology on learning and teaching was considered in some detail in Chap. 6. It is certainly changing the educational landscape, and this will only accelerate; how significant it will be in terms of employment and the subsequent societal consequences is probably still conjecture.

7.8 Summary

The macro-sociological theories of the past (e.g., Comte, Durkheim, Spencer, Mark) seem to have little generalizability in an era of post-modernity—this VUCA world. As Costa et al. (2016) wrote:

Hypercomplexity – our world is changing at an unprecedented rate, becoming more complex and globally interactive. Clear cut, unambiguous understandings are no longer an option. Thus, the capacity for self-directed and continuous learning has become the most important capabilities needed for survival in the future. Fortunately, the brain permits our ability to change, to grow, and to continuously develop our intellectual capabilities throughout a lifetime. (p. 3)

In a world of many potential threats and unpredictability, but equally much opportunity for greater understanding and progress e.g., Pinker (2019), we may just have to become more comfortable with being uncomfortable about the future. It may seem like a different mindset, but humanity has always had to deal with threats and uncertainty; I don't think the great plague was much fun for those who contracted it or lived in such localities. What we do know is that there are modern challenges to us as humans which, while different in form and require different solutions, do share a similar existential genesis. We seek to maintain a notion of the good society and well-being and stage off threats from the environment, albeit much of the real threats are within ourselves—though most do not see this—it's an evolutionary nemesis. However, on the positive side, we have made much progress as a species and with better thinking (i.e., MC) we may even call it 'global intelligence', and with better values, we may call it a 'universal morality'—and we have the capacity for both— we can use the curriculum for the twenty-first century as a vehicle to develop such capacities. In a sense, we must pitch aspects of our evolutionary human nature against other aspects of our human nature. Harris (2010), drawing from the reflections of Adam Smith captures this paradox well:

> The truth about us is plain to see: most of us are powerfully absorbed by selfish desires almost every moment of our lives; our attention to our pains could scarcely be more acute; only the most piercing cries of anonymous suffering capture our interest; then fleetingly. And yet, when we consciously reflect on what we should do, an angel of beneficence and impartiality seems to spread its wings within us: we genuinely want fair and just societies; we want others to have their hopes realized; we want to leave the world better than we found it. (pp. 58–59)

References

Brynjolfson E, McAfee A (2014) The second machine age: progress and prosperity in a time of brilliant technologies. Norton, New York

Collins A (2017) What's worth teaching: rethinking curriculum in the age of technology. Teachers College Press, New York

Costa AL et al (2016) Cognitive coaching: developing self-directed leaders and learners. Rowman & Littlefield, New York

Crawley E et al (2007) Rethinking engineering education. Springer, New York

Darwin C (1859) The origin of species. https://en.wikipedia.org/wiki/On_the_Origin_of_Species#Publication_and_subsequent_edition. Last accessed 3 Nov 2019

Diamond RM (1998) Designing & assessing courses & curricula: a practical guide. Jossey-Bass, San Francisco

Dweck CS (2006) Mindset: the new psychology of success. Ballantine, New York

Fogarty R (2009) How to integrate the curricula. Sage, London

Ford M (2015) The rise of the robots: technology and the threat of a jobless future. Basic Books, New York

Friedman TL (2006) The world is flat. Penguin, UK

Fullan M, Langworthy M (2014) A rich seam: how new pedagogies find deep learning. Pearson, London

Galbraith JK (1958) The affluent society. Houghton Mifflin Harcourt, New York

Harari YN (2016) Homo Deus: a brief history of tomorrow. Vintage, Penguin, London

Harris S (2010) The moral landscape: how science can determine human values. Free Press, New York

Hattie J (2009) Visible learning. Routledge, New York

Hattie J (2012) visible learning for teachers: maximizing impact on learning. Routledge, London

Hirst P (1974) Knowledge and the curriculum. Routledge & Kegan Paul, London

Kahneman D (2012) Thinking fast and slow. Penguin Books, London

Kelly AV (1989) The curriculum: theory and practice. Paul Chapman, London

Kurzweil R (2005) The singularity is near: when humans transcend biology. Viking, New York

Lawton D (1975) Class, culture, and the curriculum. Routledge and Kegan Paul Ltd, London

Lowe C (2017) How Latin develops the mind. https://www.memoriapress.com/articles/latin-develops-mind/. Last accessed 28 Nov 2009

McClelland DC (1973) Testing for competence rather than for intelligence. Am Psychol 28:1–14

Neary M (2014) Curriculum studies in post-compulsory and adult education: a teachers' and student teachers' study guide. Oxford University Press, Oxford

Neubauer S et al (2018) The evolution of modern human brain shape. Sci Adv 4(1):waao5961

Pellegrino JW, Hilton ML (2012) Education for life and work: developing transferable knowledge and skills in the 21st Century. National Academies Press, Washington, DC

Pinker S (2019) enlightenment now: the case for reason, science, humanism and progress. Penguin, London

Polanyi M (1966) The tacit dimension. Peter Smith, Gloucester, Mass

Reimers FM, Chung CK (eds) (2016) Teaching and learning for the twenty-first century. Harvard Education Press, Cambridge, Massachusetts

Rychen DS, Salganik LH (eds) (2003) Key competencies for a successful life and a well-functioning society. Hogrefe & Huber Publishers, Gottingen

Shulman LS (2005) Signature pedagogies in the professions. On professions & professionals (Summer, 2005). Daedalus 134(3):52–59

The National Research Council report (2012) Education for life and work: developing transferable knowledge and skills in the 21st century.

The Towards Defining 21st Century Competencies. The Foundation Document for Discussion, Ontario (2016). https://edugains.ca/resources21CL/About21stCentury/21CL_21stCenturyCompetencies.pdf. Last accessed 24 Nov 2019

Tough P (2014) How children succeed: grit, curiosity, and the hidden power of character. Arrow, London

Trilling B, Fadel C (2009) 21st century skills: learning for life in our times. John Wiley & Sons, San Francisco, CA

Wagner T (2010) The global achievement gap. Basic Books, New York

Willingham DT (2009) Why don't students like school: a cognitive scientist answers questions about how the mind works and what it means for the classroom. Jossey-Bass, San Francisco

Wringe C (1988) Understanding educational aims. HarperCollins, London

Young MFD et al (1971) Knowledge and control. Collier-Macmillan, London

Chapter 8
Assessing Twenty-First Century Competencies

Abstract This chapter focuses on the challenging area of assessing the twenty-first century competencies identified and analysed in Chap. 7. Different competencies, as well as different aspects of competence, will require different assessment methods. Assessing more complex competencies such as metacognition, critical thinking and creativity, will involve higher levels of inference in making valid and reliable assessment decisions (as greater subjectivity is inherent in assessing these skill areas), and it is both time and resource intensive. Furthermore, there is increasing pressure to ensure quality in the assessment process, to maintain standards and justify expenditure. Various strategies are explored and evaluated, but there is no easy solution to the issue of cost. Even the best combinations of methods and technology affordances may not be sufficient to meet such high expectations in a context of limited funding availability.

8.1 Introduction

The previous chapter explored key curriculum questions in the framing of twenty-first century competencies, especially what knowledge, skills and attitudes these competencies should encompass and on what basis. This chapter closes the curriculum development cycle, focusing on the big issues, questions and challenges in assessing these more complex and interrelated competencies. This is a thorny issue in education, for the simple reason that Ramsden (1992) suggests, assessment:

> …defines the actual curriculum…Assessment sends messages about the standard and amount of work required, and what aspects of the syllabus are most important. (pp. 187–188)

As a consequence, there may be little benefit in framing educational aims and outcomes that promote the cognitive, intrapersonal and interpersonal skills documented prior (e.g., critical thinking, creative thinking, metacognition, communication and collaboration), if we do not make corresponding changes in the assessment approaches and methods employed. Students, those motivated, will seek to learn what they see as the mandatory requirements of the assessments they receive, not necessarily the idealistic outcomes of being self-directed lifelong learners espoused

© Springer Nature Singapore Pte Ltd. 2020
D. Sale, *Creative Teachers*, Cognitive Science and Technology,
https://doi.org/10.1007/978-981-15-3469-0_8

by their teachers. Furthermore, to become a self-directed learner requires good thinking, volition and practice, though as humans we don't like to do too much of this stuff—for all the reasons outlined prior.

8.2   The Need for Good Assessment

Firstly, in the present context of breakneck competition for viable students, and increasing pressure on educational institutions to be accountable for their products in cost-effective ways, assessment quality is high profile. There is a need to be able to justify public expenditure in terms of value for money outputs. The quality of teaching and the cost-effective use of resources are rightly important issues in this context. However, it is the assessment credentials that largely define the value of educational programmes. If assessment practices are lacking in quality, what value can be placed on the qualifications accredited? The present situation is, as Bloxham and Boyd (2007) highlight:

> Assessment is now expected to assess subject knowledge and a wide range of intellectual, professional and generic skills in a quality assurance climate that stresses reliability with robust marking and moderation methods. (p. 4)

Secondly, and most significantly for understanding and enhancing students learning, there is increasing recognition of the important role that assessment plays in the learning process (e.g., Ramsden 1992; Boud 1995) as detailed in Chap. 2. Assessment is not simply a means to measure learning that has already occurred, it is a major facilitator in the learning process itself. As Boud (1988) illustrated:

> There have been several notable studies over the years which have demonstrated that assessment methods and requirements probably have a greater influence on how and what students learn than any other single factor. This influence may well be of greater significance than the impact of teaching or learning materials. (p. 35)

Furthermore, much research supports the view that students choose their approaches to learning rather than these being the result of innate characteristics or dispositions. For example, Prosser and Trigwell (1998) argue:

> ...approaches to learning are not stable characteristics of students. Student's approaches to learning do change with changes in perception of their learning situation and their perception of it can be changed by...teachers. (p. 83)

This is particularly significant if we accept the view that some approaches to learning are qualitatively better than others. Again, to quote Prosser and Trigwell in this context:

> ...there are better and worse ways for students to approach their learning – a deep approach being better than a surface approach. (p. 7)

What this means in practice is that how we design and conduct our assessment is fundamental to how students approach their learning, and the usefulness of that learning in terms of developing comprehensive and well-organized mental representations (e.g., neurological network in long-term memory), which are essential for conceptual understanding and transfer. Comparisons of 'Surface' and 'Deep' learning are useful frames for illustrating both a pedagogic and an assessment challenge that must be thoughtfully addressed (e.g., Biggs 2003). A surface learning approach typically involves the following characteristics and the consequent implications for learning:

- Lack of intrinsic motivation—which leads students to focus on passing the exam with minimal work; hence not developing a mindset for seeking interest in learning, making-effort and sustaining volition
- Little cognitive involvement—which leads students to fail to make sufficient connections between prior knowledge (which may be limited anyway) and new knowledge presented; hence resulting in incomplete and inaccurate mental models, limited understanding of the concepts involved, and prone to rapid forgetting.

In contrast, deep learning involves the following EBT core principles of learning:

- Intrinsically motivated to learn the topic and prepared to put in the necessary retrieval and deliberate practice to achieve mastery
- High cognitive involvement and awareness that effective learning involves the need to activate specific types of thinking (e.g., analysing, comparing, making inferences and interpretations) concerning the new knowledge presented and prior knowledge to build and extend understanding and skills.

To reinforce the point—call it retrieval practice—deep learning is a product of high MC in students. Students with high MC have a clear understanding of the learning process, how it works and what needs to be done to meet learning goals (e.g., employ appropriate task-related cognitive strategies, maintain volition and effort, and do the required practice as necessary). That's not to say that all students with metacognitive knowledge will adopt a deep learning approach; as we know motivation plays a key part in whether-or-not they will activate the necessary regulatory processes—as this takes effort and creates cognitive strain. Some may genuinely not have the time and/or not place such high value on a particular learning task.

8.3 Assessment from an Evidence-Based Approach

Assessment is more than the measurement of what has been learned (e.g., knowledge, skills, attitudes), but an essential component of the learning process, as reflected in the Core Principle of Learning: *Assessment practices are integrated into the learning design to promote desired learning outcomes and provide quality feedback.*

Irrespective of the framing of twenty-first century Competencies, this core principle, and the methods and activities that facilitate it, is central to the instructional strategies employed and applies to all contexts and modes of delivery. It is certainly the case that the use of technology will play an increasingly prominent role in doing this effectively and efficiently. I am now seeing most academic faculty using free EdTech tools such as Kahoot and Socrative to integrate formative assessment into their everyday teaching. Those who have strong pedagogic competence in applying EBT practices and principles are creating effective 'Russian Doll' strategies (remember the analogy?) in their teaching. They combine and integrate high effect methods and appropriate e-tools to create impactful instructional strategies and learning experiences tailored to student profiles and learning outcomes. For example, at the beginning of a session, they may use Kahoot as an advance organizer to check key conceptual understanding from a previous session (i.e., activating prior knowledge and stimulating thinking). From such activities, they can then address any knowledge gaps and misconceptions through two-way feedback. New knowledge can then be introduced using appropriate methods/e-tool blends. Furthermore, such technology-based assessments have the potential to provide immediate and precise descriptive feedback relating to student performance, enabling both diagnostic capability and personalization/differentiation of formative assessment. It also automates much of the assessment process, freeing up instructional time for teaching faculty. Similarly, in terms of assessing more complex technical and cognitive skills, computer-based simulations can increasingly provide data that facilitates more complex real-world performance assessment. The Towards Defining 21st Century Competencies, Foundation Document for Discussion, Ontario (2016) identified:

> Technology can support assessment *for*, *as*, and *of* learning, providing real-time assessment information that deepens our understanding of student learning gains and challenges. Technology can also support the tasks of gathering and analysing assessment information about student learning, thereby facilitating instructional decision making. (p. 35)

Assessment approaches are incorporating more performance-based/authentic assessment tasks, which focus on assessing complex performances in real work (or simulated contexts). This will often involve the assessment of a range of skills across competency areas in an integrated performance or set of performances. Assessments, particularly formative assessment, will involve more collaboration and transparency between both faculty and students, as well as among students. Peer Instruction (e.g. Mazur 1996) and Team-Based Learning (e.g., Sibley and Ostafichuk 2014) are becoming popular, as they offer many EBT affordances, especially retrieval practice and immediate, collaborative, specific two-way feedback. The learning benefits of such approaches are well summarized by Knight (1995):

> The key to the use of assessment as an engine for learning is to allow the formative function to be pre-eminent. This is achieved by ensuring that each assignment contains plenty of opportunities for learners to receive detailed, positive and timely feedback, with lots of advice on how to improve. (p. 81)

From this perspective, teaching and assessment are simply two sides of the same coin. Through collaborative formative assessment, we can get to know our students better, their learning preferences, and forge the essential communication links that foster a supportive learning relationship and build trust. However, assessment is not an exact science and the assessment of complex real-world performances, which involves the integration of a range of knowledge and skill bases, provides real challenges in terms of achieving high validity and reliability of assessment in a realistic efficiency context. As Gray (2007) pointed out:

> Finding or creating reliable, valid and appropriate assessment methods and tools matched to all learning outcomes remains a challenge. (p. 165)

Furthermore, assessment is a time-consuming process, and it is unlikely that additional resources will be available in the present context of cost-consciousness and already heavy workloads. As Boud (2000) warns:

> One of the traps in arguing for a shift in assessment practice is to propose an unrealistic ideal that can never be attained. (p. 159)

The reality, therefore, will be one of working smarter with existing resources. To achieve this teaching professionals will need to have a clear frame on what constitutes good assessment, the assessment formats, methods and strategies available, as well as the compromises that may need to be made and their impact on assessment quality. The following sections consider three essential questions which are central to achieving the quality of assessment practice:

1. *What is quality assessment?*
2. *What specific strategies can be used to enhance the effectiveness and efficiency of assessment practices?*
3. *How to develop valid and practical assessment instruments for twenty-first century Competencies.*

1. What is Quality Assessment?

Firstly, it is important to recognize that there are many different purposes for assessing student learning, and these reflect different stakeholder interests. For example, as Rowntree (1987) pointed out, the assessment may serve any of the following purposes:

- Selection and grading
- Maintaining standards
- Diagnosing learning difficulties
- Providing feedback to support the learning process
- As a source of information for evaluating the effectiveness of the teaching/learning strategy.

Furthermore, these purposes are not necessarily complementary and may conflict in practice. For example, while grades and standards may be of prime interest to employers, grading may do little to help students learn more effectively in the qualitative sense. In this context, it is important that the assessment approach effectively and efficiently addresses these stakeholder interests in a balanced manner. Courses, however, structured and delivered, must enable the validity of the summative outcome standards as well as supporting the learning process through formative assessment.

Principles of Good Assessment

In terms of quality of the assessment approach, there are certain general principles of good assessment that need to be appropriately applied in the design and conduct of assessment practices, these are:

- Validity
- Reliability
- Fairness
- Flexibility
- Sufficiency
- Authenticity.

The principles of good assessment are well documented in the literature (e.g., Haladyna 1997; Osterlind 1989; Rowntree 1987), hence only a summary reference will be made here.

Validity

This refers to the capability of a test in measuring accurately what it is we intend to measure, whether this is the recall of factual knowledge, understanding of concepts, competence in performance, or combinations of these learning outcomes. A major consideration in determining the validity of a test is the extent to which the evidence generated by the assessment items supports an accurate interpretation of the test scores in relation to desired learning outcomes. This is typically unproblematic in the assessment of factual knowledge though becomes more challenging in the case of assessing more complex performances involving the integration of a range of knowledge bases, skills and attitudinal components.

Reliability

This refers to the capability of a test to produce the same scores with different examiners, resulting in consistent and stable scoring of students over time. Fixed response items (e.g., MCQ's) are typically reliable as the answers are factual and scoring is a relatively simple process. However, in more complex assessments, reliability becomes problematic, as Banta et al. (2009) note:

> As faculty members increasingly rely on applying rubrics to student work, interrater reliability becomes another matter to address. Although multiple raters may use the same rubric to assess student work, assessment leaders should carefully determine how consistently individual assessors are judging student work. (p. 23)

Fairness

Fairness relates to a range of considerations in the assessment. However, they are all concerned with ensuring that learners, when being assessed, are provided with appropriate access to the assessment activities and are not unfairly discriminated against in the assessment process. Unfair discrimination typically means discrimination based on criteria unrelated to the assessment activity itself, for example, gender or racial characteristics. Fairness is a general concern throughout the assessment, relating as much to providing learners with the correct knowledge and time allowances for assessment, to non-discriminatory processes in marking their work.

Flexibility

Flexibility is concerned with the process of assessment, not the standard of the assessment. Learners can display their learning in a range of ways (e.g., orally, written, demonstration), provided the evidence is validly demonstrated. Flexibility typically becomes a consideration for learners with special needs (e.g., visual/auditory impairment, second language) or untypical situations (e.g., sickness on exam day). The arrangements for flexibility are usually specified by exam boards.

Sufficiency

Sufficiency is perhaps the most challenging of the principles of good assessment. In most basic terms it refers to deciding how much assessment evidence is needed to feel confident that a student is competent in an assessment task and context. This can refer to both the amount of evidence that needs to be generated or the range of applications of the competence in different performance situations/conditions For example, how many times would a student need to demonstrate mastery of a complex and critical procedure, and in what range of contexts, before we would deem him/her competent? There are no absolute answers here, and good professional judgement and collaboration with other professionals in the field are essential for establishing realistic benchmarks. I would hope that pilots flying planes do *sufficient* spaced and deliberate practice.

Authenticity

Quite simply this refers to how sure we are that the work produced by students has been done by them. In formal examination formats, we can be reasonably confident of authenticity, though with assignments being increasingly done by students in their own time, authenticity is a significant concern. Of course, internet resources further compound this concern.

Assessment Standards

Well constituted *Assessment Standards* provide a practical quality assurance framework for developing, monitoring and evaluating assessment practices and processes. It is to be noted that while such standards are generic to the overall processes, methods and procedures of assessment practices, there is always a need to contextualize them to the specific assessment context. The following are an exemplar of standards I contextualized for the adoption of the Conceive-Design-Implement-Operate (CDIO) engineering education framework in Singapore (Sale 2014). They were modelled on the well-established National Vocational Qualification standards in the UK and related to the 3 main interrelated stages of the typical competency-based assessment process:

Fig. 8.1 Summary of an
aligned curriculum

(a) Producing and reviewing an assessment plan
(b) Judging evidence and making assessment decisions
(c) Providing feedback on assessment decisions

(a) **Producing and reviewing an assessment plan**

Essentially an assessment plan identifies the why, what, when, where and how of the
assessment process for a module or unit of study. When well-constructed, it provides
a concise guide of the whole assessment process and components for both assessing
faculty and the students involved. In congruence with an aligned curriculum framework (e.g., Biggs 1996), as summarized in Fig. 8.1, it should result in assessment
methods, instruments and procedures that both effectively develop the defined student
learning outcome (formative assessment) and measure them (summative assessment)
for any specified programme.

The following criteria, which incorporate the principles of good assessment identified earlier, frame the key considerations that need to be addressed in producing a
well constituted assessment plan:

- The assessment plan specifies the assessment methods to be used, their purpose,
  the marks to be allocated, and the timing of assessments
- The selected assessment methods are valid for assessing the knowledge, skills and
  attitudinal components specified, and at the appropriate levels
- The assessment methods are well constructed and sufficiently varied to enable
  learners to display understanding/competence through different mediums
- The assessment methods are planned to make effective use of time and resources
  in producing sufficent evidence
- The assessment methods provide fair and reliable assessment opportunities
- The key aspects of the assessment plan are explained to learners
- Opportunities are provided for learners to seek clarification on assessment
  requirements
- Ways to ensure the authenticity of assessment evidence are identified
- The assessment plan is reviewed at agreed times and modified where necessary.

Of note, it is important to recognize that in practice there is often a trade-off
in terms of meeting the various principles of good assessment. For example, sufficiency is typically problematic in that what constitutes a 'sufficient' range and depth
of assessment evidence (derived from a variety of methods) is open to judgement,

as noted prior. However, addressing the sufficiency questions are crucial in areas involving safety issues. It is here that significant problems may be encountered, especially with an emphasis on more complex and interrelated competencies. Assessment is time-consuming, resource-intensive, especially for high inference assessment—where there is increasing subjectively involved in making valid and reliable assessment decisions (e.g., creativity, leadership). As Knight (2006) highlighted, reliable judgement can only be made where there have been several observations from multiple observers in a range of contexts, which is not very practical in terms of resources.

(b) **Judging evidence and making assessment decisions**

Being an assessor is in many ways akin to that of a 'caring' detective. The assessor is responsible for maintaining the standards and ensuring that assessment decisions are based on the specified performance criteria and the range of evidence sources required, but at the same time show empathy and make fair and thoughtful judgements relating to flexibility where appropriate. Flexibility can be used to accommodate individual student's special needs and circumstances through providing alternative assessment methods (e.g., oral rather than written responses) without compromising the essential knowledge and skills embedded in the standards.

One of the most significant challenges in making valid assessment decisions revolves around considerations of appropriate standard or level of proficiency. While we would all like to have clear standards from which to base assessment decisions, this is often difficult to achieve in practice. Certainly, the explicit and valid identification of performance criteria is important here. Failure to appropriately make explicit the key constructs/elements that underpin the performance areas will seriously undermine the validity of the assessment, as well as create difficulty for assessors, affecting inter-rater reliability. This, in turn, inevitably results in disagreement between assessors, as well as between assessors and those being assessed. In the worse scenario, this can lead to appeal situations and legal scenarios. In the final analysis, summative assessment involves making a judgement concerning a person's worth or capability in a certain performance area. Getting this wrong through a decision of non-competence or grading lower than the criteria indicates can have severe consequences for a learner in terms of employment or educational career access. Of note, rarely do students appeal against a top grading, suggesting that it should be lowered.

It is also the case with high inference assessment areas, that even when criteria are well derived and delineated, actual judgment in terms of how well students perform requires interpretation, and this can vary across markers. Much of the problem is identified by Knight (2006) who argues that complex learning cannot be reduced to something simple enough to measure reliably: the more complex the learning, the more we draw on *connoisseurship* (Eisner 1985) rather than measurement to make our judgment (p. 38). Connoisseurship refers to the 'art of appreciating something', and is typically used in ways to depict a person's deep awareness of, especially the subtle aspects, in an area of experience (e.g., music or art). In terms of assessment, having deep knowledge and experiences over many situations (e.g., expertise) helps to get a better holistic view of the performance; rather than just aggregating discrete elements into a final grade. In practice, breaking down complex performances into highly detailed and specific criteria can result in a level of reductionism that both fail to capture the holistic contextualized performance as well as encouraging students to focus on these more atomistic components. In consequence, this can also mitigate students adopting a deep approach to learning and the capability to transfer learning across a range of similar performance situations.

A final important consideration in this context relates to the authenticity of assessment evidence provided by students, as documented earlier. In today's globally wired world, plagiarism is a serious assessment concern.

Bearing in mind the considerations identified above, the following criteria identify the key areas of practice for making the best judgments we can:

- Learners are provided with clear access to assessment
- The assessment evidence is judged accurately against the agreed assessment criteria
- Only the criteria specified for the assessment are used to judge assessment evidence
- The assessment decisions are based on all relevant assessment evidence available
- Inconsistencies in assessment evidence are clarified and resolved
- The requirements to ensure authenticity are maintained.

(c) Providing feedback on assessment decisions

The importance of feedback is fundamental to the learning process as documented earlier—especially for formative assessment. As Gibbs (2008) highlights:

> Research in schools has identified that the way teachers provide and use feedback, and engage students with feedback, makes more difference to student performance than anything else that they can do in the classroom. (p. 6)

It is also important to ensure sound recording, collation and security procedures in conducting an assessment. The following criteria identify the key areas of practice for providing feedback and securing assessment outcomes:

- The assessment decisions are promptly communicated to learners
- Feedback to learners is clear, constructive and seeks to promote future learning
- Learners are encouraged to seek clarification and advice
- The assessment decisions are appropriately recorded to meet verification requirements
- Records are legible, accurate, stored securely and promptly passed to the next stage of the recording/certification process.

2. Strategies to enhance assessment practice

Firstly, it is important to be clear that whatever strategies are employed, the actual design of assessment items and conduct of assessment activities must be congruent with the principles and standards documented earlier. Secondly, standards are not necessarily prescriptive about the use of specific assessment methods; what is important is that the methods used should be as closely calibrated to the types of learning outcomes being assessed as is viable. The emphasis on real-world projects and tasks, which require the integration of knowledge and skills across subject and domain fields, will feature increasingly in the facilitation and assessment of twenty-first century Competencies. While students learning of key content knowledge and understanding can be effectively and efficiently assessed in written and other paper and pencil tests, this needs to be increasingly augmented by integrated real-world projects and authentic learning experiences.

All assessment methods have strengths and limitations in terms of the types of assessment evidence they can generate, and their usefulness will be largely dependent on the learning outcomes being assessed. For example, while multiple-choice items can be very effective and efficient for assessing knowledge and understanding (e.g., specific types of thinking such as analysis, comparison and contrast, inference and interpretation, and evaluation), they have little validity for assessing integrated skills in complex problem-solving activities. Similarly, performance-based items, often the most valid means for assessing more complex real-world performances, emphasized in the twenty-first century skills frameworks, are much more time and resource consuming, inevitably provide challenges both to the sufficiency of the evidence and, as they involve higher levels of inference than traditional testing methods, reliability.

Within this context, it is suggested that combinations of the following strategies can contribute to the effectiveness and efficiency of assessment practices in given assessment situations. The strategies are not meant to be exhaustive or summative, but represent practical frames from which assessment decisions can be thoughtfully made and practically customized to the range of assessment situations:

(a) Produce assessment activities that are interesting and challenging
(b) Integrate a range of learning outcomes in assessment activities
(c) Provide as much transparency as possible in the assessment process
(d) Utilize student collaboration in formative assessment.

(a) **Produce assessment activities that are interesting and challenging**

One of the central themes of enhancing learning is through learning experiences that are more intrinsically motivating for students. A major means for achieving greater engagement as well as aspects of metacognitive capability and other twenty-first century competencies is through well contextualized real-world learning tasks. Just as a well-constituted syllabus is of limited value in the hands of faculty who lack competence in pedagogic practices and the ability to create interesting and engaging learning experiences, the same logic applies to our assessment activities. If they lack interest and purpose for students and encourage rote learning, we should not be surprised to see them adopt the surface approaches to learning, as explained prior. In these situations, students will learn what is necessary for assessment but are unlikely to derive both a real understanding of the subject or a genuine interest in it. Once the assessment process is finished, much of what was learned will soon be forgotten. In contrast, where students find the assessment activities interesting and sufficiently challenging, they are more likely to develop a genuine interest in the learning involved (Struyven et al. 2002). It is motivation for mastering the tasks set that leads to a desire for understanding the important concepts and principles of a subject and makes possible the transfer of learning (e.g., McTighe and Wiggins 2000).

(b) **Integrate a range of learning outcomes in assessment activities**

There is an old English metaphor, "kill two birds with one stone". In the context of assessment, this means getting the best efficiency from assessment activities and situations. The more assessment activities enable coverage of a range of learning outcomes, especially if they integrate knowledge and skills across topic and domain areas, more efficient is the assessment task. This will be increasingly necessary for the assessment of twenty-first century competencies, as they are typically employed simultaneously in expert performance in many situations. For example, an expert in almost every field will often need to combine good thinking with good communication skills and teamwork as part of a seamless performance. In doing so, MC conducts the orchestration of skills needed for the complex work activities that we now need to enact. As emphasized prior, many twenty-first century skills are mainly 'souped-up' first century skills, and they are certainly interrelated in real-life applications. However, MC, while evolutionary hatched as a potential capability for primate mankind in terms of brain morphology, it would be far less required in yesteryear, as compared to the information overload and breakneck change of today's world. For primitive man, though with milieu variation, most aspects of life were highly predictable from year to year, and the need for thinking quite limited (at least in terms of System 2 thinking, Kahneman 2012).

The interconnectivity of twenty-first century skills, from the main competencies, will, as Lai and Viering (2012) suggests, create a need for:

> …multiple measures that either 1) represent multiple assessment modes or 2) sample from multiple content domains to permit triangulation of inferences. (p. 43)

As twenty-first century competencies have multiple sub-components of skill sets and underpinning knowledge bases—multiple measures are essential for authenticity and validity. It is for this reason that we may move more and more towards work-place assessment as a major means to attain the range of evidence-sources to make useful assessments of these competencies. For example, in assessing competence, it is essential to see how people perform in a range of work functions and situations, especially in those that are challenging, requiring high MC and technical skills applications. This can be time-consuming and requires expertise in making such assessments, both at the technical subject level as well as using cognitive, intrapersonal and interpersonal skills competently in a multi-disciplinary manner. Assessment of competence will also require high levels of skill in such areas as sensory acuity, questioning and mediating two-way feedback.

It is not surprising, therefore, that this raises significant challenges, including, as Lai and Viering (2012) note:

> the feasibility of implementing such assessments at scale in an efficient and cost-effective manner. (p. 49)

There is little doubt that this will be problematic in these times of austerity. As I remember it, *quality* was a big issue and focus in the 1980s and 1990s, whereas today, this seems to be a given, and the focus is on cost-saving as *the* competitive advantage. The question is how far does this high quality—low-cost paradigm go? This could mean more and more people working longer, harder, for less and less. In terms of assessment, we are wanting more and more quality and fidelity of measurement of highly complex interrelated skills—but attained with fewer resources.

This is an area in which technology may play a significant role in future, as it is presently providing the means for testing content knowledge and understanding (which requires types of thinking) as well as increasingly personalized specific feedback and guidance for future learning pathways and strategy use. Interactive videos, augmented reality, virtual reality, and the newer concept of mixed reality may offer useful affordances and mitigate the costing involved in expert human observation, as identified prior.

Metacognitive, cognitive and affective skills and attitudes are typically learned within the context of domain-specific knowledge. Similarly, where issues of a value-laden and ethical nature are involved, there is a need for both learning and assessment to be contextualized to the subject domain. Teaching ethics and values separate from real-life contexts is akin to teaching thinking without reference to a knowledge domain. De-contextualized knowledge is difficult to transfer and may not even be perceived as meaningful by learners. Nucci's (2001) observation about the teaching of values, is relevant in this context:

> The greatest challenge for a teacher wishing to engage in domain appropriate practice is to identify issues within the regular academic curriculum that will generate discussion and reflection around a particular value. (pp. 178–79)

Students are firstly more likely to understand and internalize learning in the affective domain when it is contextualized to specific work-related contexts and issues. Secondly, in terms of assessment, there is more likely to be authentic assessment opportunities and greater validity in the assessment of this knowledge and skills in such situations. For example, in the world of engineering practices, as in any professional domain, there are ample opportunities to naturally infuse the full range of types of thinking as well as areas of ethical concern—both in terms of personal values and wider societal issues that involve professional ethics. I have developed and conducted programmes on ethical reasoning through an exploration of specific ethical dilemmas that can manifest themselves in different work contexts. These can be structured in terms of variation of situation, emphasis, and levels of complexity. Technology can be useful here, especially short role-play videos, as suggested in Chap. 6. I produce my videos in this area, which seem to work well, and are easy to produce with free (or very reasonably priced) video editing software tools. Also, this avoids issues of plagiarism and seeking permission for reproduction, which is time-consuming. One does not need to employ 'National Geographic' video production quality—it's all about good content, and make sure the audio is clear and focused. You can also embed questions or reflection points into the video structure. When done with creativity and this is not so difficult, you are developing Creative Teaching Competence, and as you improve, it gets easier and easier.

(c) **Provide as much transparency as possible in the assessment process**

Assessment should *not* be designed to mystify students about what they are expected to learn; rather it should make explicit what is to be assessed, the specific criteria involved in making the assessment decision, and the standard of the assessment evidence required in behavioural terms—as far as is viable. For example, when doing a driving test, you know exactly what is expected both in terms of knowledge (i.e., the highway code for that country) and skills (e.g., parking the car, emergency stop, 3-point turn). You are unlikely to be asked the capital of Bulgaria in doing your test, even if you live there. There is full transparency as to what constitutes a competent driving performance. It is also equally clear as to what gaps in knowledge and competence will lead to failing the test. Other things being constant, such as assessor reliability and fairness, etc., passing or failing will depend on the performances exhibited in the test situation. However, there is disturbing evidence that factors such as mood, fatigue and hunger can influence judgements. For example, Kahneman (2012), Referring to research reported in the *Proceedings of the National Academy of Sciences*, in which parole requests were studied in terms of the time of day they were granted or not granted (the exact time of each decision was recorded, as well as the times of the judges' three food breaks), found that after each meal the proportion of requests spiked upward with 65% of requests granted. However, during the two hours or so until the judges' next feeding, the approval rate dropped steadily, to about zero just before the meal. As Kahneman concludes:

> The best possible account of the data provides bad news: tired and hungry judges tend to fall back on the easier default position of denying requests for parole. Both fatigue and hunger probably play a role. (pp. 43–44)

Similarly, the *Halo effect*, which is the tendency for positive impressions of a person in one area to positively influence one's opinion or feelings in other areas. For example, attractive-looking people are sometimes perceived as more outgoing, socially competent and powerful, sexually responsive, intelligent, and healthy (e.g., Eagly et al. 1991; Zebrowitz and Rhodes 2004). Moreover, these trait impressions are accompanied by preferential treatment of attractive people in a variety of domains, including interpersonal relations, occupational settings and the judicial system (e.g., Langlois et al. 2000; Zebrowitz 1997). The Halo effect is a cognitive bias and explains why many people try to make themselves as attractive as possible. It also explains why people try hard to make good first impressions—the Primacy effect—as a good Primacy effect, with a positive Halo effect (one can of course also suffer from a negative Halo effect)—gets one off on a 'good foot'—so to speak—with persons we are trying to impress in some way. In the context of assessment, the Halo effect can be influential in assessment decision making in 2 main ways. Firstly, for the students that we like, there may be a tendency (and this typically occurs sub/unconsciously) to be more generous or lenient in the assessment process. Secondly, if they generally produce high-quality work, and then on occasion submit lower standard work, we might still give them a higher grade than the work merits.

Given that assessments typically involve human perception and mood, and other attribution biases, you may simply just be lucky or unlucky in an assessment situation. To mitigate such systemic biases, we need to establish robust checks on the assessment process; this is typically referred to as verification. Verification, in the context of assessment, is about ensuring that the agreed processes, procedures and practices are carried out by designated personnel, as prescribed. It is not about duplicating the assessment process—through a sampling of assessment evidence and assessment decisions made is part of the overall verification role. An essential component of the verification process in the *Internal Verifier* (IV) for an educational institution. This person is typically an experienced and accredited assessor, who takes on the role of developing, monitoring and reviewing assessment practice at an institutional level, and will conduct internal quality assurance of the assessment process by:

- Carrying out and evaluating internal assessment and quality assurance systems
- Supporting assessors
- Monitoring the quality of assessors' performance
- Meeting external quality assurance requirements.

This typically involves:

- Comparing own organizations requirements with those of the external awarding body
- Identifying the outcomes needed by the agreed standards and their consequences for internal auditing
- Auditing existing administrative and recording arrangements—modify/change if necessary—to meet external audit requirements
- Carrying out assessment standardization arrangements (e.g., sampling strategy)
- Ensuring a procedure for complaints and appeals is in place (note: consider this in the local context)
- Identifying problems and developing improvement plans.

There is a range of functions, though most important is ensuring the standardization of assessment decisions and supporting teachers who have assessment roles. Key areas include:

1. Ensure that assessors consistently make valid decisions
2. Ensure that assessors make the same decision on the same evidence base
3. Ensure that all candidates are assessed fairly.

This involves conducting moderation activities (e.g., different assessors judging the same candidate's evidence and comparing their perceptions and decisions). This is an effective way to ensure that there is a common (as far as is possible) conception of standard in terms of the type, level of knowledge and skills that need to be demonstrated for an element of competence. In terms of professional development and support for assessors, the following are essential:

- Ensuring assessors have appropriate technical and vocational experience.
- Ensuring that assessors are competent to assess.
- Identifying the developmental needs of assessors and facilitating any necessary training.
- Monitoring progress and development of assessors.

The internal verification system should ensure transparency in all aspects of the assessment process, the assessment methods and tools being used, and the decisions made for all candidates. Essentially internal verifiers are the guardians of the assessment arrangements for the institution or a course of study; ensuring the principles of good assessment are applied thoughtfully from the planning of assessment approaches to recording of assessment decisions and maintaining security. Transparency is further enhanced when students are explicitly made aware of, and have opportunities to seek clarification on the syllabus learning outcomes and the assessment requirements. It also helps if the learning outcomes are written and the assessment criteria fully aligned to them. They should be if the interval verifier(s) is doing the job well.

Bloxham and Boyd (2007) summarize this in terms of a module context:

> Many researchers are now concluding that preparing students for assessment is not a distinct stage in a module but should be part of an integrated cycle of guidance and feedback, involving students in active ways at all stages. Price and O'Donovan (2006) describe a cycle commencing with providing 'explicit criteria' followed by 'active engagement with criteria', self-assessment with the submission of work' and 'active engagement with feedback'. (p. 71)

Smart students have long worked out that the secret to success in assessment boils down to basic logic, 'know what needs to be learned, learn it and know that you have learned it'. However, simply knowing this does not mean success either in learning or assessment outcomes. Within this simple maxim is an implicit essential added element, its called *effort*. Learning, at a high proficiency level typically involves considerable effort and time on task, which many students may not be prepared to do. Students who make the necessary effort and develop a high level of competence deserve to be successful—don't they? Metacognitive Capability is the best preparation we can offer our students, but, ultimately, they must take responsibility for their learning—become agentic—which is a key part of being a Self-Directed Learner.

(d) Utilize student collaboration in formative assessment

As assessment is fundamentally linked to learning and teaching, it makes sense to utilize the main stakeholders (faculty and students) to collaboratively make this 'system' work to the best advantages of both. For example, from a student's perspective, the ideal would be to use assessment to optimally support the learning process through the various processes of formative assessment. Similarly, from a faculty perspective, we would like to be able to identify student learning concerns/problems quickly and be able to effectively and efficiently deal with them—whether through instructional design or other learning support means. How then, might we create the kind of symbiosis that makes possible the best collaboration between faculty and students, without compromising the quality and credibility of final summative assessments?

Ultimately, the most fundamental way to utilize this collaboration is to help students to develop their own self-assessment capability, as emphasised prior. Students who are able, in large part, to identify what they know and don't know are already a long way to becoming independent learners and reduce the load off faculty in terms of instructional and remediation time. Time spent in developing student's capability to self-assess will result in better learning for students as well as making the instructional process more efficient.

Secondly, having students involved in peer assessment is especially important, and has a higher single Effect Size than self-assessment (0.63 as compared to 0.54, Hattie 2009). However, feedback effects from various sources (e.g., tutor–student; student–student; other credible sources) will have synergistic impacts. In summary, good feedback from good sources, and students taking an agentic approach to learning is what we want to facilitate. It must be emphasized that students will initially need direct instruction and plenty of deliberate practise to develop skills of assessment, for example:

- Analysing goals, outcomes, and tasks to ascertain what knowledge and skills are involved
- Analysing and deriving performance criteria for making judgements of worth
- Making inferences and interpretations from performance evidence to ascertain competence or otherwise, and on what basis.

If the teaching of Good Thinking has been initiated and ongoing in the earlier instruction—this should be a transfer activity for students, as this is good thinking applied to the domain of assessing performance from evidence.

3. **Develop valid and practical assessment tools**

Good design is fundamental to all assessment items, whether a fixed response, essay type or performance-tests. This is well documented in the literature (e.g., Osterlind 1989; Haladyna 1997). However, different assessment items, apart from offering different assessment evidence on students' learning, also provide different challenges in turns of making assessment decisions. For example, while multiple-choice items require skill in design, and the production of a large bank of useful items is time-consuming, marking is easy and efficient. Open response (essay-type) and performance-based items, in contrast, require a more elaborate marking system and are prone to subjectivity.

Performance-based assessments, however, are potentially the most valid forms of assessment as they provide assessment opportunities where students can display key competences in the real world or simulated activities. Such tasks provide more authentic and valid assessment opportunities, offering the following assessment advantages over more traditional pencil and paper-based approaches:

- Greater validity as the focus is on real-life performance
- Measures a range of complex skills and processes in real-world or authentically simulated contexts

- Links clearly with learning and instruction in a planned developmental manner
- Motivates students through meaningful and challenging activities.

These tasks can also encompass a wide range of activities, for example:

- Real work projects and tasks
- Simulations
- Problem-solving through case studies
- Presentations
- Any activity that essentially models what would be done by professionals in the world of work.

However, performance-based assessment poses challenges for faculty in terms of:

- More time consuming than paper and pencil type assessment
- Where courses focus on underpinning knowledge, there is less opportunity for performance-based assessment
- Subjectivity in marking, as these items often involve assessments that are high inference and therefore involve professional judgement.

The issue of subjectivity in marking is problematic, especially in project activities that integrate a range of competency areas across modules and subject domains. For example, assessing MC involves an extensive range of skills, with some being more high inference than others. Assessing student's capability for goal setting, using cognitive strategies and evaluating their learning can be assessed through a range of product evidence, focused questioning, and observation. Similarly, growth mindset and self-efficacy can be assessed through established validated psychometric tools, incorporating focused questions relating to the key underpinning constructs, self-report journals, as well as interviews exploring students' perceptions and experiences, about the work they are doing. The framework presented in Chap. 3 identifies the key components of MC, which can then be framed in terms of elements of competence, performance criteria and range statements—and subsequently packaged and contextualized to different learner contexts (e.g., levels of competence, vocational field). Invariably, there is much work involved here, and that includes time and resources, apart from expertise. As Bloxham and Boyd (2007) precisely note:

> …the research suggests that providing fairness, consistency and reliability in marking is a significant challenge caused by the inherent difficulty of reliably marking complex and subjective material combined with our marking dispositions. (p. 87)

Will technology enable the assessment of types and levels of thinking and volitional aspects of mind related to emotional regulation at the neural level through artificial intelligence networks connected to brain imaging technologies in the future? Is 'Radical Evolution' as Garreau (2005) and Kurzweil (2005) document possible, and is it likely to happen soon? Well, as indicated prior, I don't think it will happen in the next couple of decades, and it may not be as radical as being prophesied—or desired. In the absence of such technology-aided assessment tools, we must develop marking systems for performance-based assessment tasks, especially where high inference assessment is involved, through thoughtfully addressing the following assessment considerations:

- Assessment areas
- Performance criteria
- Assessment evidence
- Assessment rubrics.

8.4 Assessment Areas

Assessment areas constitute the main *performances* that are to be assessed in any performance-based activity, and typically more than one assessment area can be validly assessed. For example, in a project-based activity, there are usually opportunities to assess critical and creative thinking, teamwork and communication, ethical issues, as well as the technical subject content areas. However, just because a performance test offers such opportunities for assessment, this does not automatically mean that all possible performance areas must be assessed, especially in summative terms. What is assessed from such an activity should be considered with other assessment components for the module or unit of study. For example, if an area has been sufficiently assessed elsewhere, it may be more practical to assess other important areas that have not been previously assessed in the overall assessment scheme. However, it is, of course, useful, whenever feasible, to provide appropriate formative assessment in all the significant performance areas.

Once the summative assessment areas have been identified for the learning activity, it is then necessary to identify the marks allocation or weighting for each of the designated areas. This should reflect the learning outcomes and their relative importance within the module or unit context, as well with other components of the assessment plan. Table 8.1 summarizes the assessment areas and mark weightings.

Table 8.1 Assessment components for a design-implement project

Assessment components	Mark weighting in %
Production of car chassis components (based on practical work done in the workshop)	30
Assembly of car chassis components (based on practical work done in the workshop)	10
Performance of the Model F1 car in the racing challenge (speed and stability under test conditions)	20
Teamwork (e.g., goal setting, management of team-roles and responsibilities, dealing with conflict/challenges, etc.)	30
Oral presentation (e.g.. organization, clarity, and effectiveness of oral communication)	10
Total	100

Performance Criteria

Performance criteria are the more specific and measurable elements/behaviours that underpin the wider performance area to be assessed. For example, in assessing *Demonstrate Effective Written Communication* (a component of Communication Skills) the performance criteria may include 'write with logical organization and clear language flow', 'use concise and precise language', 'use correct grammar, spelling and punctuation', etc.

From my experience, the process of generating performance criteria for assessing the performance areas of a course syllabus can be enhanced when educational development specialists work with subject specialists, as this can ensure clarity and validity, as well as faculty buy-in. Developing highly detailed lists that have a pristine appearance but are seen as unnecessarily wieldy and obtrusive by the faculty that are required to use them are rarely used in the 'intended way' in practice.

Performance criteria should provide the necessary guidance to enable assessing faculty to make the most valid assessment decisions possible (in the context of time and resource allocation) about the learning outcomes, based on the performance evidence generated by the task activities. In designing performance-based assessment tasks, I find it useful to ask the essential question below, and answer it as best as possible:

> Will this performance task offer the student a realistic opportunity to demonstrate that he/she can meet these learning outcomes?

Table 8.2 shows how one school chose to organize their marking system for a third-year capstone project. From an analysis of the range of activities that the students were going to engage in throughout the project, and with the syllabus learning outcomes, they derived the following performance criteria that would form the basis and focus for the assessment areas.

8.5   Assessment Evidence

Assessment Evidence refers to the range of performances and products that can be validly and efficiently considered in making an assessment decision. In making assessment decisions, it is necessary to consider what the range and types of evidence are that can be generated by the various activities concerning the performance areas and criteria. For example, in assessing teamwork, a wide range of evidence sources can be generated and used to make a valid assessment decision. These could include the following:

- Feedback from students (e.g., peer assessment)
- Lecturer observation of student interactions
- Questioning
- Meeting deadlines and objectives
- Students logs/journals.

Table 8.2 Raw mark form—projects with physical deliverables

Project no. _____		Name and adm. number of students				
Assessment areas	**Performance criteria**					
1. **Conceiving** (Default 15%) [Range: 10–25%] **Selected weightage:** ____	1.1 Eliciting market needs and opportunities					
	1.2 Defining functions and concepts of the system					
	1.3 Modelling system to verify goals					
	1.4 Development of project plan					
2. **Designing** (Default 25%) [Range: 15–35%] **Selected weightage:** ____	2.1 Formulation of design plan					
	2.2 Selection of final design					
	2.3 Consideration of project costs					
	2.4 Evaluation of selected design					
3. **Implementing** (Default 15%) [Range: 5–25%] **Selected weightage:** ____	3.1 Designing the implementation process					
	3.2 Planning for hardware (or software) realisation					
	3.3 Testing, verifying, validating and certifying					
4. **Operating** (Default 5%) [Range: 5–10%] **Selected weightage:** ____	4.1 Planning training and operating procedures					
	4.2 Suggesting improvements to project					
	4.3 Planning for project disposal					
5. **Teamwork** (**10%**)	5.1 Identification of goals and work agendas					
	5.2 Utilisation of team strengths					
	5.3 Application of ground rules and management of conflict					
6. **Effective communication** (**15%**)	6.1 Logical organisation of content and language flow in the project report					
	6.2 Using correct language and grammar in the project report					
	6.3 Producing engineering drawings					
	6.4 Using effective oral communication					
7. **Personal and professional skills and attributes** (**15%**)	7.1 Using a range of critical and creative thinking skills					
	7.2 Monitoring and reviewing the quality of own thinking					

(continued)

Table 8.2 (continued)

	7.3 Managing learning					
	7.4 Acting in a manner consistent with professional codes and ethics					
CDIO skills 80% (For the above raw marks, items 1–4 = 60%, items 5–7 = 40%) **Exhibition 10% (Duty/Exhibited = 5, Bronze = 6, Silver = 8, Gold = 10)** **Deadline 10% (Project deadline—as per page 6 of logbook)** **Marks ratio between supervisor and co-examiner = 2:1**	**Score 5 to 1 in the boxes above** **Key:** 5 = Consistently met to a very high standard 4 = Mainly met to a high standard 3 = Mainly met to an acceptable standard 2 = Partially met to an acceptable standard 1 = Very poor performance					

For projects, the following generic types of assessment evidence are typically produced:

- Reports
- Progress reviews
- Logbook
- Scheduling documentation
- Engineering drawings
- Artefacts (e.g., models, prototypes, programmes, operating manuals, etc.)
- Presentations
- Responses to questions (e.g., oral, written).

Invariably, the greater the range of evidence sources that can be accessed (providing they are sufficiently valid and authentic), the more likely it is that we can make accurate assessments of performance.

Assessment Rubrics

Assessment Rubrics are rating scales in which a prepared scoring system is used for assessing learner performance for a specific task or assessment area across different levels of that performance (usually 1–5, in which 1 denotes a very poor performance and 5 denotes a very good performance). Assessment rubrics are most useful when assessing complex activities where the assessment of performance is of variation, and involving a high level of inference. For example, in assessing teamwork, it is often not a clear case of being either effective or ineffective in this performance area, but rather variation along a continuum from very effective to very ineffective. Furthermore, as there are many aspects and potentially different interpretations of what constitutes effective teamwork, it is open to different inferences by different assessors. The extent to which assessors are likely to differ in terms of assessment decisions relating to an area of performance determines the level of inference.

In performances where all assessors, assuming expertise in the area, would consistently agree on the level of performance, we can say that assessment is low inference. This would be the case in most procedural aspects of a performance in which there are clear and established, almost algorithmic, standards relating to effective performance. In these assessment situations, a checklist is a more appropriate tool and easier to use marking system. However, in areas such as creativity and aesthetics, while certain features can be identified as criteria of quality, there is still a high level of subjectivity in terms of personal interpretation of what this looks like overall. Such areas constitute high inference assessment, where assessors may have quite diverse perceptions of what is good and poor performance. In these assessment situations, the descriptors of different levels of performance are useful in mitigating the variation of assessment decisions and enhancing reliability. It only requires watching a few episodes of popular singing competitions in which one judge may see a competitor's performance as 'brilliant' but another judge (and remember these are supposed to be top experts in the field) see's it 'as self-indulgent rubbish'. I feel like it's the latter when subjected to listen (of course I try to avoid it) to much of modern music. Don Mclean who famously sang American Pie, on a visit to Singapore a decade or so back, was asked when he thought was *the day the music died* (an emphatic line in the song). He replied, in the interview, and I hope I get this right, 'some 40 years ago'. I liked his song, and his judgement on musical quality is exemplary—here's subjectivity for you. Now, would you like me assessing you're rap or techno song?

In using rubrics, decisions need to be made on whether to assess more holistically or analytically about performance areas and criteria. Essentially this relates to whether to assess the performance area overall e.g., oral communication and give a score; or break it down into key components/constructs, score these individually, then derive the overall score. There are merits in both approaches (Biggs 2003; Gosling and Moon 2003). Holistic rubrics enable a focus on the overall performance and are more economical in terms of assessment time. They are typically used for summative assessment and where some variation in reliability in parts of the assessment components can be accepted, provided the overall assessment decision has justifiable validity and reliability. In contrast, analytical rubrics enable a much greater focus on the specific elements of the areas of learning involved and make possible a much better utilization of formative assessment in the assessment process. One type of rubric is not inherently better than another (Montgomery 2001); it depends on the assessment purpose and context in which rubrics are used.

What is of primary importance is that the rubric *does not* make the assessment decision. Rubrics provide a guiding framework for focusing attention on the key elements/constructs (performance criteria) of the assessment area and summary descriptors of a range of performances. A good rubric offers sound heuristics, bringing the most important aspects/features of the task requirements to the foreground of the mind; the assessor must use expert professional judgement in making the assessment decision. Rubrics can be shared with students to further enhance transparency. While there are different opinions here, from an EBT perspective openness and transparency are consistent with sound reasoning. I don't agree that transparency of what constitutes an exemplary performance, and its components, compromise standards for assessment. I know exactly what constitutes expert tennis, and exactly what to do—I have an excellent rubric in my mind—but mediocrity still dominates my game, and there is a reason for this. With more time, good coaching, spaced and deliberate practice, and effort on my part, I would improve, and the rubric would assist in my self-regulation and self-assessment. What's wrong with that?

Designing effective and efficient rubrics can be a difficult and frustrating activity for faculty not familiar with such assessment tools. It is essential that training and support are provided by experienced educational development faculty well versed in rubric design. In my experience, educational development faculty working collaboratively with school-based faculty has proved most productive in terms of acceptance and ownership of the scoring systems developed.

There are many established texts on how to construct various rubrics (e.g., Butler and McMunn 2006; Stiggins et al. 2006). Some of the more salient considerations are summarized in Table 8.3.

In scoring student performance, it is often the case that some students do not nicely 'fit' all the behavioural indicators in any one description of performance (e.g., they may fit most indicators quite well but are better or worse on the others). In this situation, it is practical to choose the description that you feel is the most appropriate in terms of the score to be given for that performance area This can be moderated and/or adjusted holistically at the end of the assessment process for the task (especially in borderline cases). My experience, as stated earlier, is that there may be

Table 8.3 Key considerations in rubric design

Rubric design		
Identifying and writing criteria	**Writing descriptions of performance**	**Guide to scoring performance**
• Criteria identify the most important constructs/elements of the performance being assessed • Criteria are clearly aligned to the learning outcomes for the performance area • Criteria are explicitly stated and measurable based on the evidence that can be generated by the assessment method(s) employed	• Descriptors are clear and concise • Descriptors use language that is familiar and understandable by assessors and students being assessed • Descriptors provide accurate descriptions of the performance at the designated level • Qualitative terms (e.g., many, some, few) are clarified and understood by assessors and students	• Performance areas and criteria are differentiated in the weighting of marks allocated where appropriate (e.g., importance, complexity) • The scoring system makes clear how rubric scores are translated into grades

limited assessment value in highly detailed marking systems that are burdensome in practice, as faculty won't have (or make) the time for what they see as an unproductive activity—they have enough to do already. For purposes of summative assessment, a holistic rubric format usually works well, when used thoughtfully. Table 8.4 shows a typical rubric design that has been used in a range of contexts for scoring performance in Oral Presentation Skills.

This is a standard rubric design in which a performance area (in this case Oral Presentation Skills) is broken down into key behavioural indicators relating to the

Table 8.4 Rubric template for oral presentation skills

Scoring rubric for oral presentation skills
The scoring rubric provides descriptions of five levels of student performance relating to Oral Presentation (where a score of 5 represents very good performance and a score of 1 represents very poor performance)
The rubric is underpinned by specific behavioural indicators of oral presentation, these are:
• Clarity of voice, tone and modularity
• Appropriateness of presentation structure and style to specific audience
• Calibration of non-verbal communication to the spoken words (e.g., posture, eye contact and gestures)
• Answering questions in a clear, concise and focused manner
In scoring student performance, it is often the case that some students do not nicely relate to all the behavioural indicators in any one description of performance (e.g., they may fit most indicators quite well but are better or worse on the others). However, choose the description that you feel is the most appropriate in terms of the score to be given for the individual student

Score	Description of performance
5	Voice is consistently clear and effective in terms of tone and modularity Presentation structure and style fully relates to audience Non-verbal communication is highly calibrated to spoken word All questions answered in a clear, concise and focused manner
4	Voice is generally clear and effective in terms of tone and modularity Presentation structure and style mainly relates to audience Non-verbal communication is calibrated to spoken word Most questions answered in a clear, concise and focused manner
3	Voice is occasionally clear and effective in terms of tone and modularity Presentation structure and style relates to audience in part Non-verbal communication is sometimes calibrated to spoken word Some questions answered in a clear, concise and focused manner
2	Voice has limited clarity and effectiveness in terms of tone and modularity Presentation structure and style rarely relates to audience Non-verbal communication is mainly not calibrated to spoken word Few questions answered in a clear, concise and focused manner
1	A very poor performance in this area of competence

curriculum syllabus. These become the basis for the five levels of descriptive performance. Marks can then be allocated to these broadband descriptors in terms of wider assessment weightings and grading formats. Converting rubric scores into grades is more a question of logic than any specific mathematical formula. What is essential is that the marks allocation calibrates to what has been decided in the overall assessment plan for the module or unit of study.

Note: the behavioural descriptors can both reduced or added to, or further refined, depending on the learning outcome focus and level of elaboration required. Also, range statements can be specified to meet the demands of different professional contexts. In teaching, for example, I am reluctant to accredit teaching competence, based only on teaching highly motivated adult learners, unless of course, this is the specific range for a given teacher training programme.

8.6 Summary

While assessment is not an exact science, much is known about good assessment practices in terms of the principles and standards as documented in this chapter.

Hence, we can explore, from a sound evidence base, the underpinning issues in assessment quality and subsequently frame assessment approaches, strategies and methods to address the challenges of assessing twenty-first century competencies. However, as identified, developing assessment approaches and conducting a high-quality assessment of more complex interrelated real-world performance-based tasks for twenty-first century competencies, and making this cost-effective, will require much by way of teacher expertise, leadership and resource provision.

Issues and practices relating to enhancing professional development in teaching, as well as reframing aspects of educational policy and direction, will be explored in the next and last chapter of this work.

References

Banta TW, Jones EA, Black KE (2009) Designing effective assessment: principles and profiles of good practice. Jossey-Bass, San Francisco

Biggs JB (1996) Enhancing teaching through constructive alignment. High Educ 32(3):347–364

Biggs JB (2003) Teaching for quality learning at university. Open University Press/Society for Research into Higher Education, Buckingham

Bloxham S, Boyd P (2007) Developing effective assessment in higher education. Open University Press, Maidenhead

Boud D (ed) (1988) Developing student autonomy in learning. Kogan Page, London

Boud D (1995) Ensuring assessment fosters learning: meeting the challenges of more students with diverse backgrounds. In: Armstrong J, Conrad L (eds) Subject evaluation. Griffith Institute for Higher Education, Queensland

Boud D (2000) Sustainable assessment: rethinking assessment for the learning society. Stud Contin Educ 22(2):151–167

Butler SM, McMunn ND (2006) A teacher's guide to classroom assessment: understanding and using assessment to improve student learning. Jossey-Bass, San Francisco

Eagly AH et al (1991) What is beautiful is good, but…: a meta-analytic review of research on the physical attractiveness stereotype. Psychol Bull 10(1):109–128

Eisner EW (1985) The art of educational evaluation: a personal view. The Falmer Press, Basingstoke

Garreau J (2005) The promise and peril of enhancing our minds, our bodies—and what it means to be human. Doubleday, New York

Gibbs G (2008) How assessment can support or undermine learning. In: TLHE, frontiers in higher education. CDLT, Singapore

Gosling D, Moon J (2003) How to use learning outcomes and assessment criteria, 3rd edn. SEEC, London

Gray P (2007) Student learning assessment. In: Crawley E, Malmqvist J, Ostlund S, Brodeur D (2007) Rethinking engineering education. Springer, New York

Haladyna TM (1997) Writing test items to assess higher order thinking. Allyn & Bacon, Needham Heights

Hattie J (2009) Visible learning. Routledge, New York

Kahneman D (2012) Thinking fast and slow. Penguin Books, London

Knight PT (ed) (1995) Assessment for learning in higher education. Kogan Page, London

Knight PT (2006) The local practices of assessment. Assess Eval High Educ 31(4):435–452

Kurzweil R (2005) The singularity is near: when humans transcend biology. Viking, New York

Lai ER, Viering M (2012) Assessing 21st century skills: integrating research findings. National Council on Measurement in Education, Vancouver, B.C.

Langlois JH et al (2000) Maxims or myths of beauty? A meta-analytic and theoretical review. Psychol Bull 2000(126):390–423

Mazur E (1996) Peer instruction: a user's manual. Pearson, London, U.K.

McTighe J, Wiggins G (2000) Understanding by design. Association for Supervision and Curriculum Development, VA

Montgomery K (2001) Authentic assessment: a guide for elementary teachers. Longman, New York

Nucci IP (2001) Education in the moral domain. Cambridge University Press, Cambridge

Osterlind SJ (1989) Constructing test items: multiple-choice, constructed response, performance and other formats. Kluwer Academic Publishers, Dordrecht, Netherlands

Price M, O'Donovan B (2006) A scholarly approach to solving the feedback dilemma in practice. Assess & Eval High Educ, Volume 41, Issue 6, 2016

Prosser M, Trigwell K (1998) Understanding learning and teaching: the experience in higher education. Open University Press, Buckingham

Ramsden P (1992) Learning to teach in higher education. Routledge, London

Rowntree D (1987) Assessing students: how shall we know them. Harper Row, London

Sale D (2014) The challenge of reframing engineering education. Springer, New York

Sibley J, Ostafichuk P (2014) Getting started with team-based learning. Stylus Publishing, USA

Stiggins R et al (2006) Classroom assessment for student learning. Pearson, New Jersey

Struyven K, Dochy F, Janssens S (2002) Students' perceptions about assessment in higher education: a review. Paper presented at the Joint Northumbria/EARLI SG assessment and evaluation conference: learning communities and assessment cultures, University of Northumbria, 28–30 Aug

The Towards Defining 21st Century Competencies. The Foundation Document for Discussion, Ontario (2016)

Zebrowitz LA (1997) Reading faces: window to the soul?. Westview Press, Boulder, Colo

Zebrowitz LA, Rhodes G (2004) Sensitivity to "bad genes" and the anomalous face overgeneralization effect: cue validity, cue utilization, and accuracy in judging intelligence and health. J Nonverbal Behav 28(3):167–185

Chapter 9
Framing Professional Development Now

Abstract This chapter offers an evidence-based approach to framing and implementing a professional development approach to address both issues of perceived usefulness by teaching professionals themselves, as well as viability in the real world of educational institutions. While we know what knowledge, strategies and methods are most likely to enhance teaching expertise, and how to conduct the professional development process to achieve positive outcomes, good professional learning comes at a cost in terms of time and resources. Hence, both individuals and institutions have to decide what is most viable in their personal and professional contexts.

9.1 Introduction

In the preceding chapters, I framed the context and challenges that many teachers are increasingly facing now, and these will certainly dominate the educational landscape for the next decade or so. Notions of accurately predicting teaching challenges and context for the whole of the twenty-first century are highly speculative and probably erroneous.

Debates about human learning are still prevalent in terms of framing cohesive educational programmes for teacher professional development. Darling-Hammond and Rothman (2015) captured the problem clearly:

> While educators and policymakers agree that enabling teachers to improve student learning is one of the most significant ways to raise student achievement, there are heated disagreements about the most useful ways to do this. (p. 1)

Similarly, Reimers and Chung (2016) argue that:

> …teacher education programmes and educational leadership preparation programmes in many of the world's developed and emerging economies are not only based on theories of the past but are delivered in outcome ways such as rote classroom instruction. (p. 12)

Using an EBT approach I have argued that *Pedagogic Literacy* (PL), *Metacognitive Capability* (MC) and *Creative Teaching Competence* (CTC) are three major

© Springer Nature Singapore Pte Ltd. 2020
D. Sale, *Creative Teachers*, Cognitive Science and Technology,
https://doi.org/10.1007/978-981-15-3469-0_9

areas for framing teacher expertise and addressing the challenges of facilitating Self-Directed Learning (SDL)—both for students and for teachers. Certainly, teachers high in MC will have the capability to be self-directed in their professional development as they will be effective evaluators of their ability to self-regulate—it goes with the territory so to speak. As Costa and Garmston (2016) made explicit:

> Self-directed evaluators use structures and skills that engage the professional's thinking about his/her performance. In self-directed evaluation, intentionally builds the capacity of the professional for self-management, self-monitoring and self-modifying behaviors. (p. 82)

Furthermore, and equally important, they will also have the necessary competence to teach these skills to students. As Powell and Kusuma-Powell (2015) point out:

> Teachers who demonstrate self-directedness are much more likely to demonstrate emotionally intelligent classroom behaviour. They look for cause and effect relationships between their teaching and their students' learning. Their emotional intelligence contributes to the construction of powerful learning relationships with students as well as colleagues. Self-regulation is all about controlling our impulses, particularly disruptive ones, delaying gratification, thinking before acting, and suspending judgement. (p. 41)

9.2 Organizational Intelligence and Professional Capital: Two Desirable Bedfellows

MC provides the executive functions for how teachers conduct themselves as professionals and behave more intelligently. Collectively, this has the potential for developing what Powell and Kusuma-Powell (2013) refer to as *Organization Intelligence*:

> ...the emergence of understood and agreed patterns of effective interaction. (p. 22)

Schools possessing a high level of organizational intelligence—'intelligent schools'—typically have the following key defining characteristics:

> Teachers in intelligent schools are enthusiastic consumers of new knowledge. School leaders and faculty attempt to stay abreast of recent research and developments in the field. Members attend conferences, present workshops, read and discuss articles, and write for professional publications. Teachers are keen to discover what other colleagues may be engaged in as 'works in progress' and are eager to share and critique new ideas. In short, intelligent schools are inhabited by teachers who are learning and growing. (p. 24)

In many ways, organizational intelligence is very much related to the psychological climate of the school and the quality of the relationships formed across faculty members which, when highly positive, will result in not just increases in individuals' intelligence but the 'collective intelligence' of all participating. Quite simply, a positive psychological climate is equally good for both teacher and student learning as it relates to human needs and preferred norms of conduct. As Powell and Kusuma-Powell documented:

Groups that have high collective intelligence are more innovative, more likely to find creative solutions to problems, more likely to engage in reflection and therefore more likely to transfer their learning to new and novel situations. (p. 15)

However, according to the authors, high levels of organizational intelligence are not that prevalent in educational institutions. They argue:

…knowledge management in schools remains to a large extent in the Stone Age…

Schools are organizations that specialize in learning and, as such, should be very smart organizations. Frequently, however, they're not. (p. 16)

Furthermore, as we explored in some detail, these skills are extrinsically linked in that critical and creative thinking are part of Good Thinking, in which metacognition is the executive function. Equally, good thinking and the ability to self-regulate are highly linked to effective collaboration and communication, as epitomized by the old sayings (I do like these), 'think before you talk', 'think before you act'. We may say that these competencies and their underpinning skill sets and knowledge bases, in conjunction with core values (as explored prior) do provide the elements for facilitating our best educational framing for the twenty-first century at this point in time. It is then a question of having the teaching force who can do this, and the political and managerial will, policy and commitment to support them in achieving success. Reimers and Chung (2016) are correct in making what should be an obvious assertion:

…the ability of leadership to support the development of student's twenty-first-century competencies is one of the key levers to improving student learning. (p. 13)

However, they go on to argue, what seems very worrying, that:

the innovation gap in education leadership preparation is dire, and that a knowledge gap hinders educational practice and policy worldwide, as no trusted source exists of which leadership approaches are most effective. (p. 13)

Teaching professionals with high MC and CTC, as well as leadership personnel that foster organizational Intelligence, offers the possibility of developing what Hargreaves and Fullan (2012) refer to as 'Professional Capital':

… the confluence of three other kinds of capital: human, social, and decisional. It is the presence and product of these three forms of capital that is essential for transforming the teaching profession into a force for the common good. (p. 88)

Human capital in teaching is specifically about the framing and possession of the requisite knowledge and skills for the job at hand (in this case teaching). This includes knowing your subject, how to teach it, understanding students and how they learn—as well as being able to apply such knowledge effectively and efficiently in the contexts of practice. At the individual level, a high capability in human capital may constitute expertise. However, the problem here is that this is only at an individual level, and such capability may not be shared with others in the institution—not even known by others.

Social capital refers to the shared cultural aspects of human conduct and communication in the institution. The authors refer to it in terms of:

> how the quantity and quality of interactions and social relations among people affects their access to knowledge and information; their sense of expectation, obligation and trust; and how far they are likely to adhere to the same norms or codes of behaviour…Social capital increases your knowledge – it gives you access to other people's *human capital*. (p. 90)

Decisional capital can be seen in congruence to organizational intelligence as it comes about through the communication and collaboration of colleagues who are sharing their human capital, analysing and evaluating it, making refinements, modifications and improvements, which enhances the competence of all participants, especially in making more confident decisions about practice. Indeed, as Costa and Garmston (2016) hypothesize, "the basic teaching skill is decision making" (p. 147). Hargreaves and Fullan (2012) frame it in terms of:

> …the capital that professionals acquire and accumulate through structured and unstructured experience, practice and reflection—capital that enables them to make wise judgements in circumstances where there is no fixed rule or piece of incontrovertible evidence to guide them. Decisional capital is enhanced by drawing on the insights and experiences of colleagues in forming judgements over many occasions.

Invariable, these kinds of capital are mutually supporting and synergistic in that as more and more faculty work collaboratively in the development of knowledge and expertise in teaching, there are benefits to all participants, and this will have positive impacts on other stakeholders, especially the student learners.

As the authors sum up:

> collaborative cultures build social capital and therefore also *professional capital* in a school community. (p. 115)

9.3   The Pedagogy—Andragogy Debate

Firstly, here's some context on what is still an area of contention in terms of human learning, epitomised in a question I am frequently asked: "Do adults learn differently from children, and do we need to teach them differently? In Edu-speak this is sometimes referred to as the 'Pedagogy—Andragogy' debate.

Pedagogy, as identified in Chap. 2, has been contrasted with the term *Andragogy* (Knowles 1984), which focuses on the teaching of adult learners. However, knowledge from cognitive science generally throws doubt on the notion that the underlying learning processes of adults are structurally different from that of children who have attained the stage of formal operational thought (Piaget 2001), typically around 12–15 years of age. At this stage of brain maturation, children can reason logically and use a range of thinking skills (e.g., analyzing, comparing & contrasting, making inferences and interpretations, and evaluating). There is, however, evidence from cognitive neuroscience that the prefrontal cortex which, apart from other things, is responsible for the control of our impulses, complex actions, planning and organization, doesn't fully mature until a person is in his or her twenties (e.g., Swaab 2015).

This means that mature adults are likely to be more organized and determined in their learning efforts (e.g., higher MC) than younger learners who will need more structure and support in these aspects of learning. This does not limit the usefulness of developing MC in younger learners, but maybe not to expect self-directedness to be fully achieved by all students on high school graduation day. There is much to being self-directed, as Powell and Kusuma-Powell's (2015) frame implies:

> Self-direction is choosing one's own way; increasingly becoming the author of who we are coming to be; constructing an internal sense of personal meaningfulness that is founded on examined beliefs and values. Self-direction is the process by which we continually become. We perceive identity not so much as a static entity, but as a continual process of becoming. (p. 32)

Furthermore, in teaching adults, there are certainly important areas of focus that need to be attended to in qualitatively different ways. For example, there are significant differences in the level of prior experience of adults, as compared to children; the former usually having more. Adults also choose what they want to learn, and this is typically consciously directed to meet work or personal learning goals. In contrast, pupils in a school are largely told what to learn, at least in the earlier years. Knowles (1984) saw adult learners, as compared to younger learners, having the following qualitative differences:

1. Concept of the learner. Adults need to be self-directing and be treated by others as able to take responsibility. They resent others imposing their will and desire participation in decisions that affect them
2. Role of the learner's experience. Adults have a wide experience which serves to determine who they are, to create their sense of self-identity. When this experience is devalued or ignored by the teacher, this implies a rejection of the person, not just the experience
3. Readiness to learn. Adults become ready to learn when they experience a need to know or do something as a result of changes in their lives
4. Orientation to learning. Adults enter an educational activity with a life task or problem-centred orientation to learning. They are less bothered by underpinning knowledge or theory
5. Motivation to learn. Adults are motivated more by internal rather than external agents (e.g., self-esteem, recognition by peers, a better quality of life, self-actualization).

Based on these assumptions, it can be argued that there are certain key implications for teaching adults:

- Adults want more involvement in the planning/decision-making and facilitation of their learning. This may involve more flexibility in the learning arrangements to accommodate the various demands of their life contexts.
- Adults are more interested in/focused on seeing relevance/purpose in what they are learning (e.g., how it relates to a felt need or problem they are trying to address in their professional and/or personal life).

- Adults expect their adult status to be recognized and this has important human conduct issues of equity and communication style in the process of teaching and management of the learning environment.

These features of the Adult Learner, in contrast to younger learners, has been aptly captured by Rogers (1998):

> An instructor of adults is quite unlike a teacher of children or adolescents.

The person is an adult among adults. He or she cannot count on the customary advantages of age, experience, and size … Many adults will have had experiences that far surpass the background of the particular instructor. As a group, they have out-travelled, out-parented, out-worked, and out-lived any of us as individual instructors. Collectively, they have had more lovers, changed more jobs, survived more accidents, moved more households, faced more debts, achieved more successes, and overcome more failures. It is highly unlikely that we can simply impress them with our title, whether it be a trainer or professor.

> The dominant question and request of adult learners to anyone who instructs them is, "Can you help me?" (p. 27)

However, it could be argued that many of these implications apply, at least in part, to younger learners. It is unlikely than many of our students like to learn content that they see no purpose or relevance in and would resent being denied choice and some equity in human conduct issues. Similarly, the notion that adult learners prefer to learn more independently and self- discover has also been challenged by recent research. For example, Dickinson (2015) noted:

> I find that time-pressed adult workers often just want someone to tell them what to do. They have neither the time nor inclination to explore. (p. 157)

It is important therefore that teachers involved in adult education and training are aware of these different orientations and apply EBT to the adult context. Teaching adult learners requires much by way of contextualization, but the core principles of learning still apply for their learning. For example, having taught adults from many countries and cultural contexts for over 30 years, the differences in terms of motivation and orientation identified above often apply, but I am more apt to agree with Dickinson's observation, and this is important. Adult learners, and increasingly so, are doing busy workloads and often come to class tired, and their participation may be more extrinsic than intrinsic (e.g., keeping their jobs, promotion, being told they need this new accreditation as its now mandatory). Asking adult learners in the classroom environment to spend their time doing online searching on a specific topic, for purposes of later sharing and discussion may not go down well with many. Adults do like to share and learn from each other, but they could do this online research at home or in a coffee shop; they don't need to rush to class for such experience. Also, they expect, and rightly so, that you provide then with relevant new knowledge, skill applications, and facilitate their thinking in ways to build the level of understanding necessary to extend their work-related competencies. Yes, they do like spoon-feeding of this key information, but also expect skilful facilitation in making it useful.

The following sections will explore what this means for teacher professional development, both at individual and collective levels, in terms of developing Creative Teaching Competence and that superordinate competency—Metacognitive Capability. The good news is that we are becoming increasingly aware of how best to achieve this. Invariably, like the development of expertise in any domain or field, it involves much learning, persistence and hard work—but what of real value does not?

9.4 Professional Development that Does Not Work Well

There can be little argument that professional development must be central to enhancing professional practice, and much is made of the need for lifelong learning in a world of exponential knowledge production and rapidly changing occupational structures. Unfortunately, professional development in enhancing teaching quality has typically been tempered by the strong, mostly anecdotal, evidence that much professional development is not effective in terms of improving teacher practice and, most significantly, in enhancing student attainment (Timperley et al. 2008). I recall seeing a quote many years ago that is not inconsistent with the perception of many teachers that I have worked with regarding the usefulness of attending professional development workshops:

> When I die, I hope it's in a professional development workshop, as the difference will be hardly noticeable (a teachers comment, anon).

More specifically, Darling-Hammond et al. (2009), from an extensive research base, found that while 90% of teachers participated in professional development, most of those teachers also reported that it was useless. The report shows that:

> …in education, professional learning in its current state is poorly conceived and deeply flawed. Teachers lack time and opportunities to view each other's classrooms, learn from mentors, and work collaboratively. The support and training they receive are episodic, myopic, and often meaningless. (p. 2)

Research Alert (2014) posed the question:

> What form of professional learning has the most abysmal record for changing teachers' practice and student achievement? Clue: it's the kind that 90 percent of teachers normally engage in at school. If you answered, "the workshop-style training session," you're right. Despite its ineffectiveness, it still soldiers on. (p. 8)

I am guilty here, but fortunately, there will be many joining me should there be a retrospective inquisition and such practices become indictable offences. I still conduct professional development workshops in many aspects of curriculum, teaching and assessment, but I am honest with participants about what is realistically possible in such learning events. Sadly, there are still some who think that competence in a complex skill can be a realistic outcome from a short workshop involving explanation and a little 'hands-on' simulated practice. Similarly, having spent many years working out in gyms, I occasionally find it amusing when people express such disappointment that they have not achieved significant weight loss or enhancement in muscle bulk or fitness when they have only spent a few hours over a couple of weeks in the gym (and often doing very little in terms of actual exercise). There are evidence-based reasons for both of these scenarios, as we know. While I have often received positive feedback on many of my workshop programmes (and I am thankful for this), I remain reticent in believing that I have significantly influenced teaching practices to the extent that this has led to significant gains in student attainment. Over the years I have received feedback from some participants, post-workshop, claiming that they had changed aspects of practice, and this has transferred to better student learning and attainment (e.g., better student feedback relating to their teaching and improved student performance or grades). However, these are not the majority, and it seems that such individuals are doing something else, which is not the typical behaviour of participants who attend workshops without other supportive follow-up arrangements. We will explore what this is and what the implications are later in the chapter.

It seems that the most prevalent characteristic of educative professional development is that it is often met with resistance (Duffy 1993). According to Duffy et al. (2010), there are two reasons for this. First, educative professional development usually emphasizes teacher thoughtfulness, and teachers often resist being thoughtful. They do so because it is easier to operate from routines, and because their "apprenticeship of observation" (Lorte 1975) as students themselves for 13 years causes them to think that they already know how to teach (Kanfer and Kanfer 1991; Kennedy 1999). The result is often what Windshitl (2002) called "additive" change or what Huberman (1990) called "tinkering", in which teachers insert minor changes into their existing practices. Second, research indicates that learning to be thoughtful occurs in erratic spurts and not as steady growth (p. 9). Certainly, the issue of the 'lazy brain' (e.g., Kahneman 2012) and thinking being something we don't like to do too often (e.g., Willingham 2009) feature here also, but these are not peculiar to teachers per se. However, there are specific contextualized aspects to teachers in educational institutions that are both causes and the manifestation of much educational thinking (or lack of it) relating to what is good teaching and, by association, to what is good professional development. Powell and Kusuma-Powell (2015) argue that this is part of a wider systemic problem, arguing that:

> The field of education is changing with lightening-speed; schools are changing at a snail's pace. (p. 18)

They quote Caine and Caine (2001) as capturing the irony of the present situation:

> Unfortunately, many countries and cultures are employing a late 20th-century political process in an attempt to perfect an early 20th century model of schools, based on 17th-century beliefs about how people learn, to prepare children for the 21st century. (p. iv)

This may explain, at least in part, that while we are amassing strong evidence concerning what practices do work well and on what basis, it is not being widely synthesized and practised. From my experiences of working in this field for over 30 years, I would tend to agree with the findings of Timperley et al. (2008) relating to two extremes, but often used approaches to professional development:

- The first is that teachers should be treated as self-regulating professionals who, if given sufficient time and resources, can construct their own learning experiences and develop a more effective reality for their students through their collective expertise. Unfortunately, we found little evidence to support the claim that providing teachers with time and resources is effective in promoting professional learning in ways that have positive outcomes for students. (xxv)
- The alternative extreme is where outside experts develop recipes for teaching (typically based on research about what works for students) then present prescribed practices to teachers with an underpinning rationale and monitor their implementation carefully to ensure integrity. The overall evidence is that these processes can be effective in changing teaching practices, but either the changes have limited impact on student outcomes or they are not sustained once the providers withdraw. (xxvi)

From an evidence-based approach, the above scenarios may seem to suggest a parallel Educational Jurassic Park in terms of professional development to that of teaching practices. Indeed, we would expect a lack of agreement and focus on what constitutes highly effective teaching to reflect, as well as reinforce, the existing psychological and educational paradigms relating to teaching and learning. As noted previously, these have been both competing and confusing, and have done little to advance teaching as a profession guided by a strong evidence base relating to practice. Teachers are products of their biography and socialization experiences, and this applies to teacher educators also. Furthermore, once a paradigm has established itself, and members benefit from its prominence, there is little motivation to seek evidence to dispute key tenets or challenge legitimacy, especially when careers and funding opportunities are at stake.

However, apart from ideological and self-interest group influences on shaping paradigms relating to practice, other more concrete practical situated factors have contributed to teachers having negative perceptions and experiences concerning the value and usefulness of much so-called professional development activity. Firstly, given the lack of a clear evidence-based professional knowledge and practice framework, the frequent reframing of what is good teaching, and increasing workloads, it is not surprising that many teachers are reluctant to invest highly in such activities. Furthermore, teaching is somewhat unique in that career progression typically entails giving up teaching and taking on more administrative and managerial work, making highly effective and creative teaching more an act of personal choice rather than a route to genuine professional enhancement as a teacher.

Secondly, the mainstream approaches to professional development, which have proved relatively unsuccessful, are not consistent with what we know about effective learning of complex skills such as instructional strategies. Being introduced to new knowledge, even with clear explanations, opportunities to ask questions, and some simulated practice, is only effective as an advance organizer. It may provide the teacher, as a learner, with an interesting new learning opportunity and there may be intentions to try this out further down the line in classroom practice. However, several inhibitors to effective learning can typically come into play. The most basic is forgetting. Once a workshop has finished and a few days have passed by, much of the information may be lost, and this in itself may lead to it not ever being tried out. Also, preparing a new instructional strategy will most likely make some extra demands on time. Again, there may be an intention to use the new strategy, but in a busy schedule it's always easier to go with what you have done prior and it may have worked ok anyway. However, even if the strategy is tried out with good intent fairly soon after the workshop, there is every likelihood that it may not work out particularly well, and may even feel counter-productive. This is the learning process at the level of skill acquisition.

A few decades ago, I decided to learn to play the guitar as this was something I wanted to do at school but did not have the opportunity. My music lessons at school, as mentioned in Chap. 1, were far from motivating. In retrospect, the teacher probably violated every core principle of learning. I had no idea of what we were supposed to learn, saw no relevance in any of it and was occasionally caned for transgressions in reciting the musical scales. Despite my loathing of the weekly music lesson, I remained interested in playing the guitar. Sometimes in life, serendipity provides an excellent opportunity to learn something and this was the case in my mid-twenties. I met a young lady who played the guitar and sang in local venues where I lived. She was also studying for an Open University degree in psychology and asked me if I would review some of her assignments before formal submission. Jokingly I said something like, "Sure, but you will have to teach me how to play the guitar in return." This was partly in jest, but only partly. She readily agreed and the arrangement worked well. However, I soon learned why most people who are initially motivated to play the guitar typically give up within a few weeks. Learning the 'C', 'G' & 'F' chords is easy conceptually, but not at the level of skilful practice. Making a noise like a cat encased in an aluminium dustbin rolling down a steep hill, and having very sore fingers on one's left hand (I am right-handed), as well as housemates smiling quizzically as they walk past is hardly encouraging. Quite simply, knowing how to do something is far from being able to do it, when skills are involved.

In the professional development of teachers, the same scenario plays out. The real challenge is when teachers apply newly introduced instructional strategies into their classroom practise at the skill level, as they are unlikely to work on the first time, or the next time, and perhaps for many times. In this context, many teachers will give up, and this makes perfect sense. Apart from time constraints, some teachers may also be concerned about negative feedback from students in such situations, as students may experience the changes as confusing and not see the value in terms of better-quality learning at this point. Solving the gap between knowledge acquisition and understanding to one of eventual expertise, that enables transfer to other related learning contexts, is a challenging scenario for professional development. Hence, the big question, given the constraints of teacher's professional lives, is *how best to do this*?

9.5 Unpacking the Components of Effective Professional Development

An interesting anchor point in understanding what approaches and methodology of professional development works best and how stems from the extensive work of Timperley et al. (2008). The authors summarize the purpose of their work in terms of:

> …to unpack the 'black box' between the professional learning opportunities and teacher outcomes that impact positively on student outcomes. (p. 7)

In essence, there are two interrelated black boxes, as depicted in Fig. 9.1. The first concerns the necessary changes in teachers' practice, but this must be of such a nature that it brings about specific changes in student perception and behaviour related to better learning.

The approach to the professional development of creative teaching professionals outlined in this chapter is consistent with the evidence-based approach developed in the preceding chapters. The core principles of learning are equally relevant to teacher learning, as they are for student learning. While teachers, as adult learners, may have certain advantages over younger learners in that they are likely to be more focused and discerning in terms of identifying their learning goals and having greater expertise to plan, monitor and evaluate their learning (e.g., greater MC), they are subject to the same cognitive constraints of memory processing and cognitive overload as their younger counterparts.

The notion of applying the same principles of learning for students to that of teacher learning, invariably contextualized to the different tasks and contexts, is fully supported by Powell and Kusuma-Powell (2015) who argue:

> The answer to the quandary of improving learning for students lies squarely in improving learning for teachers. As a result, teachers have a sacred obligation to become architects of their own, on-going professional growth. (p. 18)

> This is what we claim to want for students: independent critical thinkers who are enthusiastic life-long learners with the capacity for healthy and accurate self-assessment and

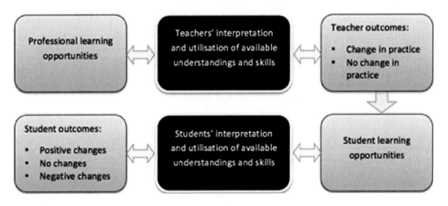

Fig. 9.1 The 'Black Boxes' of teacher professional development

self-modification. If these are desirable outcomes for students, why would we not want then for teachers as well? (p. 22)

In summary here, it is now apparent that there are many evidence-based principles and practices clearly aligned to improving teacher learning and expertise and, therefore, enhancing student learning opportunities and outcomes. This sounds good news in terms of being able to establish a solid professional development approach for enhancing teaching competence and expertise. However, the kinds of professional development activities that are effective require considerable resource time and effort, which may not be congruent with present practice in many educational institutions. For example, Gulamhussein (2013) emphasized:

The duration of professional development must be significant and ongoing to allow time for teachers to learn a new strategy and grapple with the implementation problem. (p. 3)

Similarly, Joyce and Showers (2002), from extensive research, suggest that staff development focused on student achievement must include the following essential elements:

- A community of professionals put into practice what they are learning, and share the results
- The content of staff development develops around curricular and instructional strategies selected because they have a high probability of affecting student learning—and, as important, student's ability to learn
- The magnitude of change generated is sufficient that the student's gain in knowledge and skills is palpable
- The processes of staff development enable educators to develop the skill to implement what they are learning. (p. 4)

However, as Levin (2008) noted:

To get good at a new practice takes time and effort in schools – whether as a teacher or principal or superintendent – no tradition or organization supports carefully supervised learning of this kind. (pp. 80–81)

There are no short-cuts to achieving expertise and piecemeal professional development will do little to produce the kind of teaching professionals—Expert creative teachers—needed in the present educational landscape, whether in schools or elsewhere. Furthermore, there is much that needs changing in many educational institutions in terms of relationships and structures to foster organizational intelligence and the development of heightened professional learning as framed above.

9.5.1 The Goals and Content of Professional Learning

Core Principle 2: *Learning goals, objectives and proficiency expectations are clearly visible to learners* is as applicable to teacher professional learning and development as it is for the students we teach. Timperley (2008) posed the core question:

> What do teachers need to know in order to deepen their professional understandings (e.g. pedagogical content knowledge) and extend their skills so as to have a positive impact on student outcomes? (p. 27)

I think much of the answer lies in an EBT framing of teacher expertise, which provides the underpinning Pedagogic Literacy as well as useful heuristics for guiding effective professional development. Of course, we could argue over terminology, emphasis, and valuations are always embedded. Education is essentially an experiment; one could argue, so is life. Hence, in relation to expert teaching, we can and need to make certain contextualizing statements about what Powell and Kusuma Powell (2015) refer to as "goodness" in the profession of teaching. They assert:

> As a profession, we need to work on the search for goodness. As a general rule, teachers are not skilled at deconstructing and analysing exemplary teaching and learning. And as we have seen researchers and politicians focused primarily on what isn't working in education. (p. 8)

Of interest, in this context of focusing on what's not working in education, Pentland (2014), director of MIT's Human Dynamic Laboratory, approaches the topic of effective social learning through the analysis of Big Data and through so-called Reality Mining. He points out:

> Mathematical models of learning in complex environments suggests that the best strategy for learning is to spend 90% of our efforts on exploration, i.e., finding and *copying others who appear to be doing well*. The remaining 10% should be spent on individual experimentation and thinking things through. (p. 54)

Over the chapters, I have sought to establish much by way of what constitutes 'goodness criteria' relating to teaching and when fully articulated this may underpin much of Expert Teaching now—so to speak. For example, it would be good if teachers had high MC—that 'many-headed monster of obscure parentage' and CTC as organizing superordinate competencies. The functional and generic competencies identified and discussed during the previous chapters would also be incorporated within this 'goodness framework'. Also, considerations of character traits, dispositions and values, are important here. I think it's good to have teachers with certain broad trait configurations as identified prior, have values systems that include integrity and fairness, and want to help people to learn and experience well-being.

The Gallup organization (2014), quoted in Powell and Kusama-Powell (2015), studied the characteristics of exceptional teachers for over 40 years and have identified three common attributes. Exceptionally effective teachers demonstrate:

1. *Internal achievement motivation.* These are teachers who are driven to reach higher levels of mastery and learning. They enjoy setting challenging goals for themselves, monitoring their progress and taking ownership of student achievement.
2. *Orchestration of classroom structure and flexibility.* These are teachers who balance innovation with discipline. They are structured and deliberately organized without sacrificing creativity and playfulness. They are risk-takers who view failure as an opportunity to learn. These teachers are constantly thinking about new ways to present content and to engage students in learning and discovery.

3. *Strong relationships with students, colleagues and parents.* Highly effective
   teachers understand that deep and meaningful learning takes place in a social
   setting characterized by respect and trust. These teachers deliberately set out to
   build strong learning relationships with their students and colleagues. They do
   so by supporting others to feel more efficacious and empowered as learners both
   independently and as a member of a community. (pp. 24–25)

Similarly, studies of expert teachers, while not focusing specifically on metacognition, often provide characterizations closely aligned with metacognition. Berliner
(2004), for example, offered the following description of expert teachers:

> Expert teachers are more sensitive to the task demands and social situation when solving
> pedagogical problems; expert teachers are more optimistic and flexible in their teaching than
> novices; expert teachers represent problems in qualitatively different ways than do novices;
> expert teachers have fast and accurate pattern recognition capabilities; whereas novices
> cannot always make sense of what they experience; expert teachers perceive meaningful
> patterns in the domain in which they are experienced; and although expert teachers may
> begin to solve problems slower, they bring richer and more personal sources of information
> to bear on the problem that they are trying to solve. (p. 201)

Duffy et al. (2010) suggest that observations of exemplary teachers strongly suggest that effective teachers regulate and control their thinking as they teach. However, and worryingly, they point out that while it is assumed that teachers can learn
to be metacognitive and that both cognitive and dispositional aspects of professional
development can be designed to intentionally encourage teachers to be metacognitive
professionals, the fact is that:

> current efforts are based primarily in common sense and theories rooted in constructivism
> and sociocultural thinking, and no empirical data substantiate that one or other kind of
> professional development results in teachers who are metacognitive. (p. 10)

The professional development of teaching professionals must incorporate *Pedagogic Literacy* as a primary knowledge base, and this is becoming increasingly
diverse as new knowledge relating to human learning is probably increasing exponentially. Invariably, for teaching, as in all professions, there is so much content that
may have relevance and usefulness to enhancing competency, expertise and creativity, there needs to be careful selection and prioritization of what is most relevant and
useful to the task in hand. Timperley et al. (2008) highlighted this fully:

> The world is teeming with activities variously referred to as knowledge utilization, knowl-
> edge dissemination, knowledge brokering, knowledge transfer, knowledge exchange, knowl-
> edge mobilisation, and knowledge translation. Whatever the term used, the idea is to gather
> together what the research tells us about a topic of interest and then to synthesise it into
> practical usable knowledge. (ix)

What this means is that it is not just knowledge per se, but a deep understanding
of how these knowledge bases connect to the world of practice and what pedagogic
thinking and actions are necessary for situations to create and facilitate learning
experiences that result in better learner outcomes. Hence, there must be a high level
of teacher engagement in terms of good thinking, strategic application and ongoing

collaborative evaluation of teaching for the development of deep understanding (e.g., rich integrated mental schemata in long-term memory). As Joyce and Showers (2002) made clear:

> Understanding of the theory underlying specific behaviours enables flexible and appropriate use of the behaviours in multiple situations and prevents the often-ludicrous following of "recipes" for teaching...
>
> Teachers who master the theory underlying new behaviours will implement those behaviours in greater congruence with the researched and tested ideal and are more likely to replicate results obtained in research settings with their students. (p. 81)

Hence, the importance of a sound Pedagogic Literacy in providing the theory that underpins the understanding of such behaviours. In terms of enhancing creative teaching competence, a useful comparison has been made by Gulamhussein's (2013) differentiation of teachers in terms of 'Teacher as Intellectual' as compared to 'Teacher as Technician'. The latter denotes the key functional competencies outlined prior and, at best, a 'routine expert' (in the language of Hatano and Inagaki 1986). The notion of Teacher as Intellectual seems to possess similar attributes and capabilities to a creative teacher ('adaptive expert') as framed in the context of this book. Gulamhussein suggests that this requires:

> ...time and resources which allows teachers to think through and create innovative teaching methods. (p. 23)

A crucial component in this process is practice, but not ad hoc practice. As we saw practice is most effective when it is both spaced and deliberate, which was explained in detail in Chap. 2. The same principles and procedures are appropriate in the context of the teacher's professional development. As Willingham (2009) argued:

> Teaching, like any complex skill, must be practised to be improved. (p. 147)

Also, as Petty (2015) reminds us:

> Deliberate practice is hard work, four hours a day is as much as anyone can manage. I suggest up to one hour a week for teachers.

It further helps greatly, as Petty documented (2015), if the teachers themselves respond in positive ways to the professional development experience. He specifically highlights that teachers can proactively help the learning process if they:

- Practice the use of the new methods repeatedly in a relatively short period, say five times a month
- Monitor the effects of the new methods on the learners—Did they learn? Could they cope? What did they find most difficult? ... What would help them cope better?
- Ask students for their support during these experiments, for example, ask for their opinions of the methods, and their suggestions
- Bring issues and difficulties to their peer coaching team for discussion
- Help and support the experimentation of other teachers in their team.

It is at this stage of the learning process that many institutional professional development programmes typically experience the problems of implementation identified earlier. Competence, let alone expertise, does not come easily in any area of life that requires a high skill component. One may have a good understanding of a performance area and speak very intelligently about it, but that's not competence or expertise. The world is full of expert 'armchair pundits', especially in the world of sport. Most people can watch a sport and speak convincingly on what a team or player should be doing or should stop doing. However, why are they (the actual players) earning millions of dollars a year while the rest of us are paying television fees to watch them? In all situations, early attempts at learning a new skill can be highly disheartening especially when one is doing this publicly, and teaching is very much a public situation.

It is for this reason that teachers need strong support systems if they are to do the necessary professional learning to sufficiently master the range of strategies and skills to eventually reach a stage of *adaptive expertise*. This is particularly the case when teachers are grappling with the actual implementation of new practices as they are in many ways like a novice in any performance situation, often experiencing their performance as erratic and inconsistent, even anxiety-producing. In this context, it is not surprising that many teachers stick within their comfort zone. However, in the longer term, it is no comfort zone.

9.5.2 The Process of Effective Professional Development

There is a substantial evidence base of what works and on what basis. Petty (2015), for example, drawing from the research of Joyce and Showers (2002), suggests, it is essential to first consider what the training needs should focus on. He suggests the following approach:

- Teachers democratically ascertain "their most pressing needs" and pose the question, "What do our results tell us?"
- A set of improvements is drawn up, combined and prioritized until a common goal (e.g., relating to curriculum, teaching and assessment) is identified, so as to raise attainment.

(The common goal needs to focus on a process, designed to produce better outcomes etc. It must affect the *student experience* if it is going to have an effect.)

Similarly, Timperley et al. (2008) use the term 'catalyst' as often being the originator for driving the direction for professional learning. Catalysts can take the form of:

- Information showing that current teaching practices are not achieving the desired outcomes for a group of students
- A lack of shared understanding of an area of practice
- Challenging or 'problematizing' current definitions of specific curriculum outcomes.

On this basis, training can be devised and specifically tailored to achieving a common goal. As Petty (2015) describes:

Training outcomes are agreed for knowledge, skills, and (the hardest part!) transfer to the classroom.

Once, we have established an evidence-based approach and agreed learning goal(s), we can then plan the best strategies for meeting these goals and, of course, monitoring and evaluating them—this is *Metacognitive Capability in action.* He further suggests that the training provided should be extensive (e.g., a dozen days) and conducted using the following pattern:

- **Knowledge**—explaining theory and rationale etc. using lectures, reading, video, etc.
- **Demonstration and modelling**—showing how, giving examples, seeing it done on video or live, watching simulations etc.
- **Practice** this is mainly simulated practice usually.

The importance of sustained engagement over time was extensively confirmed by Timperley et al. (2008) who concluded:

Changing practice in substantive ways is difficult. We have reached this conclusion from evidence of the length of time involved, the depth of pedagogical content and assessment knowledge typically addressed, and the multiple learning opportunities that appear to be required. (xi)

Again, certain core principles of learning are essential here. Time must be spent on the necessary retrieval, spaced and deliberate practice. Good understanding and basic competence will enable motivated teaching professionals to then take a more self-directed approach towards personal expertise. They suggest that the learning processes engaged when developing new understandings and skills involve cycles of (one or more of) the following:

- Process 1: Cueing and retrieving knowledge consolidated and/or examined.
  Outcome: Prior knowledge consolidated and/or examined.
- Process 2: Becoming aware of new information/skills and integrating them into current values and belief system.
  Outcome: New knowledge adopted or adapted.
- Process 3: Creating dissonance with current position (values and beliefs).
  Outcome: Dissonance resolved (accepted/rejected), current values and belief systems repositioned, reconstructed.

A wide range of methods, activities and resources are useful in helping teachers to go through the learning processes outlined above (e.g., Timperley et al. 2008). These include:

- Listening
- Watching
- Being observed and receiving feedback
- Reading

- Discussing practice with an expert
- Discussing own theories of practice and their limitations
- Examining student understandings and outcomes
- Analysis of current practice and reconstruction of new practice
- An authentic experience of the subject in action
- Developing self or mutually identified issues. (p. 29)

All the above activities can be aligned to core principles of learning. They involve enabling teachers to experience new concepts and practices through multiple modes and mediums, supporting the activation of prior learning and dealing with ambiguity and misconceptions, reframing on aspects of belief systems where necessary, and building rich mental schemata to ensure deep understanding through good thinking. A particularly effective technique for introducing a new concept to help teachers understand a new practice is that of 'modelling' (Gulamhussein 2013, pp. 3–4). Modelling reveals what is specifically involved in an area of learning, making the knowledge and skill components visible. This can be further reinforced through the use of video-recordings of teachers demonstrating good practices, encouraging critical discourse on what specifically is working well that positively impacts the learning process, and how best to develop these skill sets. The use of micro-teaching can take the learning process even further by bridging the gap between understanding and skilful application in practice. Microteaching typically involves teachers (usually novices, but not always) conducting short focused lessons to a small group of students (e.g., peers in training) in laboratory-type settings which are often video-recorded for later reference. After the short lesson, there is a collaborate appraisal of the lesson and its specific features. This can be done in a range of formats that usually involve combinations of the following key elements:

- The observed teacher offers his/her perception of the lesson, in term of pluses and areas for improvement, etc.
- Student participants offer their perceptions as above
- Professional development tutors and/or peer coaches offer a summary frame on what was effective and why, and what could be improved and how
- Reference to key video segments are replayed and analysed in the context of areas of practice (e.g., perceived as effective or otherwise) for illustration and reinforcement
- Situated role play may be used by a coach to demonstrate effective use of a method component or skill.

Hattie (2009) documented an effect size of 0.70 for microteaching activities conducted in this broad format. This is not surprising as microteaching encompasses many of the Core Principles of Learning in applied practice. It has also been my experience, in some 30+ years of professional development with teachers/trainers, that the microteaching activities are seen by participants as by far the most useful aspects of teacher training programmes.

9.6 Key Structures Supporting Professional Development that Does Work

As identified, certain key issues and considerations need to be effectively addressed in framing professional development that works. A failure to do this well has led to the kind of apathy that many teachers feel towards ongoing professional development, and this is hardly surprising. Furthermore, effective professional development is more than just the programme itself, but the wider context of learning at an institutional level, and beyond—as identified in the early part of this chapter.

However, there is much that can be done in terms of professional development that can lead to both better teaching capability and, in consequence, better student learning outcomes. The reasons are quite straightforward as, (1) these practices are underpinned by the core principles of learning and, (2) are more likely to be experienced as practically useful by teaching/training professionals themselves. The following are three main approaches or structures that I have used extensively, which can be highly effective in building teacher capability in most areas of pedagogic practice:

1. Reflective Practice
2. Coaching
3. Supported Experiments.

You may also be familiar with terms such as Lesson Study, Active Schemes of Work and Action Research. These are also effective approaches, and I have subsumed them within the above categorization.

9.7 Reflective Practice

Over some 30 years or so I have probably been involved in the appraisal of several thousand teaching/training professionals in a range of educational, vocational and cultural contexts, and many countries. It was once believed that goldfish, and maybe other fish also, have a memory span of only a few seconds. However, new research (not worth referencing) has challenged this and it may be more a question of months. I sometimes feel like this in recalling my teaching activities. We know that memories change in some way every time we evoke them, so I am beginning to worry about the pristineness of some of my recollections. Now that's over, catharsis not-excuses, I feel that I can authentically recall a few useful reflections on Reflective Practice. Many years back, in 1992, I completed a Master of Education degree at the University of Exeter. During this programme, I remember being introduced to, as well as doing considerable amounts of reading and thinking about reflective practice, especially its usefulness as a professional development approach for improving one's teaching. Invariably, I read the definitive texts of the time, especially the work of Schön (1983, 1987). The notion that teachers should carefully reflect on their practice, how they design, conduct and evaluate what they do in classrooms, on what basis, and the actual impact on student learning made perfect sense. After all, this is what we would expect professionals to do in any field of practice. In the literature, there are many definitions on what constitutes reflective practice, for example:

Schön (1983) defined Reflective practice as:

> thoughtfully considering your own experiences in applying knowledge to practice.

More recently, Clouder (2000) framed it broadly as:

> the critical analysis of everyday working practices to improve competence and promote professional development.

To repeat a quote used in a previous chapter, by the famous educationalist Dewey (2016):

> We do not learn from experience... We learn from reflecting on experience.

Even a short analysis and evaluation of these different definitions will inevitably lead one to make certain obvious inferences and interpretations, as common heuristics are underpinning all. Learning to do reflective practice is the same as learning to be competent at any activity, albeit this is a complex activity. One must have the knowledge bases (e.g., what is reflective practice, how it works, be knowledgeable and competent/expert in the area of practice—in this case, teaching), as well as being able to use types of thinking; especially analysis, comparison, making inferences and interpretations, and evaluation. It also helps if one can generate useful possibilities for future improvement if needed.

Schön (1983, 1987), is perhaps the person most accredited to the initial framing of Reflective Practice. Table 9.1, introduces two of his key concepts and what they entail in actual practice.

Schön saw Reflection-on-action as the more powerful of the two methods, as the practitioner can take some time recalling the various activities of the lesson, and what was noticed about the student response (e.g., attention, engagement, task completion, problems encountered). This coupled with other sources of feedback (e.g., student, peer observer) can then be subjected to scrutiny, to identify what seemed to work, and what may need some change in pedagogy for a future lesson in this area.

Table 9.1 Summary of Schön concepts of reflection-in-action & reflection-on-action

Schön on reflective practice	
Reflection-in-action	**Reflection-on-action**
• This takes place when the teacher is immersed in the learning environment, and thinking about what is really going on now • For example, a teacher may be posing his/her- self questions like, How much more discussion should I now allow on this topic area? Should I provide more examples to help students understand the concepts we are applying?	• This takes place after the teaching session, and for him, is the most critical for understanding and improving practice • The teacher may be posing questions such as, Did I manage the discussion effectively in the context of the lesson? How might I improve the level of student engagement—what strategies might I use?

Reflection-in-action can also be very useful as the teacher can make situated changes in the instructional strategy if he/she feels that the present method/activities are not working as planned; hence making possible immediate improvements to the student learning experience. Invariably, this requires a high level of sensory acuity, quick thinking and pedagogic competence to do it consistently well.

In working with teachers in training I have always encouraged them to reflect on their practice. It seemed a given method for self-evaluation and opening-up areas for future improvement. Hence, asking teachers questions relating to what they thought went well in the lesson and what areas were open to future development was standard practice. Over many years of working with different teaching/training professionals, there has been much variation in how they have responded to being asked (though more often requested) to do reflective practice, usually as part of the professional development curriculum they are undertaking. Some seem to find it useful; others seem only to pay lip service to it. Being honest, in retrospect, in the early years, I probably did not do much of it particularly well, and there were reasons for this. Firstly, teachers are busy folk, and finding time for meaningful dialogue was often difficult as they were more mindful of the next task to be done (teaching or otherwise) rather than deep meditation and exploration about what has been done (e.g., the observed lesson), Often, unless there were serious issues, it was typically the case previously that such meetings were more of an administrative activity rather than a rigorous appraisal of different practices, challenges faced and exploration of options for improvement. Furthermore, despite my background in psychology and many years of teaching, I was quite frankly, until the past 20 years or so, pedagogically illiterate in terms of possessing the kind of knowledge and understanding of human learning that is prevalent across the research fields now. Earlier, we had our tour in and out of Educational Jurassic Park, and the emergence of EBT as a means of revolutionizing professionalism. As Powell and Kusuma-Powell (2015) summarized:

> We have learned more about how the human brain learns in the last two decades than in the rest of human history put together. (p. 19)

However, from my prior experience in using reflective practise in teacher education, I did make some interesting observations, inferences and interpretations, that seemed to have a degree of generality and are worth sharing in this context. It was also a bit disturbing, but now I can use this as a resource in a new and better professional praxis. One consistent experience was that teachers, based on my observations, who were showing good skills and attitudes towards students are also the most critical of their performance in the post-lesson tutorials, much more so than those who failed to impress on those counts. The former group are typically able to identify what could be improved and why and, through discussion with them, also seemed to have employed better observational skills and empathy with what the students were experiencing, noticing changes in student's responses to different parts of the instructional strategy. They displayed enthusiasm for teaching, recognized that teaching was a complex and challenging activity, and were keen to receive feedback from me, especially feedback that identified or explored areas for improvement. It was also apparent that these teachers (those that stayed in the profession—sadly many left quite early in their careers) did develop into excellent teaching professionals—many winning teacher awards in their respective disciplines. I don't take much credit for this; they already had good communication skills and showed genuine care and concern for students. I enjoyed my work with these teachers.

In contrast, the latter group of teachers, who seemed to have much less impact in getting attention and engagement, as well as displaying a less organized and contextualized instructional strategy, we're often satisfied with what they were doing and offered relatively little in terms of thoughtful analysis and evaluation. Many examples capture the range of variation within this broad categorization, but one lecturer was particularly striking in an almost archetypal framing of a 'very poor teacher'. In terms of the core principles of learning, he managed to violate most of these in around 2-h of classroom observation. The student group comprised around 60+ students aged 16–17. Having observed this class with other teachers, I had previously noted high levels of attention and engagement, and they seemed keen to learn and fun to teach. In this situation, however, the majority were completely disengaged from any participation in the lesson content. I was able to see most of their laptops, and the screens bore no resemblance to the lesson content. The boredom was tangible, I could almost feel it. Most worrying, however, was the perceptions and interactions of the teacher in the post-observation tutorial. I always invite teachers to offer their perceptions and feelings about the lesson, in the genre of *Cognitive Coaching* (e.g., Costa and Garmston 2002). Much to my surprise, at that time (though I am not surprised now), this teacher felt that the lesson went very well and was pleased with the way he had taught it. When asked what he would change if he taught the same lesson again, he responded very clearly—"nothing in particular". In such circumstances, emotional intelligence must come quickly in play, as emotionally I felt somewhat annoyed at what seemed arrogance in that situation. I asked questions to explore what he thought went well and how he knew this, thinking this may open up an avenue for productive dialogue. This did not help the communication process and I could see he was getting more agitated by the moment. I quickly reassured him that the situation was not a problem, mentioning that this is a learning experience, though we need to have a conversation about the teaching. I did pass him my reflections and rationale on the standard teacher observation form employed at the time and offered to go through this with him now or later. He curtly replied, "not now". In summary, I asked him to read through it, have a think, and we'll go from there.

I never saw that teacher again, and he may have left the institution soon afterwards. Had I failed him in some way or was he just not suited to teaching?—I don't know. There are certainly people (for whatever reasons) who are just not suited for teaching, and for the sake of both the students and themselves, are probably best outside of the profession. In retrospect, I may have done better in terms of support. I am more aware now that such wide variations in both competences, and reflective ability, can be explained in large part by research into the differences between novice and expert teachers (e.g., Timperley et al. 2008) and between experienced and expert teachers (e.g., Hattie 2003). Timperley et al. (2008) make the following comparison between novice and expert teachers:

> The novice is someone who perceives the unfamiliar teaching situation in terms of discrete elements and, in making use of new skills and knowledge, relies on rules rather than an integrated vision of the practice. The primary focus is on the self and one's performance. As competence develops, the discrete elements become integrated into patterns, with some aspects becoming automatic and the teacher less reliant on rules. In contrast, experts have a more holistic grasp of relationships within a particular context and fluidly and efficiently solve problems as they arise. The resources on which they can draw are much richer. (p. 11)

In summary terms, expert teachers see much more of relevance in the complex dynamic situation of the classroom setting and the interactions that occur; make better inferences and interpretations of what is occurring and what specifically needs to be addressed if things don't go as initially planned. Invariably, they are also better at identifying the causes of classroom problems and, most importantly, can design learning arrangements and experiences that foster better attainment. Many of such teachers are creative teachers, within the framing heuristics of this book. Hattie (2009) uses the term "with-it-ness" to summarize much of what such teachers do differently and better than those less expert in the profession:

> Teachers need to have the skills of 'with-it-ness' – that is, the ability to identify and quickly act on potential problems and be aware of what is happening in the class (the proverbial 'eyes in the back of the head', or mindfulness). (p. 78)

Indeed, when exploring with the more open and critical faculty on the what and how they perceive in the experience of teaching it seems to be the case that some are seeing more and seeing this in better qualitative terms—more 'with-it', as Hattie described. However, this is not surprising, as when I open the bonnet of a car, I see metal, plastic and some wiring; nothing makes any sense at all. I can fondly recall a neighbour of mine, who was adamant about fixing his car, even though he had little knowledge of motor vehicle maintenance. He spent many a weekend in total frustration, often with bloody fingers, trying to fix things that he had little idea as to what was wrong and, inevitably, what to do to fix it. I have since learned not to meddle with such things and ring a local mechanic.

While discovery learning may be fun, can lead to meaningful learning and has pedagogic value for students in some learning situations, it is of limited value in a typically busy working day. The mechanic always seems to be able to fix my car, I guess he's just lucky! However, I have a deep suspicion, that when he opens the bonnet, it looks very different than what my neighbour and myself are seeing. Many factors determine such differences in perceptual acuity and capability, but most stem from prior learning and competence (or lack of it), as well as a human attribution bias whereby we think we know more than we do. In most scenarios when there is a lack of content knowledge and poor thinking, the result is a very limited framing of that domain area. For many teachers, a lack of solid evidence-based pedagogic knowledge and some 'fuzzy' thinking typically lead to both limited and inaccurate perceptions and interpretations of what is occurring in their classrooms.

Asking many novice teachers (and some experienced ones also) to do reflective practice is like asking students to do good thinking when they lack any prior useful model of what this entails and how it works. Without such knowledge bases, this is highly unlikely to occur. Several years ago, I conducted a 3-year longitudinal evaluation on the student learning experiences during the implementation of large-scale curriculum innovation in a major educational institution in Asia (Sale 2014). One of the main areas of interest in the evaluation was how effectively had students acquired key Graduate Attributes, one of which was good thinking (as framed prior). While the evaluation generally showed positive results in terms of the intended goals and objectives of the innovation, most students interviewed still had limited and idiosyncratic perceptions relating to what constituted good thinking and how to do it. This reinforces the view that good thinking will not naturally occur simply by providing tasks that involve thinking (e.g., implicit instruction). Like other complex skills, its development is subject to the same core principles of learning. Without, explicit instruction and plenty of opportunities for deliberate practice in a range of contexts, learning will be partial and fragmented. In summary, the evaluation revealed that there had been a relative failure to sufficiently infuse an explicit model of good thinking across courses as well as facilitate an effective professional development approach to address this at that time.

In reflecting on one's practice, the usefulness of the outcomes in terms of enhanced teaching proficiency and gains in student attainment depends on what constitute the content and processes that are analysed and evaluated in such reflection. The novice tennis players I view from my gym window may reflect in some way on their performance, but on what basis are they reflecting? Without a knowledge base on what constitutes key skills in tennis playing and a systematic process of deliberative practice with expert feedback, it is very unlikely that improvement will be significant—and that seems to be verified in terms of the novice tennis who have shown only minimal improvement in some 2 years of seeing them play, albeit from the gym window. In terms of reflective practice in teaching, Hattie (2009) made the key point:

> The current penchant for "reflective teaching" too often ignores that such reflection needs to be based on evidence and not post-hoc justification. (p. 241)

Earlier in the chapter, I referred to Powell and Kusuma Powell (2015) notion of "goodness" in the profession of teaching. In this context, the authors further point out that:

> …when school people do witness exemplary teaching and learning, we often tend to respond with immediate adulation and subsequent dismissal…We must learn to 'look for goodness', deconstruct it and most importantly learn from it. (p. 26)

In summary, the usefulness of reflective practice is in large part determined by 3 main (there are others) interrelated components:

1. A strong ***Pedagogic Literacy***, that encompasses the most current knowledge on how people learn and what teaching methods tend to work best, in what ways and in what context. This is the underpinning base of EBT.
2. The ability to do ***Good Thinking*** (well at least most of the time). This requires Metacognitive Capability, which was developed at some length (sorry if it got a bit boring), but that's the basis of Good Thinking from my framing and experience.

Unfortunately, I have only acquired this capability in the past two decades. As Oscar Wilde said, sadly: "Youth is wasted on the young", or the old adage, "Hindsight is 20/20 vision"—if only!

3. A *Disposition* (e.g., motivation, mindset and integrity) for authentically helping others achieve personal goals and develop the capability to be self-directed learners. This requires a high level of professional skills as well as genuine respect and concern for learners.

Much of reflective practice has failed to recognize how this works, and that's why it hasn't worked as intended. Furthermore, while institutional support and resources help considerably, these are only—albeit important—support structures. As Levin (2008) concluded:

> …resources alone, however, will not change social practices. Teachers need to see not only what they might do differently but how they could do it in the reality of their classrooms. The key to developing this understanding is ongoing work with colleagues – seeing others carry out new practices with students like yours and having others help you learn to do these new practices. (p. 86)

Asking teachers to be more creative may be of little value if they don't know specifically what this entails in the context of actual teaching practices, and how to do this effectively. Over the years I have been an observer in many workshops on creativity. Sadly, while most have been interesting in part and well-intentioned, they tend to offer generalities more than evidence-based heuristics. Having a passion and a belief that one is naturally creative and it's only a question of 'switching on the creative switch' and hey presto 'creative me' emerges, is often the message. However, this is akin to buying that expensive piece of exercise equipment, working out 10 min a day, with no pain, and within a month or two you're the 'body beautiful', or whatever. Positive beliefs and thinking are an important aspect of effective learning, but they don't inevitably lead to successful learning. Also, while notions of 'quick fixes' in terms of achieving rapid success are attractive, even seductive, it's not how successful learning and attainment of anything challenging works for some 99% of the population. Creativity, as we saw in Chap. 5, is understandable, can be adequately framed in the context of teaching, and it is a learnable capability. But it is subject to the same core principles of learning as other human skills.

Reflection on, or in practice, then, without a clear evidence-based framework will likely result only in partial and limited improvement at best. Willingham (2009) made the key point concisely:

> Education makes better minds, and knowledge of the mind can make better education. (p. 165)

9.8   How to Do Evidence-Based Reflective Practice (EBRP)

I could equally pose the question—how do you fix cars well? Firstly, you must understand how they work, know-how each part contributes to the working of the sub-systems, and how these affect the functioning of the whole (we could call this good analysis—of course, with the relevant content knowledge and understanding of car workings). Secondly, it helps if you have experience across a range of motor vehicle contexts, especially in solving mechanical problems (e.g., comparison and contrast with previous experiences of similar problem scenarios) as this will draw out prior knowledge in long-term memory, and 'hey presto', the solution comes to the conscious mind. Experts often have the answer already nicely encoded neurologically in those acquired mental models of motor vehicle defects. This helps to make correct and quick inferences and interpretations of what is wrong and how to fix it. Finally, it is just then an evaluative question—what are the ways to best fix it for the customer? (e.g., effectiveness, efficiency, cost). There's little need for creative thinking unless the customer wants it to drive like a duck. Similarly, MC may not be a needed capability in this context, providing the mechanic is well-disciplined to follow the correct procedures and pay attention throughout—we could call this autonomous learning and unconscious competence.

Hence, useful EBRP requires teaching professionals to possess Pedagogic Literacy, Metacognitive Capability and Creative Teaching Competence—to be able to frame expertise, evaluate it accurately, and improve it. In the past 5 years or so, I have developed and piloted a range of customized tools that facilitate effective reflective practice. Do note that these tools are not algorithmic in that they unproblematically capture all the components or constructs of effective teaching—or what Powell and Kusuma-Powell (2015) referred to as, "Goodness Criteria"—noted prior.

Teaching is a complex activity, and classrooms are equally complex in terms of a diverse range of student personalities, biographies and motivational status; the latter having highly changeable features in the socio-psychological dynamics of classroom interactions. My analogy with a car, whilst illustrative in context (he says hopefully), has a major flaw in that the world of engineering is a *closed system*. It can be highly complex, and when things go wrong at bad times, mechanical systems (e.g., cars) can appear to 'have minds of their own', a bit like 'Murphy's Law'—but it is still a closed system. John Cleese, a famous English comedian, who starred in the television series *Fawlty Towers*, a British television sitcom broadcast on BBC2 in 1975 & 1979, cleverly and brutally demonstrates this non-rational perceived ability of mechanical systems (in this case his car) to deliberately annoy him (i.e., break down on the road) at the most inopportune time and circumstance. Cleese firstly shouts at his car threatening it with punishment, but the car will not start on priming. He eventually gets out of his car, finds a tree branch and beats the car with it. Did the car subsequently decide to work? You may need to access the YouTube video to find out. Similarly, when we see top tennis players smash their expensive rackets on the court, it is as though they blame the racket for missing an easy shot. I have yet to meet a metacognitive tennis racket. Mine made many errors, but I think it was all down to my lack of competence. Indeed, I did not possess a high-end expensive graphite racket (or whatever they are now made of), so I have some leverage in terms of attribution of blame. The same excuse may not work so well with the top professionals who, without doubt, have access to the best tennis rackets that can be made. Teaching is heuristic and it is an *open system*, which means that its components interact and create synergistic effects, which are never completely predictable. Highly effective teams are a good example of this, in that such teams are more than the sum of the parts, and explains why teams, who are 'on-paper' not considered to be good enough to win major championships, sometimes do win these. Leicester City winning the English Premier League Championship in 2016 was a classic example. In the previous year, they just avoided relegation from the Premier League in England and were 5000-1 with bookmakers for winning this title in 2016. However, as we know, real soccer matches are not played on paper—its grass, right?

Useful heuristics are much better than fashion and fad, and the tools do not prescribe a rigid method of application—they are flexible and designed to be used for capturing interesting practice experiences—whether they are good, bad—even ugly. They are certainly not form filling-in administrative activities—which turns teachers off. However, they provide an essential structure and focus for capturing experience that can subsequently be subjected to evidence-based pedagogic analysis—with good thinking of course. Table 9.2, is an exemplar that focuses practitioners attention to the employment of the core principles of learning as the *organizing heuristics* for systematic reflection on practice.

The following example is an EBRP tool, used by 7 teacher-researchers over a 15-week module for a 2-year Ministry of Education, Singapore Tertiary Education Research Fund project, which systematically applied Evidence-Based Teaching (EBT) methods and learning principles to ascertain their impact on students' intrinsic motivation. The research used quantitative methods to measure levels of engagement (e.g., behavioural, emotional, cognitive and agentic) and self-efficacy. Most importantly, it employed students as co-participants in the learning process to gain a deep and ongoing insight into how students were experiencing their lecturers teaching over an extended time duration (Sale et al. 2018). The research was able to make qualitative comparisons between the teacher's intent and pedagogic interventions with the actual student learning experience. Table 9.3, is one used by a teacher-researcher for 2 weeks of the intervention.

The teacher-researchers in this project were all experienced teachers, including 4 who have the role of *Academic Mentors,* responsible for implementing pedagogic initiatives at school level as well as collaboration in learning units across schools. Most had received teaching awards and all were familiar with EBT. The following are samples of their reflection on conducting EBRP and using the EBRP tool:

> I have to admit that if not for the commitment to the research team, I would not be self-disciplined enough to perform my weekly post-lesson reflective log. I can firmly state that it was worth the effort and time. Because of the need to conduct ongoing reflective practice, it was not just that I had to focus on my content delivery but at the same time constantly trying to observe the responses of students at my interventions. The responses from students are it in the positive or negative served as indicators for me to continue to adapt, modify

Table 9.2 Generic EBRP tool

Evidence-based reflective practice tool

In the learning experience (e.g., lesson/session plan) was there	Evidence of effectiveness
	Evidence of effectiveness What specific *strategies*—methods/activities and resources were employed to enhance this aspect of the learning process, and how effective were they? (Based on your observation and any other feedback if available (e.g., peer observation, student feedback)
Clear communication of the learning outcomes to the students?	• What specifically is to be learned • The purpose of this learning • How this learning connects to the wider learning goals (topic areas, skills) for this module
Activation of students prior learning and connections to new knowledge presented?	• Identifying what students already know/don't know about the topic before the start of a session • Helping students to fill important knowledge gaps/clear up misconceptions • Making connections between what is to be learned now (e.g., new knowledge/skills) to what has already been learned
Instruction focusing on the key concepts and principles for understanding this topic or skill area?	• Identifying and illustrating the most fundamental concepts/principles to be learned • Explaining how these connect to the learning outcomes • Methods and activities to help students to understand these concepts/principles in real work/life contexts

(continued)

Table 9.2 (continued)

Evidence-based reflective practice tool	
Use of activities that involve good thinking to facilitate understanding?	• Enabling students to engage in the types of thinking necessary (e.g., analysis, comparison & contrast, inference & interpretation, evaluation) to connect new knowledge to what they already know • Building the necessary mental models in long-term memory
Appropriate variety in the methods, activities, media used	• Focused on the learning outcomes and the student profile • Encouraging engagement and interest • Maintaining good attention levels
Utilizing core principles of learning *In the learning experience, was there*	**Evidence of effectiveness** What specific strategies, methods and/or resources were employed to enhance this aspect of learning, and how effective were they? (Based on your observation and any other feedback if available (e.g., peer observation, student feedback)
Application of practices consistent with human memory processes?	• Chunking of content to minimize cognitive overload • Periodic recap and review of key concepts and principles • Doing Whole-Part = Whole analysis—showing how new parts of the learning connect to the wider topic or skill area
Formative assessment of student learning and provision of quality two-way feedback?	• Monitoring student learning through testing key concepts and skills • Providing clear and specific feedback (e.g., task, process, self-management) • Encouraging two-way and peer feedback where possible and useful

(continued)

Table 9.2 (continued)

Opportunities for practice to enhance understanding and/or skill acquisition?	• Retrieval practice to check key conceptual understanding • Spaced practice across sessions to build understanding and competence • Deliberate practice focused on specific skill development tailored to student's skill levels
Interactions/activities that foster a climate conducive for building rapport, encouraging success and a sense of fun?	• Use of growth mindset strategies (e.g., showing how effort impacts learning; sticking with students when they need help; mastery learning) • Use of expressive language and supporting body language in communication (e.g. expressive tone, smile, eye contact) • Allowing/facilitating humour and fun to occur in the lesson

Table 9.3 Example of a completed EBRP tool

Evidence-based reflective practice tool
Weeks 1 & 2: Oct 16–27, 2017

In the learning experience was there	**Evidence of effectiveness**
	What specific strategies, methods and/or resources were employed, and how effective were they? (Based on your observation and any other feedback if available (e.g., peer observation, student feedback)
Use of autonomy supportive style strategies? • Clear expressive facilitating language • Provides explanatory rationales • Acknowledge & accepts negative affect • Displays patience • Explores and allows student choice in the overall instructional strategy • Two-way feedback to support understanding and skill development	Use expressive facilitating language—this takes a bit of effort initially However, did use plenty (at least in my own mind) of explanatory rationales Acknowledge students feeling—they certainly do not like to read, so I elaborated on the use of jigsaw approach (see later) to spread out the workload. Also emphasized that in an increasingly complex word, it is ever more important to read and think critically to make sense of available information Patience—working on this! Set ground rules about talking in class… so far, manageable Giving students choices—did clarify that some choices are available, esp. in terms of doing extra reading and optional exercises. Emphasised if they make the effort to do, I will make the effort to mark and give them feedback 2-way feedback: Informal chit-chat with those who came to class early on how they feel about the module so far. Most said manageable—but this is perhaps still early in the semester before the hardcore stuff kicks in

(continued)

Table 9.3 (continued)

Evidence-based reflective practice tool Weeks 1 & 2: Oct 16–27, 2017	
Use of high effect strategies/methods? ● Appropriate for learning outcomes ● Appropriate for student profile ● "Russian Doll" design & facilitation (e.g., combinations of high effect methods; combinations of effective e-tools; combination of both)	Combination of strategies, Russian Doll design used—not elaborated here. See paper published on work done A challenging one is "Student Profile", as the content was initially pegged at Year 3 students, but now need to be delivered to Year 2 this academic year (AY17), who obviously had not learnt some modules yet. In fact, they are learning them simultaneously with this module. The module will revert back to Year 3 in AY18. As a result, example on more challenging applications are dropped, to make room for classroom coverage of key concepts and operation of "virtual" chemical plant (i.e. dynamic simulation of Amine Treating Unit). This topic was previously a self-study component for Year 3 students Did tell them since I had not taught them before, I have no clue who are the 'good' students or otherwise, and that IT DID NOT MATTER to me; and that any student can do well in this module, if they follow my advice and stay on top of their readings before class
Utilizing core principles of learning *In the learning experience, was there*	**Evidence of effectiveness** What specific strategies, methods and/or resources were employed to enhance this aspect of the learning process, and how effective were they? (Based on your observation and any other feedback if available (e.g., peer observation, student feedback)

(continued)

Table 9.3 (continued)

Communication to students of the learning goal/outcomes, purpose and expectations?	This was done at the beginning of a lesson, after a quick recap of previous lesson. And again, at the end of the lesson (each lesson is 2 h) In week 2, students are required to complete an (non-graded) dynamic simulation assignment for Amine Treating process, with peer marking. A rubric is shared. The assignment is to be completed by the end of Week 2, with peer marking to take place in Week 3. Explain at length the class time spent on purpose of familiarizing them with the Amine Treating process. Marking rubric given to students Outcome of this remains to be seen
Activation of prior learning and connections to new knowledge presented?	Prior knowledge on general understanding of chemical hazards are tapped into, to build a new concept of *"hazards are intrinsic to a material or its condition of use"*. Students are asked to name some chemicals they are familiar with, and the lecturer build on those chemicals named with examples that illustrate the concept. This part seemed to work Other than that, it is more of the case of giving them understanding of an accident (Bhopal Gas Disaster) and a common industry process (Amine Treating) to build up their "core knowledge" and provide context upon which new knowledge of process plant safety (such as of inherently safer design) can be added on at later weeks
Emphasis on key concepts and Principles that underpin understanding of this topic?	Week 2: Key concepts of inherently safer design (ISD) was emphasized repeatedly, in particular that it will be most effective when applied at the R&D and process development stage before the chemical plant is built. This is illustrated using the Bhopal Gas Disaster as a case study. Also the fact that the chemical MIC was stored in excessive manner (as opposed to the ISD strategy of MINIMIZE) appear to leave a strong impression among students, especially when they were informed that the storage tanks measured 40-ft in length and 8-ft in diameter—enough for an adult person to walk inside! Some are able to recall the dimensions when asked!

(continued)

Table 9.3 (continued)

Use of activities that involved good thinking to facilitate understanding?	"Slide show" of 10 real-world practices are shown one at a time, and students are to discuss in their groups and provide an answer what inherently safer design strategy is employed. In general, they did well. Some even challenged—on 2 occasions—the answer given by lecturer. A lively discussion ensued, which prompted the lecturer to consider an aspect of the answer from the students' viewpoint. Overall, they appear to develop a good understanding of the subject In the Google Slide mix-and-match case (how modifications made with good intentions ended up introducing new hazards—intended consequences) students need to identify the modifications made, link it to the rationale (at the time decision was taken), and how it led to unintended consequences. Overall, most of them got the linkages correct, with 1 or 2 groups with all correct answers
Variation in the modes and methods of information presentation and interaction?	Yes, first 2-h session: first students need to enter answers into Google Doc (what went wrong at which stage of lifecycle, and what inherently safer strategy can be used to address the issue), next the Google Slide mix-and-match. Next 2-h, involve watching a short video to understand the Amine Treating process, followed by more detailed "jigsaw" reading of more detailed write-up of the process, and answer 14 TRUE/FALSE questions on Socrative. Students are seen discussing among themselves in their respective groups. All results are viewed in real-time; and it appears that most students are able to answer most of them. Due to time running out, not able to do the MCQs part in class, and students are asked to attempt them on their own
Application of practices consistent with human memory processes (e.g., chunking of content to minimize cognitive overload; rehearsal/review activities)?	Chunking done for online video of mini-lectures (each 3–5 min) focusing on one key concept per video. Lecturer first use a video to explain overall set-up of Amine Treating process and the different units that made up the process. Jigsaw reading was then carried out in-class where each student within a group is given a unit (process description, and drawing) to understand, and collectively they work together to answer the TRUE/FALSE Socrative questions. Generally, results of class 04 better than class 03

(continued)

Table 9.3 (continued)

Incorporation of formative assessment to provide quality two-way feedback?	Feedback given on Google Doc, Google Slide activities mentioned earlier. Selected items were clarified in class, others were added to Google Doc after class, and informed students during the next class. Follow-up explains given Tue (to class 2B/03), and Wed (to class 2B/04) on queries noted on Mon (for 2B/03 and 04 present)
Use of deliberate practice to enhance understanding and/or skill acquisition?	Not able to cover a Google Doc open-ended questions in class—14 in total, designed to test/ enhance understanding of the Amine Treating Process. This was left for students to try on their own. Explained that these exercises are optional, and the benefits of trying them, i.e. able to build a stronger foundation that helps tackling later topics. Will monitor the Google Doc entry for any attempt at answering them
Interactions/activities to foster a climate conducive for building rapport, encouraging success and a sense of fun?	I think so—students sit with their groups (formed by themselves) for all lessons. These classes tend to be noisy already to begin with. FUN I think the students certainly have. Students tend to offer witty remarks all the time in the 2 weeks with them. A few attempts (covered elsewhere, or everywhere) in these submissions was aimed at encouraging success; but may not be able to quantify these—how to measure 'success' besides scoring well for tests
An aspect(s) of creativity (e.g., story, humour, activity, presentation style, example) that significantly enhanced motivation in this learning experience?	Jokes used, and funny pictures embedded into PPT slides Real world examples added as appropriate, either from case studies or from own work experience Overall class is lively, even boisterous. Thus far, able to manage 'crowd control' and students do pipe down when important matters are discussed. Not sure if these enhanced the learning experience though

or even drop certain style features or methods of teaching during this 15-week programme. It was very fulfilling and encouraging when I came to know that the students were able to understand the reasons or purposes of something I have done or created for them during lessons. I also feel much more capable and confident in my role as Academic Mentor, in that I can better understand more specifically what teaching challenges some of our faculty are personally facing in their teaching and can provide specific and useful feedback on what might help them meet these. It enables me to go beyond personal experience and have a sense of scientific backing to what I am doing in a coaching/mentoring context.

In using the Evidence-Based Reflective Practice Tool, this provided a clear structure for evaluating my teaching in very specific terms and being able to change/modify aspects of strategy on the spot in the lesson context, if I felt it needed such action. I noticed that when I taught the same topic 2–3 times in a week or so to different groups, I felt that I am improving session by session. It made me very metacognitive to what I was doing, and over time it becomes somewhat automatic.

The EBRP tool, which involved a bit more work initially, helped to focus my planning and post-lesson reflection. It will be useful for teachers who genuinely want to improve their practice. After a few sessions, it became less time consuming, and as it is a flexible tool, not a form-filling in exercise – it was not burdensome.

Reflective practice using the EBRP tool kept me checking and thoughtfully planning the lesson structure and sequences and ensuring that I was both focused on the learning goals and the students' needs weekly. Over time, this became easier and easier, as I become increasingly familiar with the design process that is the basis of EBT – although it is still a challenge to make all lesson parts interesting for students.

Doing reflective practice using the EBRPT was both easy to use and helped me to think more carefully about the structure of my lessons and how I might make the learning both easier for the students to deal with the content and get them more engaged and interested. I also noticed that by using EBT the students were giving more comprehensive answers to questions. Overall, I felt that I was able to get a good relationship with the students and that they were learning better.

9.9 An Evidence-Based Frame on EBRP

A little knowledge can save one's life or it can be a perpetual nemesis. For folk who had prior knowledge, and especially experience, when the signs of a tsunami become apparent, could take the quick evasive action of going upland and probably be safe. In contrast, the unfortunate folk who had no knowledge (or did not believe those who told them to run and go upland), especially if they went down to the beach to explore this strange occurrence, would perish for the lack of this small bit of factual knowledge. Similarly, parenting, which is a much-contested issue, raises questions about useful and not-so-useful—even dangerous—knowledge. The so-called 'snowflake' generation (e.g., Winget 2017; Piper 2017) are not the result of brain mutation; numerous factors are impactful here, but certain parenting practices are likely to contribute to such dispositions. I won't pursue this further here, but it will have implications for teaching, learning and well-being, especially for teachers seeking to develop self-directed learners. SDL requires much in the way of volition, persistence, grit, or whatever is the vogue term at the time. Snowflakes don't have much of this if I read the analogy correctly.

The use of the EBRP tool, as noted, cannot capture all the important interactional features in a classroom lesson. The teacher, even with the help of a trained observer, may miss certain events that may be significant for enhancing understanding of the learning experience for students. However, there is much of merit in expert peer appraisal, in doing reflective practice, especially from an EBT perspective. There is always subjectively, and the teacher must recognize this. Systemic perceptual and attribution biases, as reinforced extensively in this work, abound in the case of human apprehension (e.g., Kahneman 2012). There is a tendency to attribute blame to external factors in the environment, including other people. Let's be honest, how often do we see people, when something has gone wrong, immediately stand up and say, "it's my mistake, I am responsible for this mess." Most people will acknowledge some personal blame if the evidence is striking, but the tendency to attribute some of it elsewhere is prevalent. In 3 decades of coaching underperforming teachers (defined in terms of consistently low feedback scores), few acknowledged their role in producing such results. Only a handful said something akin to 'Maybe I need to improve my teaching'. In most cases, a range of external factors was identified as the basis of their feedback. These included the students they had to teach (e.g., these are 'poor' students); the curriculum they had to teach (e.g., not well written, obsolete content); and time constraints (e.g., too much other work to do), etc. Changing their attributions, apart from building trust and some rapport, was always the initial challenging task.

EBRP does provide an approach and a methodology to both diagnose lesson effectiveness and predict teaching and learning possibilities through a systematic, focused and more objective base for understanding the experience of learning. It applies both *Reason* and *Science* (e.g. Pinker 2019), as best as we can presently frame it. As Treadwell (2017) summarizes in the wider context:

> Our ability to reflect on our practice underpins all professional associations. If the medical profession stopped looking for best practice, then the role of technologies and the realignment of how hospitals met their purpose in the 21st century would not have resulted in the transformation of hospital services over the last 20 years.

> Education has just begun this process and it will take the same passionate determination to ensure that every school is developing each learner's competencies, enabling their ability to learn – anything, anywhere, with anyone, at any time. (p. 195)

In the context of teacher education, EBRP makes possible more meaningful and productive dialogues that can authentically change teachers' mindsets—essentially beliefs and perception, and eventually feelings. In using the EBRP tools, it is essential that teachers who are asked/told to do reflective practise, for whatever reason (e.g., part of a formal course on teaching; been referred for mentoring/coaching due to low student feedback scores; a new institutional requirement relating to performance management goals) are inducted into the main tenets and practices underpinning EBT. Without a strong knowledge base relating to what constitutes effective practice and the underpinning learning science, there is likely to be a very limited analysis of the lesson, invalid inferences and interpretations of what is being experienced by students in a class, and subsequent poor evaluation of events.

EBRP also enables a more focused and meaningful dialogue about practice. Instead of a dialogue which may comprise vagueness and misconceptions of terminology and high inference judgement, it provides a structure—a set of guiding heuristics—for focusing on specific aspects of the learning process and/or method employment. This has face validity and there is nothing more valid than direct evidence when dealing with teachers who are asked/told to change their perception and practices (including the underlying beliefs) that they previously held 'dearly' for many years. For example, showing teachers examples and explaining how they work in the context of a teacher's subject context (especially how they can meet key learning outcomes effectively and efficiently) is the most authentic and influential strategy. I have observed that some teacher educators make judgements of worth about practice, but lack the capability to offer practical specific instructional options for improving what has been noted as 'an area for development'. Giving feedback to a teacher that the lesson needs to be more engaging is likely to be of no value if the teacher knows little about types of engagement, and what are useful cognitive and/or motivational strategies to achieve this pedagogic goal. Even worse, is when the teacher educator cannot provide clear examples of what may work, and on what basis, and how to do this.

9.10   EBRP, Lesson Study and Active Schemes of Work

Teachers using EBRP will be able to both better predict the likely outcomes from their lesson planning, as well as make a more accurate diagnosis of them after the event. They will be doing what Schön referred to as *reflection on action*; the difference now is that such reflection will not be "post hoc Justification" as Hattie (2009) alluded. This approach has certain similarities with *Lesson Study* (e.g., Stigler and Hiebert 1999), a Japanese innovation that has proved highly successful in terms of effective, efficient and engaging lesson creation. The approach starts with identifying teaching methods/activities that seem to work best with students (based on observations and student feedback), followed by the design of instructional strategies incorporating these proposed methods/activities. The lessons are then piloted with various teachers over time with peer appraisal. The idea is that the best methods are scrutinized over time, with changes/modifications, and improvements being made. This process typically continues until the teachers involved feel that they have sufficiently validated the lesson in terms of its potential for being an effective, efficient and engaging instructional strategy for that topic area, student profile, and the desired learning outcomes to be met.

Even a basic pedagogic analysis of Lesson Study, reveals that it involves much by way of EBT practices and principles of learning:

- Researching the behaviour, interests and experiences of the learning group, finding out how well they are learning (as well as not learning), activating prior knowledge, and asking their learning preferences in terms of instructional methods (e.g., Autonomy Supporting style)
- Identifying levels of attainment and engagement (e.g., behavioural, emotional, cognitive and agentic) as a basis for instructional design
- Designing instructional strategies for topics/learning outcomes—based on this research
- Using spaced and deliberate practice with expert feedback for evaluating the various components of the lesson in terms of student attainment and engagement
- Encouraging metacognition through all stages of the implementation process (e.g., setting clear challenging goals, planning strategy, monitoring and reviewing strategy, as well as evaluation and extending learning).

A similar pedagogic structure is that of *Active Schemes of Work* where participating teachers collectively produce a plan of a wider structure of learning for a programme (e.g., module or unit) and continually improve the instructional design of the various lesson components (e.g., method blends, activities, examples, technology supports) over time. Typically, a scheme of work is a structured summary breakdown of the whole module or unit, focusing on:

- The specific learning outcomes relating to the key learning goals or topic areas for the particular curriculum programme. These are organized in terms of the best sequencing and appropriately allocated to each of the lessons comprising the programme
- A preferred instructional strategy (e.g., methods, activities, resources) as well as other supporting data for each of the lessons (e.g., details and timelines for completion of assignments, notes relating to infusing related process skills such as thinking, learning strategies)
- Details and necessary guidance on the assessment to be used at specific stages and activities in the programme (both summative and formative).

Petty's (2015) framing of Active Schemes of Work advocates that these essential planning organizers should be used dynamically by all the teaching team not only to initially identify best evidence-based instructional strategies but continually improve them, similar to that of the Lesson Study approach. By explicitly making the schemes of work dynamic and part of a collaborative improvement process, this energizes all the teaching team to actively input into the lesson design, especially activities likely to be most effective for enhancing engagement and attainment for the student groups. This is an ongoing process, and when driven by an EBT approach, reviewed by ongoing feedback and peer discussion, can significantly improve the learning experience for students. Active Schemes of Work provide the perfect context for faculty skill development and creativity, They also provide a great way to foster relationship building and motivation for the faculty involved. Petty summarizes the rationale and range of benefits as follows:

- **Active learning works**. Research shows that active learning is by far the best for recall, student enjoyment, deep learning (full understanding), and for correcting the learners' misunderstandings.
- **It improves results**. School improvement research shows that teachers have about three times the effect on achievement as their managers. So, achievement and students' life chances can only be improved if teaching is improved.
- **It is likely to get a commitment to improvement**. Subject centred discussion on how to teach well is at the heart of a teacher's role, teachers usually enjoy being involved in practical development in their subject area.
- **Teams share best practice**. So, the best teaching methods are available to all.
- **It raises expectations of teaching quality**. Active schemes of work can raise expectations of what it means to teach well, as well as showing how this can be done.
- **It 'stores' best practice**. Good teachers who leave the college leave behind their methods for others to benefit from and enjoy.
- **It supports beginning teachers**. Novice teachers are given effective methods to adopt and to learn from.
- **It promotes professional development**. Writing the scheme promotes subject centred discussion on effective teaching and so develops staff.

9.11   Coaching

Teachers as learners are subject to the same learning principles and constraints as their students, especially when they are learning new knowledge and skills. In this situation they have to deal with acquiring new knowledge, building understanding, and doing the necessary practice, as well as dealing with their existing beliefs and emotional responses to change. While they may have greater experience and maturity in self-regulation and metacognitive strategies than many of their students, there is still much cognitive and emotional effort needed to build the necessary understanding and competence. The importance of being in a learning relationship and context that is congruent with the Core Principle of Learning: *A psychological climate is created which is both success-orientated and fun* is equally applicable to teachers as learners.

Coaching has become an increasingly popular term in the language of training and professional development. Previously, the term was typically used in other fields, especially in sport. We rarely hear terms like a teacher of soccer; soccer coach seems to be the norm. However, does not a soccer coach also teach soccer? Definitely. I have been coached in soccer and have coached soccer for many years. In both situations, there is an emphasis on improving performance, which entails aspects of learning. Furthermore, this learning is not just knowledge and skill-based, but often involves attitudinal components such as states of mind, dispositions, emotions and how to manage them. These affective components, such as emotional and belief management (e.g., mindset formation and maintenance) may be as important as the skill components. How many times have we seen professional sportspeople massively underperform (e.g., appear to freeze) at the most crucial stages in major events (e.g., championship point—even breakpoints—in a tennis match; miskick a penalty in a world cup soccer tournament). Hence, coaching is more than skill development but developing competence/expertise, and this is more than knowledge and skills per se, as we explored in some detail in Chap. 5.

There are many terms in the literature relating to 'helping people to learn' and 'enhancing performance', one could easily write a significant text on this alone—there are probably some already. For example, there is sometimes confusion over the terms of mentoring and coaching. In practice, mentoring is a wide-ranging term, which while focusing on helping a person to understand themselves better, usually in relation to their professional field (e.g., dealing with a significant challenge in a present role or decision-making on changing roles or career), it can involve a range of sub-roles (e.g., model, sponsor, adviser, teacher, counsellor)—even coach. There are similar points of comparison and contrast with notions of teacher, trainer—even facilitator. Does not a teacher train, and a trainer teach? Much is an issue of framing preferences, though the historical 'baggage' seems to have perceived training as focusing more on specific skills development, whereas teaching is wider and more holistic. There is a present saying, and I have no idea of its source, but it has become somewhat of a vogue term, "The teacher is no longer the sage on the stage but the guide in the guide". Simplistic, naïve, and certainly not useful. Expert teachers can train people, facilitate learning and mentor and coach—that's how I see it. For example, in the modern context, the full range of facilitation skills should be part of the skill repertoire of expert teachers. For example, they should be competent in using such technical skills as:

- Listening actively
- Observing carefully
- Using active learning methods and process tools
- Asking and responding to questions
- Paraphrasing
- Giving & receiving feedback
- Staying neutral
- Testing assumptions
- Staying on track
- Synthesizing ideas
- Providing summaries
- Collecting and organizing ideas/knowledge generated.

There are many definitions of coaching in the literature. I particularly like that of Costa and Garmston (2002) as it offers a process orientated frame and is consistent with my experiences over a few decades:

> Coaching serves as a foundation for continuous learning by mediating another's capacity to reflect before, during and after practice. (p. 23)

Of specific importance in the professional development context, good coaching enables the clarification of important knowledge, stimulates good thinking, structures focused deliberate practice and provides quality feedback. Coaching acts as a key organizing catalyst for facilitating deep understanding and sustainable competence to enable learners (in this case teachers) to transfer learning across a range of related teaching contexts in a fluent and contextualized way. For example, Joyce and Showers (2002) found that:

> A large and dramatic increase in the transfer of training – effect size of 1.42 – occurs when coaching is added to an initial training experience comprised of theory explanation, demonstrations, and practice. (p. 77)

Furthermore, their research revealed that coaching appeared to contribute to the transfer of training in five ways, in that coached teachers:

- practised the new strategies more frequently and developed greater skill
- used the newly acquired strategies more appropriately in terms of curriculum alignment
- exhibited greater long-term retention of knowledge and skill use with the strategies
- were more likely than uncoached teachers to explain new models of teaching to their students, ensuring that students understood the purpose of the strategy and the behaviours expected of them when using the strategy
- exhibited clearer cognition regarding the purposes and uses of the new strategies.

In coaching, as in teaching, language (the jargon) that is specific to what is being learned must be made explicit and clear so that both coach and those being coached have as accurate an internal representation of the *thing* (understanding, skill, attitude) that is to be developed/improved. For example, in tennis, putting 'slice on the ball' refers to a specific technique that creates deception on the movement of the ball when bouncing. This requires understanding the technique, its purpose in different game situations and, of course, the skilful application of putting slice on the ball. Achieving a high level of congruence of metal maps between coaches and those being coached can be a massive challenge as tasks become more complex. It is for this reason that coaching people to develop MC is difficult, challenging and takes lots of practice. It takes skilful coaching and motivated learners. The reason is that there is a lot to learn, and much thinking and volition are required—as I have emphasized across the chapters. However, it is as it is, and we are helping no-one by pretending it can be achieved by clicking some magic learning or creativity switch—which is no more than either ignorance or lies. Expert teachers must be evidence-based on matters of cognition, but equally on values and the behaviour that communicates those values. If we genuinely want to promote SDL, we must be role models, and this involves core values that underpin effective learning and well-being.

The framing of Pedagogic Literacy is the foundation of a *language of learning*, derived from EBT, just as evidence-based practice in medicine derives from knowledge domains in the biomedical sciences, and results in medical terminologies. This is why when doctors talk to each other, I often don't understand much of it—it is their language for communicating accurately about medical conditions and how to treat then (call it their jargon if you like). Now don't knock jargon, it may mean that the operation your need will, in practice, be done properly. Somehow in teaching, even using the term pedagogy scares some teachers, let alone that many-headed monster 'metacognition'. Of course, we don't want jargon for jargon-sake, but essential terminology that captures key aspects of the learning process will facilitate rather than hinder collaborative teacher learning and professional development.

For example, when I coach (train or mentor) teachers, I ensure that we have a common understanding of such terms as, activating prior knowledge, cognitive overload, task and process feedback, retrieval, spaced and deliberate practice, interleaving, mental models, specific types of thinking, etc.—and what is important about these terms in the context of teaching practices and student learning.

The importance of an explicit content-specific language is what makes possible the diagnosis and predictive capability of learning events from a more objective base. In the context of coaching, as Downey (2003) summarized:

> …language is what allows the client to be self-generating, and the practice that makes it possible for the client to be a long-term excellent performer. (p. 9)

The goals of coaching can also be framed in either very specific terms or in wider developmental terms, which depend on the goals/needs of the person being coached and the context of learning. Using a personal example may illustrate the very specific end of a continuum. As I do many overseas workshops often involving several days, there is usually a closing ceremony that has a social/fun component to it—typically a meal, speeches and some games, etc. I am often asked to do something as the trainer/facilitator. Now doing a speech is not difficult, and I can tell jokes and stories for some time—too long according to my wife. However, when asked to sing, and this often happens in Asian countries, the alarm bells go off in my head—this scares me. The reason is simple, I am a novice singer—ok, I'll be honest—I am a dreadful singer. Hence, I decided, over a decade ago, to have some singing lessons, but with one specific objective: to sing one song not too badly. Let's not get into what level of competence this is, but I would not get through any first round of a singing contest in this sector of the universe. In summary, I spent around $400 on being coached to sing The *Green, Green Grass of Home*, a song made famous by Tom Jones. I had no interest in developing self-directedness in learning or music appreciation, but simply to get through any song without total embarrassment. Well, I did learn to sing the song a bit better than prior, through some development in breathing and voice control techniques, as well as spaced and deliberate practice. However, I have yet to be offered a singing contract, and according to my wife, it still sounds "quite dreadful". I don't lose any sleep over this, and my overall self-efficacy has not been impaired. I accept that this is not a skill set that I have much by way of innate aptitude. For me to develop even a foundational level competence (however defined) would involve more practice and volition than I am prepared to give this goal. In terms of intrinsic motivation and cost-benefit analysis—it's not worth it to me. If pushed into singing a song somewhere, I do my limited version of The Green, Green Grass of Home, and generally, get away with a modicum of embarrassment. Positive comparisons with Tom Jones are unusual.

In contrast to my singing development, working with teaching professionals who are struggling with important aspects of practice and feeling personal stress, the coaching role and process takes on a different level of learning considerations that may significantly impact their future and well-being as a person. Teachers being identified as underperforming and being sent to a supposed expert teacher is both potentially daunting and stressful. While there is no one correct way to conduct coaching activity, as in the case of teaching, there are useful evidence-based heuristics and better practices in terms of helping people to achieve better outcomes, though these may be defined in different stakeholders-terms. For persons being coached, an immediate concern is often one of how to enhance their feedback scores and not have to go through this 'remediation process'. For management, there may be multiple concerns; for the persons under their purview, key performance indicators, complaints from students, parents, employers, etc. For faculty being coached, there are serious concerns about negative impacts on their learning and well-being.

9.12   The Structure of Subjective Experience: A Neurolinguistic-Programming (NLP) Approach

Before exploring the stages, strategies and techniques for achieving a productive outcome in coaching relationships, it is useful to explore some fundamental underpinning assumptions about human interaction and communication. Coaching, like teaching, involves an investment in terms of attaining a trusting relationship that supports good rapport. All approaches to coaching place an important emphasis on understanding the reality of the situation as perceived and experienced by the client. This applies to teaching, mentoring and especially psychotherapy. The basic assumption is that human behaviour is in no small part determined by a person's perceptions and that a change in perception and thought is a prerequisite to a change in behaviour. Hence, in coaching teachers, it is essential to get a clear understanding of their perceptions, thoughts and feelings about the learning situation they are involved in. This is especially the case where teachers are referred to a coach for improvement, as this can involve emotive aspects.

The field of neuro-linguistic programming (NLP) has provided much insight into 'the structure of subjective experience' (e.g., Bandler and Grinder 1990; Dilts 1980), and how individuals make sense of their experience and interact with others, much of this occurring subliminally. While neuroscience focuses on brain functioning at the level of neuronal networks and neurotransmitters in the brain in response to different environmental stimuli, NLP has sought to explain aspects of how the mind works at the level of subjective experience. This is particularly interesting as there is often the assumption that the mind is simply the result of what the brain is doing. This is far from the case in terms of present understandings. Neuroscientists are increasingly identifying correlations and causal links between electrical currents in the brain and various subjective experiences, but how these billions of electrical signals result in a mind that experiences a stream of consciousness that has specific feelings such as I am angry or I am happy, is not known (Harari 2016). In his words:

As of 2016, we have absolutely no idea. (p. 128).

However, he goes on to argue that what we can be sure of is that the stream of consciousness we directly experience in every moment is the surest thing in the world. As he illustrates:

You cannot doubt its existence. Even when we are consumed by doubt and ask ourselves: 'Do subjective experiences exist?' we can be certain that we are experiencing doubt. (p. 123)

Hence, the study of subjective experience is of primary importance at the level of human interaction and the construction of meaning. Coaching is very much engaged in the meeting of two persons, each with their own 'streams of consciousness'. How these play out in terms of client perception and feeling are crucial in terms of the learning that may result from such encounters—either positive or negative. In most basic terms, if the resulting subjective experience for the client does not result in perceptions and feelings favourable to learning (e.g., trust, understanding of what needs to be done, self-efficacy for the task in question), outcomes may not be conducive to learning and well-being.

NLP can be broken down to a relationship between 3 interrelated components of human psychological functioning that affects the way we perceive the world around us and subsequently interact in it. The essential components are:

Neuro	This refers to the neurological processing of senses—seeing, hearing, feeling, tasting and smelling—and how they shape our perceptions and thinking about things in our world
Linguistic	This refers to the language patterns (verbal and non-verbal), which affect our understanding of the world and how we communicate with others
Programming	This refers to how we organize and orientate our thoughts, feelings and beliefs to interact with the world in our personalized way.

A significant aspect of NLP is concerned with understanding how these components interrelate internally to shape—even determine—the structure of subjective experience. Dilts (1980) captures this important framing of NLP when he referred to it as:

...the study of the components of perception and behaviour which makes our experience possible. (p. 1)

While each person has his/her individualized inner view of reality (Map in NLP terms), the essential processes of the way that the map is constructed is generic and understandable. Dilts went so far as to argue that:

When the confusions and complexities of life experiences are examined, sorted and untangled, what remains is a set of behavioural elements and rules that aren't too difficult to understand at all. (p. 5)

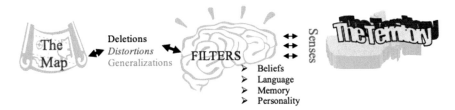

Fig. 9.2 Summary of the structure of subjective experience

For NLP then, a major focus is on understanding how the brain structures the inner world of subjective experience and, from such understanding, how we might influence human perception and action in more productive ways. The structure of subjective experience from an NLP perspective is summarized in Fig. 9.2.

For purposes of translation, *The Territory* represents the external world, which is experienced through our senses. However, the external world (The territory) does not get represented as it is in our internal world—*The Map*. The external world cannot be understood separate from the interpretations we put on it, as each individuals' interpretation of the external world is always mediated by existential *Filters*, which process the sensory information from the territory—the main ones being our belief systems, language, memory and personality traits. The result of this process of interpretation is that we construct our inner world (map) along with the deletions, distortions and generalizations that result from this filtering process. What is most significant here is that the way we experience the world, and our actions about that experience, gives us the feeling of objectivity (e.g., this is the real world), but in fact, it is a very personalized subjective framing of the world. In NLP terms, the *Map is not the Territory*, only our representation of it.

From an NLP perspective, beliefs are the most significant neurological filter determining how we perceive and experience external reality. Furthermore, our perceptions of reality will determine both our thinking and behaviour. In Chap. 3, we considered in some detail how beliefs impact student learning, motivation and well-being. The same theoretical perspective applies to the learning, motivation and well-being of teachers in a coaching situation.

Indeed, differences in beliefs may represent man's biggest threat to survival. In the context of teaching and learning, a major barrier to successful learning for many students (and some teachers) are limiting beliefs about intelligence, self and the nature of learning itself. It is to be emphasized that maps are not *just* theoretical constructs, they are the internal realities of all of us as we try to make sense of the world we live in and find personal meaning for living *a life*. Furthermore, maps can both assist us in our search for personal success and meaning as well as constitute the biggest barrier to such fulfilment and well-being. In a nutshell, some maps are better than others—much better. As Hall (2001) argues:

> The richer our map, the more accurate, adequate, and useful our menu, the more choices. The more impoverished our model, the fewer choices. The richer and fuller our linguistic map, the richer our mind…Maps induce states, and states govern perception and behaviour. (pp. 26–27)

In Costa and Garmstons' (2002) framework of Cognitive Coaching, working with a persons' subjective experience is fundamental, as they point out:

> It mediates invisible, internal mental resources and intellectual functions. These resources and functions include perceptions, cognitive processes, values, and internal resources.
>
> Cognitive coaches focus on the thought processes, values and beliefs that motivate, guide, influence, and give rise to the overt behaviours. (p. 13)

Costa and Garmston (2016) summarize the role of a cognitive coach:

> As cognitive coaches, therefore, we are interested in operating on the inner thought processes – in "coaching cognition". Teachers possess wide and expanding bodies of information and skills and they make decisions about when to use what from the extensive range of their repertoire. Cognitive coaching assists teachers in becoming more conscious, efficacious, precise, flexible, informed, and skilful decision-makers. Together, teachers and coaches create greater student learning. (p. 147)

Hence, the coach is involved in helping clients better understand how they are thinking, feeling and acting, to help them to enhance the capacity for better thinking, decision-making and self-regulation. The effective coach is the mediator in this process, helping to "promote behavioural changes towards more effective practice" (Costa and Garmston 2016, p. 9).

It is not surprising that human communication, thinking and behaviour is often messy, confusing and not productive to effective learning and well-being. Rationality seems to be largely trumped by unconscious biological processes that challenge notions of free will (e.g., Harris 2010). However, while humans are individually unique in terms of their personality configurations, biology and experiences, there is much similarity, which is to be expected given our universal brain morphology, human needs, and the way the mind works in terms of the dynamics of subjective experience. Hence, while the content of streams of consciousness can vary greatly, and they may be framed as productive or unproductive for learning and well-being, there will be streams of consciousness that can be understood and dealt with. For Costa and Garmston (2002) a cognitive coach can play the important role of helping another person to take appropriate action toward his or her goals, while simultaneously helping that person to develop expertise in planning, reflecting, problem-solving, and decision making. (p. 13).

In coaching, as in all human interaction situations, there are key touchpoints in which perceptions and feelings can be most poignant. Given the many ways in which biases can affect perception, feelings and behaviour, awareness of such touchpoints, and skillful communication, can mediate perception more favourably towards building trust and rapport—essential for getting productive results with other people. For example, in customer service, a key touchpoint is the first encounter—the Primacy effect—and we know what can happen here. Being ignored or having a curt tone of voice, or 'that look' (I can't be bothered-look) can frame the nature and form of what is to follow. I have lived and worked in Singapore for over 24 years, and I am still amazed at the courtesy and friendliness of most locals. It is encouraged in schools and the local community, and cynics may call it social engineering. All of culture and socialization is social engineering, and I would argue that some memes are better than others—much, much better.

In coaching teachers, there are key touchpoints, which include the first meeting, dialogues post-practice observations, helping clients solve problems and framing future scenarios. How one conducts oneself is fundamental to how effective the mediation of coaching is. The following are some reflections on key *touchpoints* in the coaching process as well as key skills and techniques for building trust and rapport in facilitating self-directedness in clients (teachers or others).

9.13   The Stages of Coaching

The terminology may vary, but the following represent broad interrelated stages that may be reiterated over time, depending on outcomes and perceptions:

1. Planning Stage
2. Action Stage
3. Reflection Stage.

The Planning and Action Stages

This can vary in how it arises. For example, in coaching clients, you may have had some prior experience with them, which may have created some preconceptions about you, which may or may not be favourable, and also based on very idiosyncratic data. I still find it amazing how people talk about celebrities, that they have never met, and make detailed evaluative judgements about what they are like as people— whether good, bad or ugly. It's as though they don't know that many celebrities hire public relation professionals to manage their media image presentation. Terms like 'fake-news' are now part of media-speak. On what basis would the general public know what is true and false on things that they don't have any direct contact with. It's hard enough to know what is true and false even in one's territory, as much will be determined by our neurological filters and the various cognitive biases that are systemic to human brain functioning. To push the point, but not too far in this context, how well do we know ourselves, and what exactly is the self? Kahneman (2012) puts this in stark perspective when he asserts that:

> The notion that we have limited access to the workings of our minds is difficult to accept because, naturally it is alien to our experience, but it is true: you know far less about yourself than you feel you do. (p. 52)

Equally, they may know very little about you and will go through the Primacy effect of first impressions, and a myriad of factors—both conscious and unconscious—will come in to play in terms of how they perceive you at this point—also potentially positive or negative.

As a teacher, coach, or mentor, the same processes play out; hence I am especially mindful of my presence, voice, tone and manner to mitigate negative perceptions, as the initial focus is always on helping clients to feel as comfortable as possible in being in this situation with me. The goal is to establish rapport and trust, explore through mediation their framing of the situation—their map of the territory. Once this is established, if it is established, then I consider what information and examples—often stories—that may help them reframe in ways that might produce action and resources for meeting their goals, as well as personal self-directedness in their future learning. If this is attained, helping them develop ways to enhance practice becomes more of a technical issue rather than an interpersonal one. If a client is not comfortable with you, and trust is not there, there will not be rapport, and the coaching encounter may be benign and unproductive.

It trust and rapport are established, a meaningful continuation of the coaching—client relationship can now work towards identifying personal goals that clients frame as meaningful and useful to their professional learning and development. As their coach, you can assist and facilitate the framing of these goals, helping them to choose challenging but achievable targets. Clients must understand what is involved, and what resources they may need, and finally must commit to the actions they have identified as necessary for goal attainment. I ensure that clients can clearly articulate what they need to do and how they intend to go about this action. Of course, I will support and provide feedback as is necessary, but we must have clarity and commitment on both sides.

If there are goals and there is clarity of strategy (always recognizing that these can be modified based on new information and understandings), it is necessary to identity specify key success indicators for the action to be taken, and the evidence sources to be collected, to support its success or otherwise. For example, in the cases of teachers who have been sent to me for consistent poor feedback scores, I avoid any judgements on their teaching but ask if I can observe one of their lessons as a starting point. As this is direct authentic evidence from which to have a meaningful conversation and share perceptions of teaching in order to do useful analysis of practices in terms of some criteria of effectiveness (e.g., engagement, attainment, core principles of learning), it typically becomes the process, and the evidence base, from which to work for future progress.

Finally, as this stage is about establishing a process for monitoring and review, especially for supporting self-assessment, I use an EBT approach with the situated use of the EBRP tool. This has been effective and productive as it provides both a rationale and methodology for improvement and has credibility with clients. I have noted, especially for those who genuinely want to improve, that many clients experience quick positive results in terms of student engagement in lessons, which fosters the development of a positive belief in the approach being employed. Also, they usually feel a better sense of mastery in using effective strategies, which further enhances confidence, self-efficacy and a better sense of well-being in their practice.

Reflection Stage

It is useful to conduct this stage as soon as possible after the teaching observation and any other aspects of the action stage (e.g., collection and analysis of student feedback, peer observation data) but giving sufficient time for clients to do their evidence-based reflective practice as documented earlier—what Schön referred to as *reflection on action*, but from the standpoint of a sound pedagogic literacy and good thinking. In this way, clients can reflect on their practice using a systematic diagnostic approach, valid data sources, and a tool to focus their pedagogic analysis.

If the planning stage has been effective, both in terms of understanding at the technical level (e.g., pedagogic literacy) and in building trust and rapport, this stage should prove less challenging and problematic (but not always).

Using the same basic model of understanding the client's maps, and their experience, perception and feelings about the territory (the reality of teaching), I invite their analysis, inferences and interpretations, evaluations of this teaching event, both in terms of their cognition and feelings. In this situation, and this applies throughout the coaching process (and most human interaction contexts), certain skills are fundamental for facilitating successful outcomes. These are outlined in the next section.

Key Coaching Skills

Effective coaching requires an extensive range of skills, both technical to the professional field, as well as interpersonal and intrapersonal.

Firstly, in many coaching situations, coaches are likely to possess high-level competence in the technical areas they are coaching, whether it be soccer, singing, or teaching. However, they may not have been the very best, and in coaching the more psychological aspects of human performance, may have little knowledge of the technical field. For example, the famous success coach Anthony Robbins coached Andre Agassi, who subsequently achieved major success in the grand slam tournaments. Robbins is no great tennis player, but it wasn't Andre's tennis that needed the coaching, but other psychological attributes. In the context of teaching, however, it is unlikely that a coach is not competent in teaching, especially if they are tasked with improving teaching practices. The following are some key areas that I focus on:

Sensory Acuity and Mindfulness

Sensory Acuity refers to the ability to notice, monitor and make sense of the external cues provided by other people's communication style. This involves both good observation skills, especially of non-verbal behaviour (often referred to as paralanguage) and empathic listening (putting oneself in the position of the other to extract the essential meaning of what they are trying to communicate in their words, tone and body language). Skill in recognising patterns in body language and voice characteristics helps to understand clients states of mind, which provides essential feedback in terms of how we communicate, and the questions we might ask, to help them clarify their thoughts and get into a more productive state of mind—if this is needed. When communicating with others, this means noticing the small but crucial signals (e.g., voice, tone, gesture, eye movement, colour changes that indicate state and mood changes), which provide a window on their perceptions and feeling, and the basis of how they are responding.

Mindfulness is a quality of consciousness in which humans are openly and non-defensively aware of what is truly taking place. As Kabat-Zin (2003) summarizes:

> It is very much an 'allowing' and receptive form of experiencing, such that when people are more mindful, they are more accepting of what they experience without focusing, resisting, or manipulating it. (p. 257)

Expertise in both sensory acuity and mindfulness (essentially a component of MC) requires deliberate practice over time. Many humans tend to focus their listening more on what they are going to say next, rather than on a deep understanding of what the other is trying to communicate. If not convinced, do a simple social psychology experiment of people watching in respect to this behavioural set over the next few days. You may be doing it yourself—I do it sometimes.

Calibration of Communication Style

There is an old saying, "people like people like themselves" and there is a reason for this; it's because they have a good rapport. They have things in common like interests, views on key aspects of reality, what constitutes fun, similar rules of engagement—in the language of NLP, 'similar personal maps about the territory'. Neuroscience is increasing showing how different interpersonal experience trigger physical and emotional changes in the brain, either promoting or inhibiting open and trusting interactions with others (e.g., Glaser 2014). Costa and Garmston (2016) go as far as saying:

> In rapport, brains synchronize, crossing the skull and skin barriers to beat as one. (p. 46)

In terms of friendships, rapport typically occurs (or does not occur) naturally and largely unconsciously. I have not seen (or should I say heard) people say, "Hey we are building rapport here, and that's because we are doing x, y, & z". Quite simply, they are sharing streams of consciousness from their experiences and their maps about things in the territory and are being stimulated in some positive way cognitively and/or emotionally—the latter often being it is fun and includes some humour.

In a coaching situation, rapport needs to be coaxed somewhat, and this takes skill. As coaches, we want to get rapport, as we know we can be more useful to clients' needs. However, clients must perceive us as trustworthy and an ally in helping them meet their goals. Building trust and rapport can take time, though losing it can be immediate and based on what may appear a trivial act or a misperception. There is an analogy, that it only takes one cockroach in a bowl of beautiful cherries to ruin its appeal; however, one beautiful cherry does nothing to reduce the disgust of a bowel of cockroaches.

Another tenet of NLP that I have found useful relates to linguistic preferences, and how one can understand aspects of a person's subjective experience to create synchronicity that facilitates communication ease and rapport building subliminally. For example, it postulates that people tend to have preferred *Representational Systems* (See Table 9.4) which incorporate specific Linguistic Predicates in their conversational language to convey meaning.

Skill in recognising patterns in linguistic terminology and body language helps to understand people's maps and state of mind. For example, people with a strong visual preference will tend to use terms like, "let's image", "we are drawing a blank

Table 9.4 Examples of NLP linguistic predicates

Visual predicates	Auditory predicates	Kinesthetic predicates
Imagine	Talk through	Hold on
Focus	Tune in	Put finger on
Look at	Listen to	Strikes me
Point out	Rings a bell	Get a grip of
Seeing it	Explains it	Close fisted
Show it	Deaf to	Tingling
Blind to	Crashing down	No stomach for it
In a flash	Hear me out	Hanging on
An eyeful	A little voice	In touch with
Bright as day	Lowering the tone	A handful
Dark as night	Harmony	Touched me
Drawing a blank		

here", etc. In contrast, more kinesthetically orientated people are more likely to use terms like, "let's get a grip of", "it strikes me", etc. In a nutshell—am I revealing a linguistic cue here? —people respond better to those who "speak their language". Also, this typically involves good sensory acuity and mindfulness.

It is through understanding clients' representational system preferences and the linguistic predicates used, that the coach can calibrate his or her communication style in ways to facilitate better understanding. This is often referred to as *Mirroring* the behaviour of the other person. By using similar linguistic predicates, tone of voice, and calibrating body language (e.g., gestures and body language of the other) providing it is not obvious imitation, helps in the rapport-building process. This has neurological support in the notion that humans have certain types of neurons termed *Mirror Neurons* (e.g., Rizzolatti and Sinigaglia 2008). The authors discovered unique neurons in the frontal and premotor cortex while researching the neural representation of motor movements in monkeys. Unlike other motor neurons, these neurons not only fired when engaged in planning a motor movement, but also through the observation of a related movement in another person or other monkey. Such neurons exist in humans, which can easily be understood in terms of how we respond to the experiences of characters in films. Do you cry in sad movies? I do, even though I know this is Hollywood fiction and the actors are being paid millions of dollars. Whether we like it or not, our minds play over the actions we see around us, and we feel the same or at least similar emotions. It's also why laughter is often referred to as contagious; once a person starts to laugh, others typically follow. In my experience some join in who probably did not experience the basis for the original laughter, it's just how the mind works. Smiling works the same way, as we explored prior. It also provides important feedback to others about your mood and approachability, both at conscious and subconscious levels, and has contagion effects. For example, if you smile at a student, he or she is likely to smile back at you and this can quickly spread to his or her classmates. The skillful and natural use of the smile helps in building rapport, and when used appropriately and calibrated to the communication style of the client, it works especially well.

Questioning

The skillful use of questioning in promoting good thinking has been illustrated prior, though Robbins (2001) view on the importance of questions is worth restating in this context:

> Thinking itself is nothing but the process of asking and answering questions
> Questions immediately change what we focus on and, therefore, how we feel (pp. 179–8)

Good questioning, using the other skills of sensory acuity, linguistic predicates, voice tone and paralanguage calibration, facilitates rapport and the mediation of experiences towards the goals of helping teachers to be self-directed and capable of their evaluation and professional development directions. NLP also uses a *Meta-Model of Language*, that helps to clarify meaning and reduce miscommunications, which Dilts (1980) refers to as:

> an explicit set of questions as well as a model for asking questions. (pp 77–79)

This specifically focuses on how language works in and affects our neurological states of mind, emotions, perceptions, relationships, and skills. This makes us aware of how the Words we use, our Tone of voice and supporting Body Language makes the difference in building rapport with other people. I have found this a useful set of heuristics in terms of framing and using questions to clarify the meaning, uncovering distortions, deletions and generalizations that are typically embedded in peoples' maps that may be seriously impacting the quality of their cognition, and subsequent action in the territory.

By asking certain questions, we can gain insight—*enter-into*—another person's model of the world (Map) and understand the world from their point of view. Questions focus on:

- What
- How
- Who.

Such questioning structures facilitate the other person's active involvement in examining his/her maps and the mapping process to run *Quality Control* checks on the mapping—enabling them to reframe and change meaning in the light of new evidence. In many ways, this process has similarities with the Gallwey's (1987) famous work on the *Inner Game*. His pioneering work was in understanding what goes on in the heads of tennis players that enhance or inhibit their performance. This applies to all people in situations where performance is involved (especially where it is high stakes and under observation from others) and many factors can come into play in affecting it. Invariably, this is very noticeable in top-level sport, for example, epitomized by a world-class soccer player failing to get the ball on target when taking a penalty kick or a golfer missing a 6-inch putt. However, it can also happen when actors/comedians get what is referred to as 'stage fright' or when people doing a public speech get 'lost for words'. Gallwey's key point was that a person's performance at any time has what he referred to as 'Potential'—their level of actual capability/skills. Hence, if they truly played to their potential that would be their best performance at this level of capability. For, example, have you ever done something (e.g., taught a lesson, given a talk, played a sport) when you felt that this was your best effort. Well, that would be near your potential. However, as human beings, in our minds (hence the term inner game) there is much that can significantly impact our thinking and feelings- both positively or negatively. Negative aspects that can undermine performance he refers to as 'Interference'. Interference is anything that enters our mind (and sub/unconscious stuff is involved here) that creates negative disturbance and, consequently, mitigate performance. Most documented are feelings of anxiety and fear that seriously affect concentration and even bodily control. However, it is our thinking and what beliefs and images flit through the mind that are the typical causes of interference at the psychological level, which quickly impact emotions, brain behaviour and psychomotor aspects, often epitomized in the phrase 'I feel like jelly'. For Gallwey, much of coaching is about improving Potential and reducing Interference to increase Performance. In doing this, there is also a synergistic effect, as increases in performance enhance belief and confidence, which in turn builds a resource for managing interference.

I remember when Garbine Muguruza won the ladies Wimbledon Title in 1917, referring back to her defeat in the same final two years prior, in the post-match interview, saying something like she was a 'different player now than then', and this wasn't about the tennis, but her resources to manage the interference in her head (Note: she did not use such terms, but the intended meaning seemed congruent with such affect). Similarly, she reflected on how she played in this final, and her thinking when faced with set-points that would have resulted in the loss of the first set:

> When I had those set points against me, I'm like, 'Hey, it's normal. I'm playing Venus here.'
> It's so I just keep fighting. And I knew that if I was playing like I was playing during the two weeks, I was going to have eventually an opportunity. So I was, like, calm. If I lose the first set, I still have two more. Let's not make a drama.

An important aspect of this process is generating the right questions for clients needs at the right time, as they can use their metacognitive thinking to explore their beliefs and feelings and get into a more productive state, focusing on what can be done, rather than what negative outcomes could occur; this is using one's inner voice, so to speak (again no pun intended). Treadwell (2017) framed it succinctly:

> Our unique capability to be able to talk to ourselves and use our 'inner voice' to question and interrogate our world, is essential to our learning. (p. 39)

Muguruza executed it perfectly, as she won the final in straight sets, winning the second set 6-0, against Venus Williams—which is no mean feat.

There is no use for the *Why* question in the meta-model. From an NLP perspective why questions, at best, get justifications and do nothing to change the situation. I am often amused when I see interviewers asking candidates why they want to teach. Do they expect anyone to say, "I like the long holidays"? I have even heard of approaches that advocate asking 5 why questions to probe deeply. Quite frankly after a second or third why question, I think I might just become a tad annoyed. The skilful use of what and how questions are particularly effective as it opens up the opportunity to unpack aspects of practice and how it works or don't work. Ask clients how they felt things went, what specifically, how they know this, etc. The manner, tone and calibration are of course crucial—I make it a conversation with a purpose, informal but focused. We may even have a joke in between the dialogue. I usually do most of the reflection stage in an informal setting over coffee if clients feel comfortable with this. It has worked well for me over the years, though the cost of buying coffees has put a few years on my working life. It has much in common with the open and flexible approach of creative interviewing (Douglas 1984), which involves:

> ...the use of many strategies and tactics of interaction, largely based on an understanding of friendly feelings and intimacy, to optimize cooperative, mutual disclosure and a creative search for mutual understanding. (p. 24)

Creating such a relationship, has strong neurological correlates in terms of brain responses, For example, Costa and Garmston (2016) referring to the work of Glaser (2014) note that conversations trigger physical and emotional changes in our brains and bodies, releasing either oxytocin, which fosters bonding and collaboration, or cortisol, a hormone that evokes stress and fear (p. 41).

In using questions it is important to recognize that before people can make a response, they need to process what has been asked and what it means; consider whether or not they have a response that they are prepared to make; then actually make the response. This can take some time. I have often observed teachers in many countries and contexts, after a period of exposition, ask the class "any questions", and often within 1–2 s, say something akin to "no, good, we'll move on". Now this is a double negative, as students will get used to such a communication set, and very soon not even bother to go through the cognitive strain of thinking about having questions, and will surely pick up on the verbal cue, and likely associated paralanguage, that asking questions in some way is 'not good'. Hence, we need to be mindful in question use.

Another important skill in the process of questioning and related mediation is the technique of *Paraphrasing*. Paraphrasing is seeking clarity on how a person feels in response to a situation or a prior question, by communicating back to them what you think they intended to communicate. This requires sensory acuity, quick thinking and being able to say the right words in the right way at the right time. It's intended to make the person feel comfortable in making sense of the situation or event in their terms and giving them the time, psychological space and necessary support to get better clarity on their thinking. In this way, issues and concerns can be made explicit, clear and tangible, which keep the conversation moving productively. In meeting clients in a post-observation context, I typically make explicit that understanding what is going on in classrooms often takes a bit of cognitive work in order to process all the information and make meaning, and, therefore, it helps if we may both work together to check our mutual understanding; I may say something like, "Let's support each other on this". In other words, I make paraphrasing an explicit technique that we can collaboratively utilize as we mediate the teaching experiences for better clarity and outcomes. Once there is rapport and trust, this is not difficult to introduce in most cases.

The importance of feedback in learning has been explored in detail. What is important, apart from the communication manner, is that it is data-driven and focused on the area(s) that may be high leverage in terms of improvement (e.g., task, process). Furthermore, simply giving what you think is useful and clear feedback does not mean that it will be interpreted in such terms. Bandler and Grinder (1990) make a poignant statement in this context:

> The meaning of your communication is the response you get. (p. 61)

9.14 Supported Experiments

Petty (2015) describes a Supported Experiment as "…a pilot or trial of a teaching strategy new to that teacher". Essentially, the teacher will use a strategy (ideally based on evidence of what methods work) for a given period to adapt it where necessary to the student group(s) and develop the necessary skills to use it effectively and fluently. In this process, the teacher will have the support of other teachers, who will be reviewing the experiment and its impact on student learning. As Petty summarizes:

> This might include discussions with peers, advanced practitioners, mentors, managers, trainers, or some combination of these… As a rule, experiments do not work well-first time, and that's fine if we learn from them!
>
> At a designated point, the experimenter will decide whether the experiment has worked or not, in their particular context. This is reported back to other teachers who can also learn from the experiment.

Supported experiments provide a clear structure for conducting professional development activities. Furthermore, as an increasing number of teachers embark on conducting supported experiments, openly sharing and thoughtfully appraising each other's work, there is a building of professional knowledge on effective teaching customized to the situated context of the school and its learners (e.g., Communities of Practice, Lave and Wenger 1993; Professional Capital, Hargreaves and Fullan 2012). Petty (2015) suggests that there are many benefits in using supported experiments to enhance professional development, as they:

- model and develop a culture of continuous practice
- include all teachers in continuous improvement
- provide a blame-free culture needed to encourage and support risk-taking and development

- prevent teaching skills from 'plateauing' and becoming stale
- provide the blame-free support needed to <u>really</u> change classroom practice
- encourage the development of teaching strategies that respond to known difficulties
- are inspiring for staff and can even reinvigorate quite jaded teachers.

As in all learning, whether for students or teachers, there needs to be the necessary time and support for competence to develop. Academic faculty will likely need to use the new instructional strategies several times before they reach levels of proficiency that achieve the high impact potential in terms of student attainment for particular method use. Secondly, and equally important, students need to become comfortable and see the relevance of the methods to their learning, which will also take some time. As Petty emphasizes:

Students also need to learn how to respond to the new methods, as effective methods are always more demanding of students than conventional teaching. They need to know why these new methods are being used, what it demands of them, and how to respond.

In my experience, this has been fully borne out over many years. During the past 5 years, I have been using supported experiments with several academic faculty on the use of flipped classroom learning (as documented in Chap. 6). The initial supported experiments were primarily used to identify useful strategies in terms of enhancing student engagement and attainment within the flipped classroom format, and at a school-based level. However, the success of these supported experiments led to an increasing number of academic faculty joining what is now an institution-wide project in designing and facilitating flipped classroom learning using an EBT approach. The interest continued and resulted in a two-year research project, with the work being presented at the Redesigning Pedagogy International Conference, National Institute of Education, Singapore (Sale et al. 2018).

In terms of involving students in the implementation process, as Petty suggested, two students from each class were invited (not conscripted) to be "co-participants" (Lincoln 1990, p. 78) in the research project to add a more authentic ethnographic component. These students chose to participate and knew that the teaching faculty were genuinely attempting to improve their learning experiences and attainment opportunities at the institution. They were given a full briefing on the research purpose and their role and responsibilities in participating in the research. For the flipped classroom implementation, they were specifically required to:

- Communicate with classmates to identify significant experiences relating to the new teaching approaches used
- Make personal notes and/or blog their experiences with both structured and open questions in the designated student blog
- Meet with the researchers at least once a semester for group sharing.

Informing and involving students from the onset of the implementation of the flipped classroom experiments provided many valuable insights into their learning experiences, as well as the essential 'buy-in' for the important changes that were being made in their classrooms.

Supported Experiments or Action Research?

In essence, supported experiments can be equally framed as mini action research projects in that they are directed at understanding and/or improving an aspect(s) of practice. One may ask, when does a mini-research project cease to be mini and become large-scale research? For me, this doesn't matter, it is the outcomes that count, and as the saying goes, "start small", and there's a reason for this. However, having identified specific pedagogic methods that might be of use and need some implementation practice, a supported experiment provides the affordances identified above without extensive time and resource commitments, which are often potential barriers to such work. However, if this proves of interest and is seen as beneficial to an important aspect of learning, especially if supported by management, a research project can be framed to extend the work in a more formal way. Stringer's (2004) frame on action research captures the approach in practical and viable terms:

> Action research differs quite significantly from the highly objective and generalizable exper-
> imental and survey studies that continue to provide significant information about schools
> and classrooms. It does, however, encapsulate the systematic qualitative research routines
> now becoming commonplace in the educational arena and increasingly applied by teachers
> and administrators as part of their work in schools. (p. 6)

What this means is action research is less likely to scare busy teachers away from doing this essential activity, but at the same time necessitates the systematic application of research methodologies used in qualitative research. As outlined above, it is possible to start small in terms of specific practice in one's own classroom, develop strategy possibilities, try them out and share this with colleagues, and eventually develop validated instructional approaches that impact beyond the classroom, the institution, the local community, and even to the wider global educational community. This potentially extending process of action research is summarized in Fig. 9.3.

Finally, I want to emphasize the motivated teacher's capability to do this work well. Given time and support, teachers with their specific professional training, MC

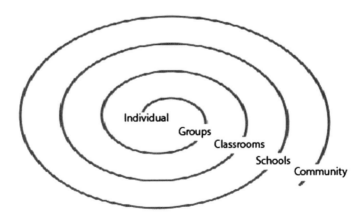

Fig. 9.3 Action research spiral of influence

and CTC are well equipped to excel as researchers investigating their practices. We know that setting challenging goals leads to better learning for students; hence why do we not see this as important for our teachers? Equally of note, such an approach involves both teachers and students as collaborators involved in examining ways to improve learning—now that's a real focus on learning, not a debate about what are teacher-centred or student-centred methods. All methods are centred on learning for both students and teachers, and this would enable, as Hattie (2009) claims:

> The ultimate requirement is for teachers to develop the skill of evaluating the effect that they have on their students. (p. 36)

9.15 Institutional and Societal Features that Facilitate Good Professional Development

The main components, activities and processes for effective professional development have been identified and explained in the previous sections. This section summarizes institutional and societal features that are likely to be powerful enablers in both developing high impact professional development for teachers and, of course, the best learning opportunities for our students in terms of attainment and well-being.

Valuing Good Teaching

Firstly, it is important to bear in mind that professional development approaches will be largely ineffective without a strong motivational base and commitment from teachers themselves. As Hargreaves and Evans (1997) stated:

> …where educational change is concerned, if a teacher can't or won't do it, it simply can't be done. (p. 3)

Secondly, educational institutions seeking to build and retain a high performing teaching force must create the conditions, platforms and support structures to bring this about. As noted earlier in the chapter, institutions and other societal conditions are often not always conducive to professional development that works (Powell and Kusuma-Powell 2015). They further observed that even if such good practice comes to some public attention in the institution, that:

> …when school people do witness exemplary teaching and learning, we often tend to respond with immediate adulation and subsequent dismissal. (p. 26)

They also quote Costa et al. (2014) in this context, who wrote:

> While our media-rich culture places a high value on talent, the irony is that in most schools, talent is underrated and often teachers remain silent about their own beliefs about talent. (p. 75)

It is not surprising. Therefore, that Darling-Hammond and Rothman (2015) note that:

While educators and policymakers agree that enabling teachers to improve student learning is one of the most significant ways to raise achievement, there are heated disagreements about the most useful ways to do this. (p. 1)

Hargreaves and Fullans' (2012) framing of building Professional Capital (e.g., institution-wide expert capability) is interesting in this context:

If you concentrate your efforts on increasing individual talent, you will have a devil of a job producing social capital. There is just no mechanism or motivation to bring all that talent together. The reverse is not true. High social capital does generate increased human capital. Individuals get confidence, learning, and feedback from having the right kind of people and the right kinds of interactions and relationships around them. (p. 4)

The main issue, as the authors point out, is:

…good learning comes from good teaching…So, let's concentrate our efforts not on bigger budgets, smaller classes, changing the curriculum, or altering the size of schools – but on procuring and producing the best teachers we can get. (p. 13)

This makes perfect sense from the perspective taken here. Firstly, the teaching force is the most single important factor in terms of educational quality. After all, they are the front line in the teaching and learning process, just as the team players are the front line in professional team sports. Rarely do poor teams win major championships in any professional sporting arena. Also, it helps if there is good leadership, and the same comparisons between school leaders and professional team coaches equally apply. Having worked extensively in educational development in Singapore for 24 years, as advisor, researcher and conducting workshops across the educational sectors, I was not surprised when the Economist Magazine (2018) referred to the Singapore Education System as the "best in the world", noting that the major contributing factors included a focus on quality teaching and pedagogy being based on educational research. What was increasing becoming 'face-validity' for me, has taken on a wider empirical frame.

Darling-Hammond and Rothamn's (2015) extensive analysis and evaluation of high performing educational systems (e.g., Finland, Ontario and Singapore) have spelt out the main features that seem to make the difference:

- Deep respect for the profession from the top levels of government and throughout society
- Strong common training for all teachers and leaders around these shared goals
- Systemic mentoring and induction for new teachers by trained senior teachers
- Continual development of educational knowledge, skills, and talent through extensive, governmentally subsidized professional development opportunities and a career ladder offering roles that expand and share experience
- Significant scheduled time for teachers to collaborate and learn together through lesson study, action research, and other reflections on practice.

Societal Commitment to the Importance of Learning

Referring specifically to Singapore, as I know it well, there are deep interrelated systemic factors that have contributed to Singapore's success, which is not only globally noticed in terms of its educational system but also in its economy, recently reported to be the most competitive in the world (Straits Times, October 2019). Firstly, Singapore as long sought to develop a workforce that not only can learn faster but also better (my interpretation). The development of personal attributes relating to what are now referred to as twenty-first-century competencies, as long been in the Singaporean educational and cultural landscape. For example, in 1995, the Education Minister, Lee Yock Suan, highlighted the need for students to be able to:

> ...learn to think independently and solve unexpected problems to survive and prosper in the years ahead, when knowledge and skills will become obsolete faster than before. (Conference to top civil servants on June 30)

A particularly salient landmark was the then Prime Minister of Singapore, Mr. Goh Chok Tong's (1997) framing of Singapore's educational system in terms of "Thinking Schools, Learning Nation" at the opening of the 7th International Conference on Thinking. He stated:

> We must get away from the idea that it is only the people at the top who should be thinking, and the job of everybody else is to do as told. Instead, we want to bring about a spirit of innovation, of learning by doing, of everybody each at his level all the time asking how he can do his job better. (p. 1)

Educational systems, philosophy and practices inevitably reflect the societal context in which they prevail. They are also likely to incorporate the interests and concerns of dominant decision-making groups in that society. In Singapore, there is a heavy reliance on the continuous development of its human resources to sustain and enhance competitive advantage, which will only be possible in the future with a workforce capable of responding to the enormity and complexity of economic and technological change with both productiveness and creativity. As the Prime Minister of Singapore, Mr. Goh Chok Tong (1997) stated:

> The old formulae for success are unlikely to prepare our young for the new circumstances and new problems they will face. We do not even know what these problems will be, let alone be able to provide the answers and solutions to them. But we must ensure that our young can think for themselves so that the next generation can find their solutions to whatever new problems they may face. (p. 3)

The Thinking Schools, Learning Nation program was launched in 1997, with quality in schools being highlighted as a key to ensuring student success. This is often referred to as the *Ability-Driven Phase*, which focused on "the development of every child to maximize his or her full potential through an education system tailored to that purpose" (Tan and Low 2016, p. 31). The concept of thinking schools was based on the notion that:

> ...the development of thinking and committed citizens would be crucial in ensuring that future challenges would be confidently dealt with, ensuring the continued success of Singapore. Learning was promoted as a national culture by encouraging creativity at every level of society. The role of teachers was also redefined, so that each school would be perceived as a model learning organisation. (Singapore Infopedia)

A Learning Nation envisions a national culture and social environment that promotes lifelong learning in our people. The capacity of Singaporeans to continually learn, both for professional development and for personal enrichment, will determine our collective tolerance for change. (Singapore Ministry of Education 2014, paras. 2–4).

Schools were encouraged to develop a spirit of creativity and innovation, in that both students and teachers were to be involved in action research, scientific investigations, and entrepreneurial activities. Well-prepared and well-supported teachers were seen as central to these aims.

Central to the vision was a comprehensive review of the curriculum in educational institutions, undertaken by the Ministry of Education, to promote more creative and critical thinking. As Tan and Gopinathan (2000) have commented:

It focuses on developing all students into active learners with critical thinking skills and on developing a creative and critical thinking culture within schools. Its key strategies include: (1) the explicit teaching of critical and creative thinking skills; (2) the reduction of subject content; (3) revision of assessment modes… (p. 7)

At that time, I was Education Advisor at Singapore Polytechnic and was tasked with developing a whole curriculum approach to promoting thinking, which I euphemistically referred to as "The Thinking Curriculum: A Response to Thinking Schools, Learning Nation" (Sale 2004). Furthermore, while Singapore maintains a strong adherence to developing cognitive capabilities and high educational attainment levels, as necessary goals for its educational direction, it is also committed to a holistic education that incorporates interpersonal and intrapersonal competencies in the wider educational aims and curriculum goals. In 2011, what is referred to as the *Values-Driven Phase*, sent a clear signal that the holistic education of individuals was essential to survive in the twenty-first-century workplace and society? Mr. Heng Swee Keat, Minister for Education, explained that values and character development had to be placed at the core of the education system because parents and educators alike called for schools to develop students holistically in response to changing demands in the global environment. As Tan and Low (2016) summarize:

The goals of this phase are "every school a good school," "every student an engaged learner," "every teacher a caring educator," and "every parent a supportive partner." (p. 31) This values-driven phase goes hand in hand with the ability-driven phase, where schools not only teach academic and life skills but also help instil values and build character in students. (p. 31)

In this student-centric phase, clear desired goals and outcomes of schooling and education were spelt out. The goal of the Singaporean education system is to nurture every child, regardless of his or her ability or achievement level. The ecology of educational reform is seen as resting on a set of shared values, *The Desired Outcomes of Education*, which are attributes that educators aspire for every Singaporean to have by the completion of formal education. These outcomes establish a common purpose for educators, drive policies and programmes, and provide a means of determining how well the education system is doing. They seek to develop:

- a **confident person** who has a strong sense of right and wrong, is adaptable and resilient, knows himself, is discerning in judgment, thinks independently and critically, and communicates effectively;

- a **self-directed learner** who takes responsibility for his learning, who questions, reflects and perseveres in the pursuit of learning;
- an **active contributor** who can work effectively in teams, exercises initiative, takes calculated risks, is innovative and strives for excellence; and,
- a **concerned citizen** who is rooted to Singapore, has a strong civic consciousness, is informed, and takes an active role in bettering the lives of others around him.

How these outcomes are developed through key educational stages is summarized in Table 9.5.

Most significant, in Singapore, policy quickly becomes active in practice, and new thrusts are supported with comprehensive professional development and support for teachers. I was astonished and delighted to have so much opportunity for professional learning. It bore no resemblance to what I had experienced prior, and this is not in any way critical of my prior workplaces. Teachers here have about 20 h a week built into their schedule for shared planning and learning, including visits to one another's classrooms, as well as 100 h per year of state-supported professional development outside of their school time. Furthermore, to create the space for critical thinking in the classroom, the content of all subjects has been significantly reduced (by up to 30%). Testing and assessments are increasingly being redesigned to encompass critical thinking skills. Darling-Hammond and Rothman (2015) point out that the purpose of the curriculum reduction was to free up space and time to focus on

Table 9.5 Key stage outcomes of education

At the end of primary school, pupils should	At the end of secondary school, students should	At the end of post-secondary education, students should
be able to distinguish right from wrong	have moral integrity	have moral courage to stand up for what is right
know their strengths and areas for growth	believe in their abilities and be able to adapt to change	be resilient in the face of adversity
be able to cooperate, share and care for others	be able to work in teams and show empathy for others	be able to collaborate across cultures and be socially responsible
have a lively curiosity about things	be creative and have an inquiring mind	be innovative and enterprising
be able to think for and express themselves confidently	be able to appreciate diverse views and communicate effectively	be able to think critically and communicate persuasively
take pride in their work	take responsibility for their own learning	be purposeful in pursuit of excellence
have healthy habits and an awareness of the arts	enjoy physical activities and appreciate the arts	pursue a healthy lifestyle and have an appreciation for aesthetics
know and love Singapore	believe in Singapore and understand what matters to Singapore	be proud to be Singaporeans and understand Singapore in relation to the world

promoting thinking and self-directed learning, which are recognized as important skills required for the global economy (p. 45). Furthermore, from their analysis of high performing educational systems, notably Singapore and Finland, they identify that while successful systems may differ in several ways, there are common systemic features—in that they are systems for teacher and leader development. Key features include:

> Multiple components, not just a single policy, and these components are intended to be coherent and complementary, to support the overall goal of ensuring that each school in each jurisdiction is filled with highly effective teachers and is led by a highly effective principal. (pp. 76–77)

> The evidence shows that school leadership is second only to teaching in its effects on student learning. About a fourth of the school-related variation in student achievement can be explained by school leadership (Leithwood et al. 2004). (p. 88)

Similarly, Ryan and Deci (2017) concur with the above analysis of these systems as well as capturing the essence and purpose of this chapter.

> Despite massive differences in curricular approaches, they have one important thing in common: they treat and train their teachers as professionals. These nations have invested in higher salaries and higher-quality training, to recruit the best and the brightest and help them internalize and develop effective classroom practices. In turn, the more competent and professional the population of teachers, the more they can be expected to benefit from, and make good use of professional autonomy. (p. 378)

> The primary focus is student flourishing – that is, not only growing in cognitive skills and knowledge but also developing and strengthening personal and social skills and experiencing psychological health and well-being in the process. (p. 380)

9.16 Epilogue

Teaching expertise (in terms of Martin's (2009) *Knowledge Funnel*) is now less of a *Mystery* and more understandable in terms of useful *Heuristics*. As a consequence, we can now develop tools (e.g., highly effective, efficient and creative instructional strategies) that will enhance competence and performance in teaching, of which Drucker framed as only possessed by the 'naturals' who somehow know how to teach. Invariably, teaching itself needs significant reframing. As Treadwell (2017) points out:

> The role of the educator is now far more dynamic and requires a deeper professionalism and rigour in our understanding of the Learning Process, the concept frameworks within the domains and the competencies in which we are operating, as well as the thoughtful and reflective practices we are encouraging. (p. 156)

For me, the aim is to improve teaching quality globally, as I feel that's something worthwhile in terms of a *moral landscape* (e.g., Harris 2010). I also think that a global teaching force, skilled in Metacognitive Capability and Creative Teaching Competence, would help a generation of learners to be able to better understand the world (the one inside their heads, and that external to self). It may also contribute to better decision making for mankind's future. This might represent a major existential feature of human progress, akin to Pinker's (2019) "Enlightenment Now".

References

Bandler R, Grinder J (1990) Frogs into princes: the introduction to neuro-linguistic programming. Eden Grove Editions, Middlesex

Berliner DC (2004) Describing the behavior and documenting the accomplishments of expert teachers. Bull Sci Technol Soc 24(3):200–212

Caine G, Caine R (2001) The brain education and the competitive edge. Scarecrow Press, Lanham, MD

Clouder DL (2000) Reflective practice: realising its potential. Physiotherapy 86(10):517–522

Costa AL, Garmston RJ (2002) Cognitive coaching: a foundation for renaissance schools. Christopher-Gordon Publishers Inc, Massachusetts

Costa AL, Garmston RJ, Zimmerman DP (2014) Investing in teacher quality. Teacher College Press, New York

Costa AL, Garmston RJ (2016) Cognitive coaching: developing self-directed leaders and learners. Rowman & Littlefield, New York

Darling-Hammond L, Rothman R (2015) Teaching in the flat world: learning from high-performing systems. Teachers College Press, New York

Dewey J (2016) The University of Hong Kong, e-learning blog. https://tl.hku.hk/2016/03/reflection-how-do-i-do-it-john-dewey-we-do-not-learn-from-experience-we-learn-from-reflecting-on-experience/. Last accessed 29 Nov 2019

Dickinson M (2015) Evolution of an e-learning developer guide instructional design and project management. In: Waldrop JB, Bowden MA (eds) Best practices for flipping the college classroom. Routledge, New York

Dilts R et al (1980) Neuro-linguistic programming Vol. 1: the study of the structure of subjective experience. Meta Publications, California

Douglas JD (1984) Creative interviewing. Sage Publications, Beverly Hills

Downey M (2003) Effective coaching: lessons from the coaches' coach. Texere Publishing, New York

Duffy GG (1993) Teachers' progress toward becoming expert strategy teachers. Elementary Sch J 94:109–220

Duffy GG et al (2010) Teacher as metacognitive professionals. In: Hacker D, Dunlosky J, Graesser A (eds) Handbook of metacognition in education. Lawrence Erlbaum and Associates, New Jersey

Economist Magazine (2018) What other countries and learn from Singapore's schools. Retrieved from https://www.economist.com/leaders/2018/08/30/what-other-countries-can-learn-from-singapores-schools

Gallup Pool Organization (2014) The state of American schools

Gallwey TW (1987) The inner game of tennis. Jonathan Cape, London

Glaser J (2014) Conversational intelligence: how great leaders build trust and get extraordinary results. Bibliomotion, Brookline, MA

Goh CT (1997) Speech by Prime Minister Goh Chook Tong at the opening of the 7th international conference on thinking on Monday, 2 June at The Suntec City Convention Centre, Singapore

Gulamhussein A (2013) Teaching the teachers: effective professional development in an era of high stakes accountability. National School Boards Association, Centre for Public Education

Hall LM (2001) The spirit of NLP. Crown House Publishing, Bethel

Harari YN (2016) Homo Deus: a brief history of tomorrow. Vintage, Penguin, London

Hargreaves A, Evans R (1997) Beyond educational reform: bringing teachers back in. Open University Press, Buckingham

Hargreaves A, Fullan M (2012) Professional capital: transforming teaching in every school. Teachers College Press, New York

Harris S (2010) The moral landscape: how science can determine human values. Free Press, New York

Hatano G, Inagaki K (1986) Two courses of expertise. Child development and education in Japan, pp 262–272

Hattie J (2003) Teachers make a difference: what is the research evidence? Distinguishing expert teachers from novice and experienced teachers. Australian Council for Educational Research

Hattie J (2009) Visible learning. Routledge, New York

Huberman M (1990) Linkage between researchers and practitioners: a qualitative study. Am Educ Res J 27:363–391

Joyce BR, Showers B (2002) Student achievement through staff development, 3rd edn. ASCD, Alexandria, CA

Kabat-Zin J (2003) Mindfulness-based interventions in context: past, present, and future. Clin Psychol Sci Pract 10(2):144–156

Kahneman D (2012) Thinking fast and slow. Penguin Books, London

Kanfer R, Kanfer F (1991) Goals and self-regulation: applications of theory to work settings. In: Maehr M, Pintrich PR (eds) Advances in metacognition and achievement, vol 7. JAI Press, Greenwich, CN, pp 287–326

Kennedy M (1999) The role of preservice teacher education. In Darling-Hammond L, Sykes G (eds) Teaching as the learning profession: handbook of policy and practice. Jossey-Bass, San Francisco

Knowles M (1984) Andragogy in action. Jossey-Bass, San Francisco

Darling-Hammond et al. (2009) Professional learning in the learning profession: a status report on teacher development in the United States and abroad. National Staff Development Council

Lave J, Wenger E (1993) Situated learning: legitimate peripheral engagement. Cambridge University Press, Cambridge

Lee YS (1995) The Straits Times, July 1

Leithwood et al (2004) How leadership influences student learning. University of Minnesota Center for Applied Research and Educational Improvement

Levin B (2008) How to change 5000 schools. Harvard Education Press, Cambridge

Lincoln YS (1990) The making of a constructivist: a remembrance of transformations past. In: Guba EG (ed) The paradigm dialog. Sage, London

Lortie DL (1975) Schoolteacher. University of Chicago, Chicago, IL

Martin R (2009) The design of business. Harvard Business Press, Massachusetts

Pentland A (2014) Social physics: how good ideas spread—the lessons from a new science. Scribe Publications, London

Petty G (2015) Resources provided on *Geoff Petty Homepage*. Can be accessed at https://geoffpetty. com/

Piaget J (2001) The psychology of intelligence. Routledge, New York

Pinker S (2019) Enlightenment now. Penguin, U.K.

Piper E (2017) Not a day care: the devastating consequences of abandoning truth. Salem, New Jersey

Powell W, Kusuma-Powell O (2013) The OIQ factor: raising your schools organizational intelligence. John Catt Educational Ltd, Melton, Woodbridge

Powell W, Kusuma-Powell O (2015) Teacher self-supervision: why teacher evaluation has failed and what we can do about it. John Catt, Suffolk, UK

Reimers FM, Chung CK (2016) Teaching and learning for the twenty-first century. Harvard Education Press, Cambridge, Massachusetts

Research Alert (2014) Educational Leadership May 2014, vol 71, no 8. Professional learning reimagined. (https://www.ascd.org/publications/educational-leadership/may14/vol71/num08/Double-Take.aspx). Last accessed 15 Nov 2019

Rizzolatti G, Sinigaglia C (2008) Mirrors in the mind: how we share our actions and emotions. Oxford University Press, New York

Robbins A (2001) Unlimited power. Pocket books, London

Rogers A (1998) Teaching adults. Open University Press, Buckingham

Ryan RM, Deci EL (2017) Self determination theory: basic needs in motivation, development, and wellness. The Guilford Press, New York

Sale D (2004) Promoting thinking: a Singaporean response. SPGG J

Sale D (2014) The challenge of reframing engineering education. Springer, New York

Sale D, Cheah SM, Wan M (2018) 3 paper symposia on enhancing students' intrinsic motivation: an evidence-based approach. In: ERAS-APERA international conference, Singapore, 12–14 Nov 2018

Schön DA (1983) The reflective practitioner: how professionals think in action. Temple Smith, London

Schön DA (1987) Educating the reflective practitioner: toward a new design for teaching and learning in the professions. Jossey-Bass, San Francisco

Singapore Infopedia. https://eresources.nlb.gov.sg/infopedia/. Last accessed 7 Nov 2019

Singapore Ministry of Education (2014) https://www.moe.gov.sg/education/education-system/desired-outcomes-of-education

Stigler JW, Hiebert J (1999) The teaching gap. Free Press, New York

Stringer E (2004) Action research in education. Pearson, New Jersey

Swaab D (2015) We are our brains. Penguin, London

Tan J, Gopinathan S (2000) Education reform in Singapore: towards greater creativity and innovation. NIRA Rev 7(3):5–10

Tan OS, Low EL (2016). Singapore's systemic approach to teaching and learning twenty-first century competencies. In: Reimers FM, Chung CK (eds) Teaching and learning for the twenty-first century: educational goals, policies, and curricula from six nations. Harvard Education Press, Cambridge, Massachusetts

The Straits Times (October, 2019) https://www.straitstimes.com/business/economy/singapore-is-worlds-most-competitive-economy-world-economic-forum. Last accessed 22 Nov 2019

Timperley H et al (2008) Best evidence synthesis on professional learning and development. Report to the Ministry of Education. Wellington, New Zealand

Treadwell M (2017) The future of learning. The Global Curriculum Project, Mount Maunganui, NZ

Willingham DT (2009) Why don't students like school: a cognitive scientist answers questions about how the mind works and what it means for the classroom. Jossey-Bass, San Francisco

Windschitl M (2002) Framing constructivism in practice as the negotiation of dilemmas: an analysis of the conceptual, pedagogical, cultural, and political challenges facing teachers. Rev Educ Res 72:131–175

Winget L (2017) What's wrong with damn near everything!: how the collapse of core values is destroying us and how to fix it. Wiley, New Jersey

Printed in the United States
By Bookmasters